SALT WEATHERING OF BUILDINGS
AND STONE SCULPTURES

**SWBSS
2017**

20 – 22 September 2017

FH|P UNIVERSITY OF APPLIED SCIENCES POTSDAM

Proceedings of SWBSS 2017

Fourth International Conference on
Salt Weathering of Buildings and Stone
Sculptures

University of Applied Sciences
Potsdam, Germany

20-22 September 2017

Edited by Steffen Laue

Verlag der
Fachhochschule
Potsdam

Impressum

Bibliografische Information der Deutschen Nationalbibliothek:
Die Deutsche Nationalbibliothek verzeichnet diese Publikation in der Deutschen Nationalbibliografie; detaillierte bibliografische Daten sind im Internet über http://dnb.d-nb.de abrufbar.

Dieses Buch ist auch als freie Onlineversion über die Homepage des Verlags sowie über den OPUS-Publikationsserver der Fachhochschule Potsdam verfügbar.
http://nbn-resolving.de/urn/resolver.pl?urn:nbn:de:kobv:525-15913

Steffen Laue (Hrsg.)
Proceedings of SWBSS 2017
Fourth International Conference on Salt Weathering of
Buildings and Stone Sculptures
Potsdamer Beiträge zur Konservierung und Restaurierung Band 6

Verlag der Fachhochschule Potsdam
www.fh-potsdam.de/verlag

© 2017 Fachhochschule Potsdam
Texte und Abbildungen in Verantwortung der Autorinnen und Autoren der Beiträge

| ISBN | 978-3-934329-93-5 | (Druckausgabe) |
| URN | urn:nbn:de:kobv:525-15913 | (elektronische Ausgabe) |

Layout: Semler Grafik oHG, Berlin
Herstellung und Vertrieb: tredition GmbH, Hamburg
Gesetzt in TheAntiquaSun

Cover image: Photo: Hans-Jürgen Schwarz, Design: Benita Lohse

Previous SWBSS CONFERENCES

SWBSS 2008, organized by L.M. Ottosen and colleagues, The National Museum Copenhagen, Denmark, 22-24 October 2008

SWBSS 2011, organized by I. Ioannou & M. Theodoridou, University of Cyprus, Limassol, Cyprus, 19-22 October 2011

SWBSS 2014, organized by H. De Clercq and colleagues, Royal Institute for Cultural Heritage, Brussels, Belgium, 14-16 October 2014

SWBSS 2017
Scientific Committee

De Clercq, Hilde – Royal Institute for Cultural Heritage, Brussels, Belgium
Diaz Gonçalves, Teresa – Laboratório National de Engenharia Civil, Lisboa, Portugal
Espinosa-Marzal, Rosa M. – University of Illinois at Urbana-Champaign, USA
Flatt, Robert – ETH Zürich, Switzerland
Hamilton, Andrea – University of Strathclyde, Glasgow, UK
Ioannou, Ioannis – University of Cyprus, Nicosia, Cyprus
Laue, Steffen – University of Applied Science Potsdam, Germany
Lubelli, Barbara – Delft University of Technology, The Netherlands
Ottosen, Lisbeth – Technical University of Denmark, Lyngby, Denmark
Siedel, Heiner – Dresden University of Technology, Germany
Steiger, Michael – University of Hamburg, Germany
Török, Ákos – Budapest University of Technology and Economics, Hungary
Vergès-Belmin, Véronique – Laboratoires de Recherche des Monuments Historiques, Champs-sur-Marne, France

Organizing committee

Laue, Steffen – University of Applied Science Potsdam and Salze im Kulturgut e.V.
Steiger, Michael – University of Hamburg and Salze im Kulturgut e.V.
De Clercq, Hilde – Royal Institute for Cultural Heritage, Brussels, Belgium

Sponsored by

DBU

The Deutsche Bundesstiftung Umwelt DBU
(German Federal Environmental Foundation)

Preface

Welcome at the University of Applied Sciences in Potsdam!

The first SWBSS event was held in Copenhagen, Denmark in 2008, with Lisbeth Ottosen as driving force, the second was arranged by Ioannis Ioannou and Magdalini Theodoridou in Limassol, Cyprus, in 2011, and the third SWBSS conference was organized by Hilde De Clercq and colleagues at the Royal Institute for Cultural Heritage in Brussels, Belgium.

This is the fourth conference in the series. It is a great pleasure to see that SWBSS 2017 has been met with interest from so many different countries both in Europe and outside of Europe.

We received around 50 contributions, consisting of papers and abstracts and bringing together conservators, restorers, engineers, scientists, young students and experienced researchers.

The success confirms the significance of the SWBSS agenda for the scientific community. Being aware that salt crystallization is of major importance in the weathering of porous building materials, I do hope that this conference contribute to an exchange of knowledge, create new solutions for the preservation of our cultural heritage and stimulate new research ideas and future collaborations within this interesting field.

On behalf of the Conference Organizing Committee and the participants I would like to take the opportunity to thank the many reviewers of the Scientific Committee who so generously gave their time to ensure that the papers accepted are of a high quality and represent a true advance in the state-of-the-art.

Furthermore, I would like to thank Hilde De Clercq, Royal Institute for Cultural Heritage, Brussels, and my colleagues of Salze im Kulturgut e. V., especially Mike Steiger and Hans-Jürgen Schwarz, for co-organizing this SWBSS conference. Thanks also to the Prussian Palaces and Gardens Foundation Berlin-Brandenburg, particularly to Kathrin Lange and Roland Will, for organizing the excursion. And without the help of my colleagues and the students of our University we could not have organized this international conference: thank you all!

Last but not least, I would like to express my deepest gratitude to our sponsor The Deutsche Bundesstiftung Umwelt DBU (German Federal Environmental Foundation) for supporting the conference.

I sincerely hope that you will enjoy SWBSS 2017 and wish you a pleasant stay in Potsdam!

Steffen Laue

Conference Chair
University of Applied Sciences Potsdam
Department for Conservation and Restoration

Table of contents

Salt sources, transport and crystallization

Deicing Salts: An Overview
A. E. Charola, B. Rousset and C. Bläuer (Washington, DC, US) 16

Traffic-induced salt deposition on facades
M. Auras (Mainz, Germany) 24

Wick action in cultural heritage
L. Pel and R. Pishkari (Eindhoven, The Netherlands) 33

Measurement techniques and experimental studies

A preliminary study on dynamic measurement of salt crystallization and
deliquescence on a porous material surface using optical microscope
M. Abuku, S. Hokoi and D. Ogura (Higashi-Osaka, Japan) 42

Diagnostics and monitoring of moisture and salt in porous materials by
evanescent field dielectrometry
C. Riminesi and R. Olmi (Florence, Italy) 49

Determination of the water uptake and drying behavior of masonry using a
non-destructive method
A. Stahlbuhk, M. Niermann and M. Steiger (Hamburg, Germany) 57

Measurement of salt solution uptake in fired clay brick and identification of
solution diffusivity
E. Mizutani, D. Ogura, T. Ishizaki, M. Abuku and J. Sasaki (Kyoto, Japan) 65

Local strain measurements during water imbibition in tuffeau polluted by gypsum
M.A. Hassine, K. Beck, X. Brunetaud and M. Al-Mukhtar (Orleans, France) 74

Assessment of the durability of lime renders with Phase Change Material (PCM)
additives against salt crystallization
L. Kyriakou, M. Theodoridou and I. Ioannou (Nicosia, Cyprus) 81

Salt crystallization tests

Salt crystallization tests: Focus on their objective
A. E. Charola, I. Rörig-Dalgaard, J. Chwast and J. Elsen (Washington, DC, US) 92

Mitigating salt damage in lime-based mortars with mixed-in crystallization modifiers
S. J. C. Granneman, B. Lubelli and R. P. J. van Hees (Delft, The Netherlands) 100

Studies for conservation issues

- Efficiency of laboratory produced water repellent treatments on limestone
  C. Charalambous and I. Ioannou (Nicosia, Cyprus) 110

- Environmental control for mitigating salt deterioration by sodium sulfate on
  Motomachi Stone Buddha in Oita prefecture, Japan
  K. Kiriyama, S. Wakiya, N. Takatori, D. Ogura, M. Abuku and Y. Kohdzuma
  (Kyoto, Japan) 118

- Numerical analysis on salt damage suppression of the Buddha statue carved into
  the cliff by controlling the room temperature and humidity in the shelter
  N. Takatori, D. Ogura, S. Wakiya, M. Abuku, K. Kiriyama and Y. Kohdzuma
  (Kyoto, Japan) 125

- Management of sodium sulfate damage to polychrome stone and buildings
  D. Thickett and B. Stanley (English Heritage, UK) 135

- Conservation of marble artifacts by phosphate treatments: influence of
  gypsum contamination
  E. Sassoni, G. Graziani, E. Franzoni and G. W. Scherer (Bologna, Italy) 143

- Electrode placement during electro-desalination of NaCl contaminated
  sandstone – simulating treatment of carved stones
  L. M. Ottosen and L. Andersson (Lyngby, Denmark) 150

- How *not to bother* soluble salts while grouting
  C. Pasian, F. Piqué, C. Riminesi and A. Jornet (Lugano, Switzerland) 158

- Moisture transport during poultice application
  C. Franzen, M. Aulitzky and S. Pfefferkorn (Dresden, Germany) 168

- The application of hydroxyapatite-based treatments to salt-bearing porous
  limestones: A study on sodium sulphate-contaminated Lecce Stone
  G. Graziani, E. Sassoni, G. W. Scherer and E. Franzoni (Bologna, Italy) 176

Salt reduction

- Evaluation of desalination and restoration methods applied in Petra (Jordan)
  W. Wedekind and H. Fischer (Göttingen/Berlin, Germany) 190

- Salt extraction by poulticing – results of a pre-investigation at the archeological
  site of Coudenberg
  S. Godts, H. De Clercq and M. Rich (Brussels, Belgium) 200

· Desalination of Cotta type Elbe sandstone with adapted poultices: Optimization of poultice mixtures and application terms
J. Maitschke and H. Siedel (Potsdam/Dresden, Germany) 208

· Tracing back the origins of sodium sulfate formation on limestone as a consequence of a cleaning campaign: the case study on Charité and Espérance sculptures of Chartres cathedral
S. Benkhalifa, V. Vergès-Belmin, O. Rolland and L. Leroux (France) 219

· Granite and schist masonry desalination by poultices at Jacobine Church in Morlaix, France
B. Brunet-Imbault, B. Reidiboym and C. Guinamard (Studiolo, Paris, France) 229

Case studies

· Salt content of dust and its impact on the wall paintings of the church St. Georg at the UNESCO World Heritage site Monastic Island of Reichenau in Germany
J. Frick, M. Reichert and H. Garrecht (Stuttgart, Germany) 242

· Investigation of salts sources at the Karadjordje's Gate on the Belgrade fortress
M. Franković, N. Novaković, S. Erić, P. Vulić and V. Matović (Belgrade, Serbia) 252

· Investigation and examination of a degraded Egyptian painted limestone relief from Tell Hebua (Sinai)
E. Mertah, M. Othman, M. Abdelrahman, M. Fatoh and S. Connor (Cairo, Egypt) 261

· Technique for transportation of stone sculptures damaged by salt crystallization
Vinka Marinković (Zagreb, Croatia) 271

· Investigation and conservation concept of salt damaged epitaphs in the church of Werben (Saxony-Anhalt)
S. Laue, D. Poerschke and B. Hübner (Potsdam, Germany) 275

· Salt-induced flaking of wall paintings at the Mogao Grottoes, China
L. Wong, S Bomin, W. Xiaowei, A. Rava and N. Agnew
(The Getty Conservation Institute, L.A., US) 285

· Development of a network-based climate monitoring system for climate assessment and regulation
C. Leonhardt, S. Leonhardt and J. Heller (Kiel, Germany) 297

Salt sources, transport and crystallization

Deicing Salts: An Overview

A. Elena Charola[1]*, Bénédicte Rousset[2] and Christine Bläuer[2]
[1] Museum Conservation Institute, Smithsonian Institution, Washington, D.C., USA
[2] CSC Sàrl, Conservation Science Consulting, Fribourg, Switzerland
* charolaa@si.edu

Abstract

The world production of salt (NaCl) was over two hundred million tons in 2015. The US is the second larger producer of salt after China, produced over four million tons of which 43% were consumed in highway deicing. While NaCl is the most commonly used salt, other salts are added to it to improve its performance, such as $CaCl_2$, $MgCl_2$. To reduce the use of the deteriorating NaCl, other salts are also used, such as magnesium acetate, calcium magnesium acetate or potassium and magnesium formate. The addition of sand and other inorganic insoluble compounds to aid in making surfaces less slippery is discussed, as well as the recent use of organic deicers and the problems that these can induce.

The paper aims to present an overview of deicing salts, and the differences with anti-icing or antifreeze solutions. It also discusses the problems they induce to vehicles, buildings and constructions, while also considering the negative aspect they have for the environment as well as their contribution to air pollution. Some examples are presented to illustrate the problem and less aggressive alternatives are discussed, especially with regard to the conservation of valuable architectural heritage.

Keywords: deicing salts, building deterioration, environmental pollution

1. Introduction

The world production of salt (NaCl) was over two hundred million tons in 2015. The US, the second largest producer of salt after China, generated over four million tons of which 43% were consumed in highway deicing. While NaCl is the most commonly used salt, other salts are added to it to improve its performance, such as $CaCl_2$, $MgCl_2$. To reduce the use of the metal corrosive NaCl, other salts are also used, such as magnesium acetate, calcium magnesium acetate or potassium and magnesium formate, urea, and even sugar containing solutions from either sugar processing or equivalent procedures.[1]

While the use of deicing salts is necessary, they do have a negative impact on the environment, such as: soil contamination, negative effect on plants and trees near the highways/streets, contamination of water courses and eventual drinking water, air contamination by powdered salts, corrosion of reinforced concrete in bridges and structures, as well as of cars and trucks. For example, in Austria it has been estimated that half the induced vehicle corrosion could be attributed to de-icing salt.[2] As deicing salts are distributed, the finer particles (<10 µm, usually referred to as PM10) can remain suspended in the air, thus contaminating it.

Differentiation between deicing and anti-icing or antifreeze should be made. Deicing salts are applied after snow events, their effectiveness being based on lowering the freezing point of water. Antifreeze solutions of glycerol or various glycols are applied prior to the event to prevent a strong bond between the pavement surface and frost by applying a freezing point depressant. These are mainly used on aircraft, machinery and vehicles as they are non-corrosive, however, most of them are toxic.[3] Other solutions have been developed based on special coatings.[4]

2. Deicing salt varieties

Deicing salts can be roughly divided into inorganic salts, such as sodium chloride (NaCl), organic salts, such as magnesium calcium acetate (CaMg(CH$_3$COO)$_4$) and organic compounds such as urea (CO(NH$_2$)$_2$). Salts can be used in various mixtures, and other substances added, such as anticaking agents, e.g., potassium ferrocyanide[5] or anticorrosion agents such as ammonium phosphate or sodium hypochlorite.[6] Also, they can be spread directly in granulated form, or as a solution, i.e., brine. With the former, the mixture with sand (e.g., 75% sand-25% NaCl), or other equivalent materials such as fine gravel or expanded clay pellets contributes to decrease the slippery surface of compacted snow[2]; however, they do increase small particulates in air by about 45%.[7] Recently, potassium carbonate (K$_2$CO$_3$) has been studied in comparison to NaCl, and it was found that while it was more adsorbed to soil colloids, the pH was elevated more than for NaCl, and the species composition of the area where it had been applied changed significantly.[8]

2.1. Chloride based deicers

Chloride ions from deicing salts will mobilize and increase soil salinity near the roadways where they are applied. While magnesium and calcium ions in-crease the stability and permeability of the soil, sodium ions will decrease them. Furthermore, sodium, magnesium and calcium chlorides may contribute to the mobilization of trace metals from the soil to surface and groundwater. The solid chloride deicers, i.e., NaCl, may contribute to air pollution through particulates released into the air.[2,7]

2.2. Acetate based deicers and others

Soil microorganisms will break down acetate ions resulting in oxygen depletion of the soil, which can impact vegetation. A similar oxygen depletion is most likely to occur in slow flowing streams and small ponds into which these ions migrate.[7] While the toxicity of calcium magnesium acetate (CMA) to fish and invertebrates is low, when also containing potassium, CMAK (50% CMA-50% KA), they have higher toxicity. Acetate deicers will result in the decrease of air pollution as sand use can be reduced; however, the solid deicers, CMA and sodium acetate, NAAC, may contribute fine particulates to the air increasing its pollution. These deicers are mainly approved for use at airfields and aircraft, as they are less corrosive, as is the case for potassium formate, in either liquid or solid form.

2.3. Urea

Urea [CO(NH$_2$)$_2$] is used as a deicing agent for airport runways[9] though it has been mostly discontinued in larger US airports.[10] The main reason is that as a fertilizer (46% by weight nitrogen content) it contributes to environmental pollution, e.g., acute toxicity to aquatic invertebrates and plants, as well as some fish.[11] Several soil bacteria contain the urease enzyme that catalyzes the decomposition of urea into NH$_4^+$ and HCO$_3^-$. Furthermore, NH$_4^+$ (or NH$_3$) is oxidized by nitrifying bacteria, Nitrosomonas and subsequently by Nitrobacter, in a two-step process to NO_3^-, an ion that is regularly found on building façades.

Urea forms an eutectic mixture with water (at ~33% by weight) with the eutectic point at 11.5°C. Solubility is about 1Kg/L at 20°C, the dissolution being endothermic, and the equilibrium RH is 76.5% at 25°C. In dilute solutions (not specified but probably below 5%), urea decomposes to NH_3 and CO_2 (the formation of isocyanic acid occurs upon heating, temperature not specified). The most common impurity in synthetic urea results from the condensation of two molecules to form biuret ($C_2H_5N_3O_2$) or carbamylurea, a compound that interferes with plant growth. As a deicer, urea proves practically useful, i. e., deicing within 15-20 minutes, at temperatures below -9.4°C taking into account that its dissolution is endothermic.[12] Many studies have addressed the decomposition of urea in aqueous solutions[13], while others address its use to decrease vehicular emissions of NOx which contribute to the formation of nitrates or nitrites in buildings along the streets.[14]

2.4. Glycols and other alcohols

Methanol was used as antifreeze in windshield fluids, but because of health concerns the amount added is restricted. Ethylene glycol, commonly referred to as "glycol" is used as engine cooling antifreeze. The freezing point of ethylene glycol is about –12°C, however, mixed with water, this is depressed, e. g., a mixture of 60% EG-40% water freezes at –45°C. Propylene glycol has replaced ethylene glycol in many uses because of its lower toxicity. These products are used for aircraft deicing fluids (heated aqueous solution of ethylene glycol), and as antifreeze, as undiluted, thickened propylene glycol.

2.5. Other organic deicers

In the USA, the Minnesota Department of Transportation claims to have pioneered the use of sugar beet juice based on the huge sugar beet industry in the Red River Valley of Minnesota/North Dakota, and the massive need for re-use of sugar beet waste helped create a market for it, and the fact that these states get a lot of snow and ice contributed to the testing.[15] The sugar beet syrup is mixed in with traditional salt, sand or chloride brines to improve performance and reduce the impact on the environment.[16] Not only sugar beet syrup is used, but other residues of distilled or fermented agricultural products[7] such as corn, barley and even pickle brines. The addition of syrup from sugar processing to brines has been shown to improve their effectiveness and has been approved in Switzerland since 2015.[17]

3. Impact of deicing salts on buildings

When considering the effect of deicing salts on buildings and constructions the immediate image that comes to mind is the damage at the foot of walls, resulting from the rising damp from the solution of the melted snow and salts, as shown in *Figure 1*.

Experience has shown that to this deterioration mechanism two other direct contamination processes have to be added. The first one occurs in damp winter conditions and affects buildings located along high traffic roads, where topography contributes to accumulate the salt containing melted snow and that vehicular traffic and snow clearance vehicles splash on to the building walls or disperse into the air so that they enter directly at a certain height (*Figure 2* left). The second process occurs during dryer winter periods when the excess deicing salts applied recrystallize and accumulate at the base of buildings (*Figure 2* right).

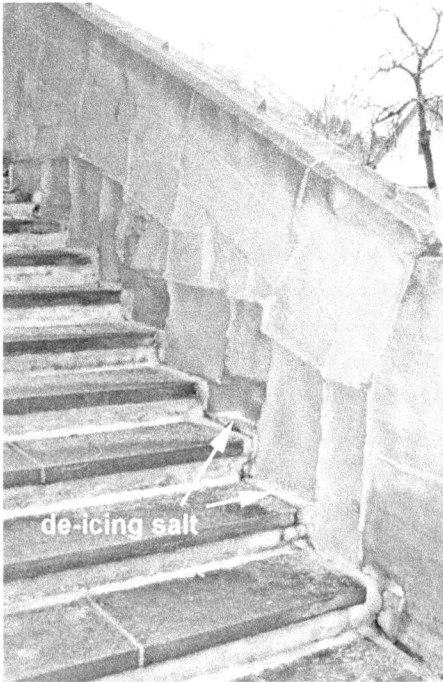

Figure 1: Lausanne (canton de Vaud, CH), Pierre Viret stairs (20.01.2010). Deicing salts are spread lavishly on the practically non-porous gneiss, and the salt and melted snow mixture accumulates at the base of the porous molasse-sandstone bridge wall that promptly powders and disaggregates.

In certain streets, where regular and intense winds are prevalent, the salt grains on the ground can be mobilized and suspended in the air, as well as thrown against the façades by whirl winds. If the building surfaces are moist or rough, the salts will be "attached" to them and deterioration will eventually occur *(Figure 3)*.

Following the above mentioned processes, it is logical to ask what will be the eventual impact of the presence of these deicing salts in the air on the conservation of buildings and monuments. Recent studies regarding the composition of fine particular matter suspended in urban air have shown the recurrent presence of NaCl.[7,18,19] In sea-side areas, it is obvious that most of the salt present can be attributed to marine spray and fogs, however, this cannot be applicable to inland areas. For example, in Putaud et al.[20] (see figure 3, p. 2584), the measured annual average values of PM10 for NaCl (from January 1998 to March 1999), is referred to as "sea salt", in both rural areas, such as Chaumont, and urban areas, such as Basel and Zurich, as well as at street level (kerbside)

Figure 2: Left: Belfaux (canton de Fribourg, CH), north façade of the Lanthen-Heid manor (1526).The damages that affect the render and the underlying molasse-sandstone are mainly the result of splashing traffic along this street (3.09.2008). Right: Lausanne (canton de Vaud, CH), rue Saint-Etienne (20.01.2010). During dry winter cycles the residues of recrystallized deicing salts can be seen at the foot of façades that can either be mobilized and suspended in the air or enter the material when dissolved in water via hygroscopicity, rain, or more snow.

Figure 3: Lausanne (canton de Vaud, CH), rue de la Barre, Château Saint-Maire (10.04.2012). The sanding and disaggregation that affects the base of the building and up to 2 m in some areas is mainly due to the presence of NaCl.

in Bern. The PM10 immission of this city is strongly influenced by traffic and showed a value of about 1.5 µg/m³ while the other sites range between 0.5 µg/m³ for the urban sites to 0.2 µg/m³ for the rural site of Chaumont

The chemical composition and the quantity of the coarser fractions in air pollution reflect the contribution of mineral dusts put into suspension by vehicular traffic. These include remaining deicing salts and explain the relatively high concentration of NaCl, as clearly stated by Gianini et al.[21, p. 104]

4. Discussion

There is no question as to the contribution of deicing salts to the deterioration of our architectural heritage as well as to the environment. But it is also clear that deicing salts, as well as antifreezing formulations for airports and aircraft are necessary and have to be used to avoid traffic accidents, disruptions in the economy, and taking into account that the negative impact of closing roads far exceeds the cost of snow and ice removal.[22,23] While airport and aircraft are allowed the use of some formulations that cannot be used freely elsewhere, no such proviso has been considered for our cultural heritage.

Another problem is the introduction of additional hygroscopic salts into salt-containing structures, as the hygroscopic salts will mobilize and activate those present within the structure so that they will migrate and eventually crystallize in other areas. This has been observed on monuments, where salt efflorescence changes places after events introducing moisture and new salts, as described by Rolland et al.[24] who called it the "transporting brine hypothesis". But the problem is now increased with the use of products from the food industry[25] that are added to deicing salts. No one has as yet raised questions regarding their long term effect when retained together with the deicing salts in structures. For example, sugar beet syrup is hygroscopic, with a DRH ~60%[26], so supposing that this, together with NaCl is taken up into the structure, in theory, less salt should crystallize out. Of course, other problems could set in with the introduction of organic materials, such as biocolonization, and deterioration of the stone matrix, should it have some solubility in water. But this is an area that still needs to be studied formally.

Another topic that requires evaluation is whether applying a polymer surface overlay system, such as SafeLane® having an epoxy bonding agent, a special aggregate capable of storing a deicer that automatically releases before frost and ice can adhere to it[4]. The question is how fast is this overlay system worn down by the traffic and how much epoxy is released into the environment.

Conclusions

It is clear that to maintain the economic system that has been developed since the industrial revolution, many changes have been made in the environment and its ecology. But these changes take time to implement, so it is important that, for those of us concerned with the conservation of our architectural heritage, we should point out the problems that threaten it to raise the awareness of the general public so eventually some actions can be taken to protect them. As every building and its situation are different, solutions adapted to the individual case have to be found. Often it is possible to use gravel and no deicing salt at the proximity of the walls of historic buildings, or it is possible to cover the basis of the walls, e.g., with boards to keep salt containing snow away from the walls, similar to the covers used to protect statues during winter. In some instances, a French-drain could be installed by the wall where deicing salts are applied, so that the melted snow and salt will get trapped in it, or as in Switzerland, many staircases are half closed and only a small part kept free of ice or snow. Ingenuity has been the mark of humankind, and it is time for it to come to the rescue should we want to preserve our architectural heritage.

References

[1] Chemikalien-Risikoreduktions-Verordnung, Anhang 2.7, 2017. https://www.admin.ch/opc/de/classified-compilation/20021520/index.html#app27.

[2] Heisses Thema Eis und Schnee, Die Umweltberatung, 2011. http://images.umweltberatung.at/htm/eis-und-schnee-ratgeber-wasser.pdf.

[3] Brunning A., Periodic Graphics: Deicers and Antifreeze. C&E News, (93:2), (2015), 30. http://cen.acs.org/articles/93/i2/Periodic-Graphics-Deicers-Antifreeze.html.

[4] Cargill, Safelane Surface Overlay, 2017. http://safecote.com/project/anti-icing-surface-overlay-safelane/, https://www.cargill.com/industrial/winter-road-maintenance/safelane-surface-overlay.

[5] Mansfeldt T., Rennert T., Götzfried F., Eisencyankomplex-Gehalte in nordrhein-westfälischen Strassenrandböden nach dem schnee-reichen Winter 2009-10 Strasse und Autobahn, (6), (2011), 389-393.

[6] Wrochna M., Malecka-Przybysz M., Gawronska H., Effect of road de-icing salts with anti-corrosion agents on selected plant species, Acta Sci. Pol. Hortorum Cultus, (9:4), (2010), 171-182.

[7] Fischel, M., Evaluation of selected deicers based on a review of the literature, Report No. CDOT-DTD-R-2001-15, Colorado Department of Transportation Research Branch, 2001. https://www.codot.gov/programs/research/pdfs/2001/deicers.pdf.

[8] Erhart E., Hart W., Effects of potassium carbonate as an alternative de-icer on ground vegetation and soil. Annals of Applied Biology, (136:3), (2000), 281-289.

[9] Meesen J. H., Urea, Ullmann's Encyclopedia of Industrial Chemistry, Vol. 37, Wiley-VCH Verlag GmbH & Co. KGaA, Weinheim, 2012, 657-695.

[10] Ground Support Magazine, Snow and Ice Control Chemicals for Airports Operations, 2005. http://www.cryotech.com/snow-and-ice-control-chemicals-for-airports-operations.

[11] MeltSnow.com, Urea, Material Safety Data Sheet, 2010. http://meltsnow.com/development2010/wp-content/uploads/2010/08/UREA_MSDS_MSWS.pdf.

[12] Peters Chemical Company, Breaking the Ice, 2006. http://www.peterschemical.com/break-the-ice-comparison-of-ice-melting-chemicals/.

[13] Alexandrova A.N., Jorgensen W.L., Why urea eliminates ammonia rather than hydrolyzes in aqueous solution, J. Phys. Chem. B, (111:4), (2007), 720–730.

[14] Durickovic I., Thiébault L., Bourson P., Kauffmann T., Marchetti M., Spectroscopic characterization of urea aqueous solutions: Experimental phase diagram of urea-water binary system. Applied Spectroscopy, (67:10), (2013), 1205-1209.

[15] MacDonagh L.P., Deicing with beet juice, Green Infrastructure, Deeproot, January 29, 2014. http://www.deeproot.com/blog/blog-entries/deicing-with-beet-juice.

[16] Kinney T., Cities, states testing beet juice mixture on roadways, USA today, February 21, 2008. http://usatoday30.usatoday.com/weather/research/2008-02-21-beeting-ice_n.htm.

[17] Schweizerische Eidgenossenschaft, Bundesamt für Umwelt, Auftaumitteln, 2015. https://www.bafu.admin.ch/bafu/de/home/themen/chemikalien/fachinformationen/chemikalien--bestimmungen-und-verfahren/auftaumittel.html.

[18] Van Dingenen, R., Raes F., Putaud J.-P., et al., A European aerosol phenomenology - 1: Physical characteristics of particulate matter at kerbside, urban, rural and background sites in Europe, Atmospheric Environment, (38), (2004), 2561–2577.

[19] Putaud, J.-P., Van Dingenen R., Alastuey A., et al., A European aerosol phenomenology - 3: Physical and chemical characteristics of particulate matter from 60 rural, urban, and kerbside sites across Europe, Atmospheric Environment, (44), (2010), 1308-1320.

[20] Putaud, J.-P., Raes F., Van Dingenen R., et al., A European aerosol phenomenology - 2: chemical characteristics of particulate matter at kerbside, urban, rural and background sites in Europe, Atmospheric Environment, (38), (2004), 2579-2595.

[21] Gianini, M.F.D., Fischer, A., Gehrig, R., et al., Comparative source apportionment of PM10 in Switzerland for 2008/2009 and 1998/1999 by Positive Matrix Factorisation, Atmospheric Environment, (54), (2012), 149-158 doi:10.1016/j.atmosenv.2012.02.036.

[11] Kuemmel D.A., Hanbali R.M., Accident analysis of ice control operations, Report, Marquette University, 1992. http://www.trc.marquette.edu/publications/IceControl/ice-control-1992.pdf.

[23] HIS Global Insight, American Highways Users Alliance, The economic costs of disruption from a snowstorm, 2014. http://www.highways.org/wp-content/uploads/2014/02/economic-costs-of-snowstorms.pdf.

[24] Rolland O., Vergès-Belmin V., Etienne M., et al., Desalinating the Asyut dog in the Musée du Louvre, Science and art: a future for stone, Proc. 13th Int. Congress on the Deterioration and Conservation of Stone, Hughes J. and Howind T. (eds), University of the West of Scotland, Paisley, 2016, vol. II, 1247-1253. http://research- portal.uws.ac.uk/portal/files/397446/13th_ICDCS_Paisley_2016_VOL_II.pdf.

[25] Boller M., Bryner, A., Questions fréquentes sur le salage des routes, Eawag, Swiss Federal Institute of Aquatic Science and Technology, 28.03.2017.http://www.eawag.ch/fileadmin/Domain1/Forschung/Oekosysteme/Oekosysteme/FAQs_Salage_Eawag-update2016.pdf.

[26] Maudru J.E., Paxson, T.E., The Relationship of Sugar Moisture to Relative Humidity. American Society of Sugar Beet Technologists, 6th Biennial Meeting, Detroit, Michigan, 1950, 538-540. http://assbt-proceedings.org/1950Proceedings.htm.

Traffic-induced salt deposition on facades

Michael Auras*
Institut für Steinkonservierung e. V., Mainz, Germany
* auras@ifs-mainz.de

Abstract

Air pollution has been reduced significantly in Germany during the last decades. Nevertheless high pollution levels are recorded in urban environment along heavy-trafficked roads. Relevant portions of the emissions of nitrogen oxides and fine particulate matter are caused by car traffic. The change of environmental conditions has led to a change of salt deposition on facades. Actual research results show high concentrations of nitrogen oxides at historical facades but contradictory data for the deposition of nitrates. Additionally, the redispersion of de-icing salt by car traffic leads to a deposition of chlorides on facades not only in the base zone.

Keywords: air pollution, traffic emissions, salt input

1. Introduction

In the last decades air pollution in Germany has been reduced significantly. Most notably the emission of sulphur dioxide and its secondary products are reduced now to a level of less than 10 %, compared to the amount emitted in the late 1980ies. Consequently the pH value of rain increased to a level being inherent by the saturation with carbon dioxide. Also the emissions of nitrogen oxides and particulate matter could be reduced, but they still are at problematic levels, regarding the actual threshold values. High proportions of these pollutants originate from road traffic. Besides their negative health effects, their impact on materials has to be considered. This is particularly true for historic buildings, because they have been exposed to air pollutants for decades or even centuries and they also will be affected by future emissions.

The paper will report some results from recent research on the impact of the actual pollution regime on historic stone buildings.[1-3] Especially results regarding the input of salts and salt-forming substances are presented.

2. Methods

In the research project presented here, the impact of traffic-related immissions on the building materials of historic monuments in several German cities was studied. Data on traffic volume and air quality, numerical modelling of the dispersion of air pollutants in the surrounding of selected monuments, exposure experiments, as well as laboratory and onsite measurements were combined to allow for an evaluation of the traffic-related pollution and its effects on stone buildings. More details are given in a specific paper.[2]

Data from five cities, Bamberg, Würzburg, Mainz, Essen, and Munich were used to obtain an overview of the situation in these cities being characterized by different climate and traffic volume (Tab. 1).

Maps of traffic volume were drawn and superimposed to maps of the stock of cultural heritage buildings. Then the

City	Area (km2)	Inhabitants	Motor vehicles
Bamberg	54.62	70,863	44,168
Würzburg	87.63	124,577	71,342
Mainz	97.74	202,756	107,004
Essen	210.30	566,862	307,943
Munich	310.70	1,388,308	771,625

Table 1: Size, inhabitants and number of registered motor vehicles of the selected cities. Data from 2011, from [1]

Rack code	City	Building	Average traffic volume (vehicles/d)	Rack: Horizontal distance from street	Rack: Height above street
BA	Bamberg	Nonnenbrücke 1	15,000	1 m	5 m
WÜ, up	Würzburg	Residenz	15,000	4 m	up: 23 m
WÜ, low	-	-	-	-	low: 4 m
MZ, up	Mainz	-	-	up: 35 m	up: 31 m
MZ, low	-	Christuskirche	27,000	low: 24 m	low: 13 m
E	Essen	Wasserturm	30,000	2 m	6 m
M	Munich	Bayer. National-museum	52,000	5 m	3 m

Table 2: Exposure sites (2 racks at Mainz and Würzburg - up: upper rack, low: lower rack; from [1]

number of buildings positioned along main road and thus exposed to increased traffic emissions was determined.

Using data from air quality survey and meteorological stations, the dose-response functions from the MULTI ASSESS program[4] were calculated for the recession of the reference material Portland Limestone.

In each of the five cities one historic building in a heavy-trafficked road was chosen for the exposure of several pas-sive samplers *(Tab. 2)*. Special exposure racks were constructed and fixed at the facades of the buildings in traffic-near positions. In Würzburg and Mainz, racks were mounted at two different heights of the buildings. Then slabs of three stone varieties, Portland limestone, Baumber-ger sandstone, and Carrara marble were exposed in sheltered and unsheltered positions. Only in Mainz a Mank's carousel was additionally installed to allow for a comparison with literature data.

Annual emissions (total)

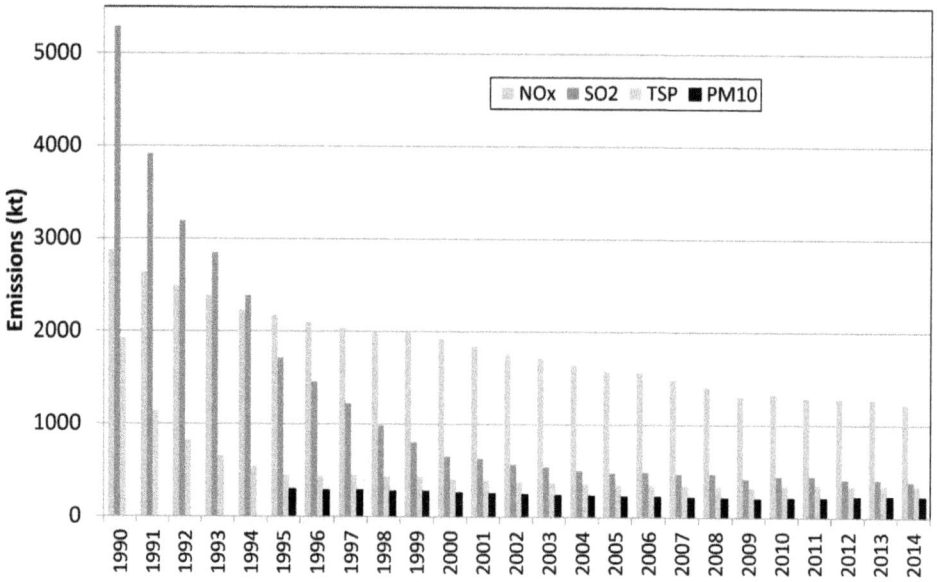

Figure 1: Development of pollutant emission from 1990 to 2010. Data: Umweltbundesamt [8] TSP: Total suspended particles.

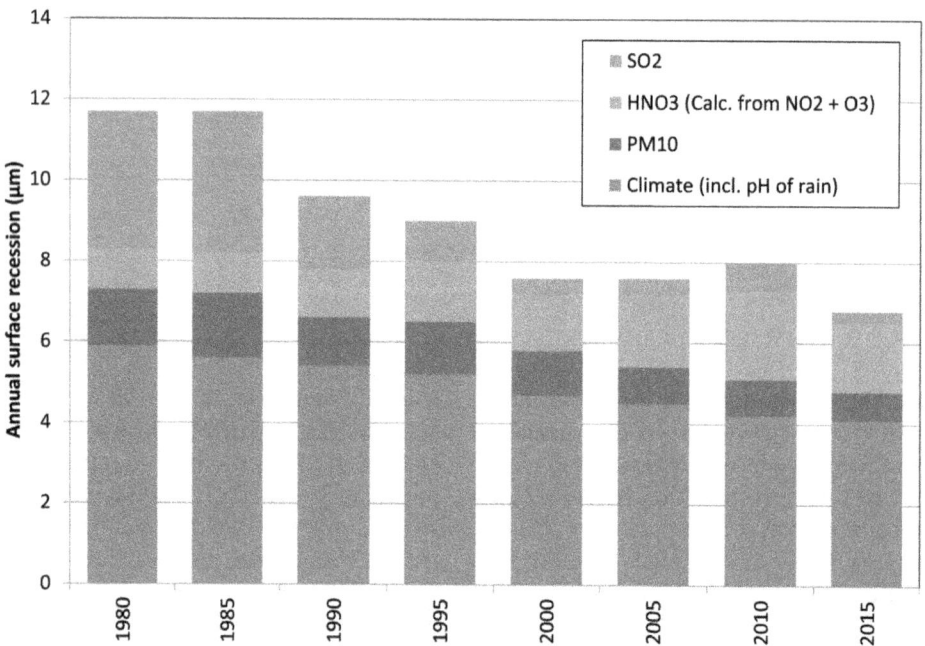

Figure 2: Calculated recession rates of Portland limestone under environmental conditions of the Karlsplatz at Munich during the last 35 years. Modified from [9]

On all racks horizontally oriented rain-sheltered passive samplers made of a boron substrate were used to collect particulate matter for ESEM studies and EDX-analyses. The samplers were replaced bi-monthly during one year. The collected particles were characterized morphologically and chemically by semi-automated individual-particle analysis with a Scanning Electron Microscope (FEI Quanta 200 Feg) combined with energy-dispersive microanalysis (EDX). Due to the large spectrum of particle sizes, 2,000 particles from each sample were analysed within two size classes (300 nm to 3 μm and > 3 μm). Image analysis showed that the coverage of the substrates was mainly due to the deposition of particles of the > 3 μm fraction. Therefore so far only the larger fraction is evaluated. Altogether approximately 42,000 particles were analysed. Each particle was checked visually and after elimination of artefacts and particles with poor images about 31,000 particles were evaluated. Based on chemical composition and morphology, the particles were classified into 14 groups, and the relative particle number, as well as the relative and absolute area coverage of each group was calculated.

Surface Active Monitors (SAM) according to Rumpel[5] are filter papers impregnated with a solution of alkali carbonate. They were used to sample sulphate, nitrate and chloride and were replaced bi-monthly.

The contents of chloride, sulphate and nitrate of the SAM filters and of the stone slabs that were exposed under sheltered conditions were analysed by ion chromatography (Dionex ICS 1000).

Diffusive samplers NO_2 and HNO_3[6,7] were provided and analysed by courtesy of the IVL Swedish Environmental Research Institute.

3. Results

Air pollution by sulphur dioxide has been diminished drastically. The emissions of nitrogen oxides and fine particulate matter were also reduced, but to a much lower extent *(Fig. 1)*.

Calculation of recession rates of the reference material Portland limestone via the MULTI ASSESS formulae[5] show a decline of recession caused by the lowering of pollutant concentrations with time. These calculations need an input of data of air pollution (SO_2, HNO_3, pH of precipitation) and climate (relative humidity). Missing HNO_3 data are calculated from the concentrations of NO_x (calculated as NO_2) and ozone, and relative humidity[5]. By setting one of these factors after another to zero, the influence of the single pollutants on the calculated recession rate can be evaluated. *Fig. 2* shows the result of such calculations with environmental and meteorological data from Munich. Although the recession rate decreases continually with time, an increasing impact of HNO_3 (more precisely NO_2 and O_3) on the weathering of limestone is observed.

The question of the role of the actual pollutant regime on historic buildings is not an academic one. In several German cities the stock of historic buildings being positioned along main roads and thus being exposed to increased traffic emissions caused by more than 5,000 vehicles daily is considerable *(Tab. 3)*.

Single particle analysis of deposited particulate matter shows a distinct increase of chloride particles in the winter months at several buildings *(Fig. 3)*. This proves, that the deposition of chloride particles is mainly due to the redispersion of de-icing salt by car traffic. Calculating annual surface coverage rates from the bi-monthly samples, it can be shown, that at Munich and at Bamberg about 10 % of a horizontal, rain-sheltered surfa-

	Listed monuments	Therefrom at main roads	Number of historic buildings along main roads				Exposed to traffic volume > 5,000 vehic./d	
Traffic volume (x10³ vehicles/d)			< 5	5-20	20-40	>40	Total	%
Bamberg (city)	1348	507	163	344	-	-	344	26
Würzburg (city)	802	318	222	87	9	-	96	12
Mainz (part of city)	866	328	167	94	57	10	161	19
Munich (part of city)	4882	1212	90	702	344	76	1122	23

Table 3: Stock of listed historical buildings and monuments exposed to elevated traffic volume. Data from [10]

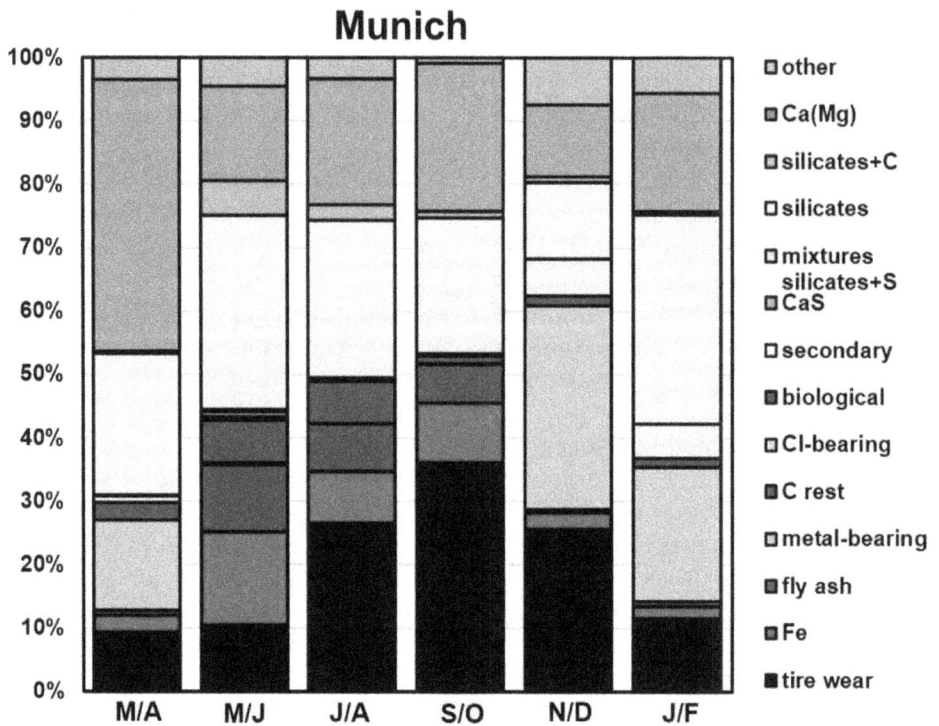

Figure 3: Variation of the composition of deposited particles > 3 μm at Munich (relative surface coverage). Letters on x-axis indicate bi-monthly sampling. From [11]

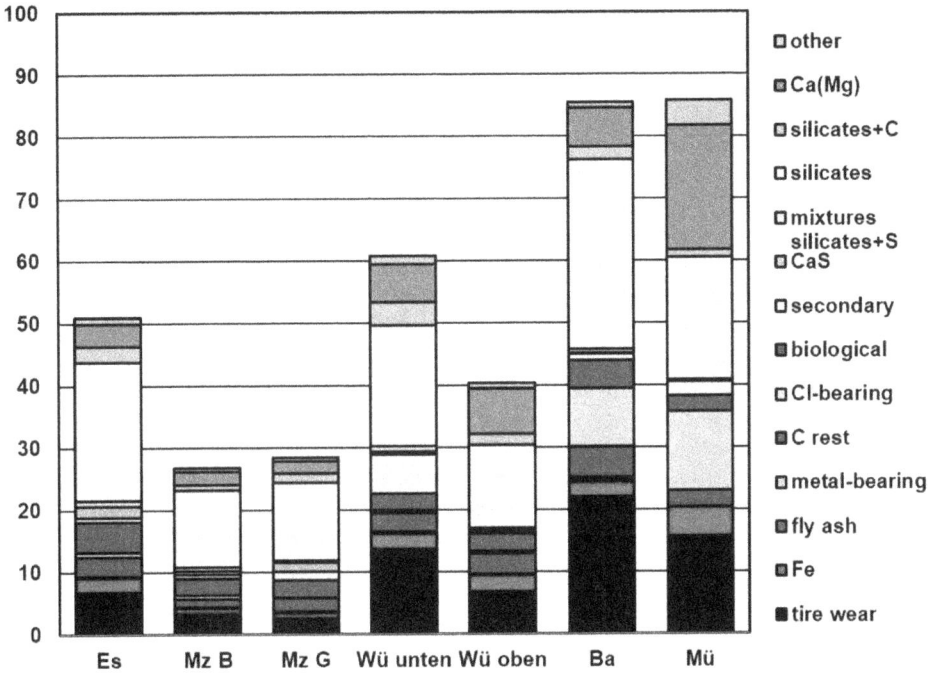

Figure 4: Calculated annual area coverage by particle deposition. From[11]

ce might be covered by chloride particles after one year *(Fig. 4)*. The deposition of redispersed chloride particles predominantly affects the lower parts of buidlings, but in two cases it was found also in remarkable height (MZ, low and WU, up). Using surface activated monitors (SAM), the deposition rates are higher for nitrate compared to sulphate. Using however natural stone samples, more sulphate than nitrate is deposited *(Fig. 5)*. It is supposed, that this is caused by the ability of the SAM-samplers to bind nitrogen oxide gases chemically, while at stone surfaces gaseous substances are only bound physically by weak adhesion forces. So the input on SAM-filters seems to depend on the concentration of NOx-gases, while the input on stone samples is assumed to be determined by the formation of HNO_3.

The measurements of HNO_3-concentrations in air directly at the facades show two trends: First there is a distinct seaso-

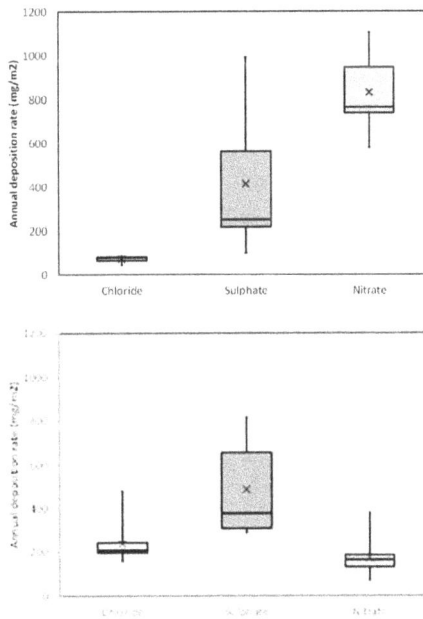

Figure 5: Annual deposition of salts on passive samplers from historic buildings situated at traffic-rich urban main roads. Left: Results for SAM-samplers. Right: Results for stone samples.

HNO₃ - Passivsammler

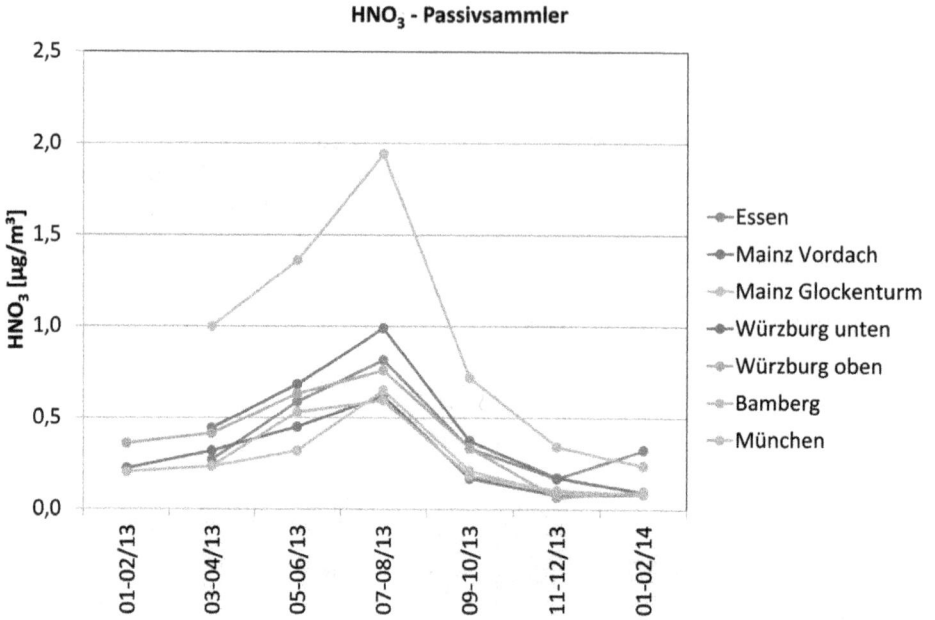

Figure 6: Results from passive sampling of HNO$_3$ at various facades. Data by courtesy of the Swedish Corrosion Institute, from [12]

HNO$_3$ (IVL)

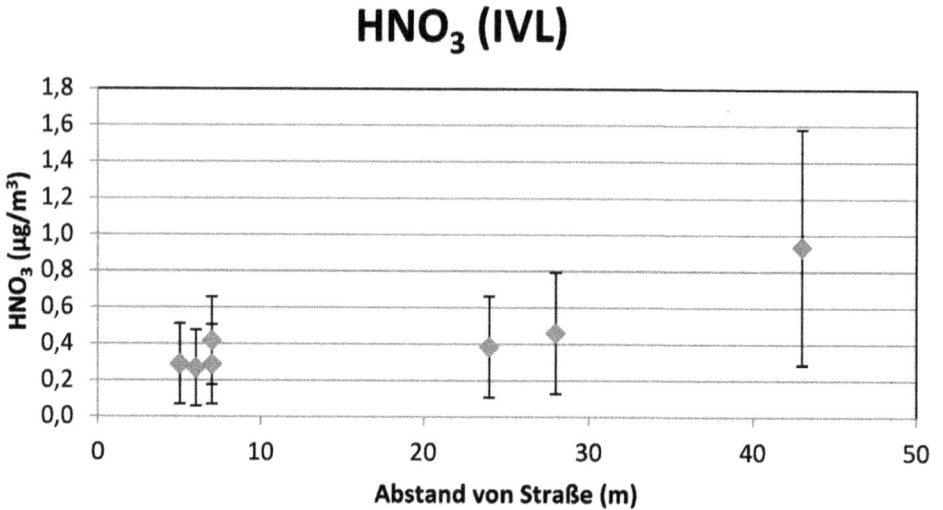

Figure 7: Mean values of the data from figure 6, ordered by the distance from the emission source. Data: see fig. 6

nal dependency with maximum values in summer *(Fig. 6)* and secondly a positive correlation with the height of the exposure racks at the building is indicated *(Fig. 7)*.

4. Discussion

While the observation from *figure 6* is due to the intensity of solar insolation, the dependency on building height could indicate that the formation of HNO_3 from NO_2 takes some time. HNO_3 seems not to be formed immediately at the source of the exhaust emissions, but in some distance. This could indicate that the reaction of nitrogen oxides with water and oxygen might to be rather slow. A consequence of this time-dependency might be a formation of HNO_3 not only at the higher parts of the buildings but also in the side roads, where it is superimposed by the dilution effect due to the minor NO_x emission rates.

5. Conclusions

Although the environmental conditions have been significantly improved over the last decades, there is still an impact of pollutants on many historic buildings. Besides accelerated soiling by tire wear and others the deposition of salts takes place on facades. Chloride and sulphate is deposited in form of fine particulate matter. Chlorides originate mainly from the redispersion of de-icing salt.

The role of nitrogen oxides and HNO_3 is not very clear up to now. The contradictionary results obtained by using different sampling materials need further research on the formation and deposition of nitrates.

6. Acknowledgements

Many thanks to Silvia Beer, Stefan Brüggerhoff, Petra Bundschuh, Thomas Dirsch, Joachim Eichhorn, Martin Ferm, Anette Hornschuch, Dirk Kirchner, Inga Kraftczyk, Karin Kraus, Martin Mach, Wolfgang Mühlschwein, Dirk Scheuvens, Klemens Seelos, Björn Seewald, Rolf Snethlage, Roger Thamm, Stefan Weinbruch and Tim Yates for their cooperation and assistance. Thanks also to all the involved authorities for the supply of data. Financial support by the Deutsche Bundesstiftung Umwelt is gratefully acknowledged.

References

[1] Auras, M., Beer, S., Bundschuh, P., Eichhorn, J., Mach, M., Scheuvens, D., Schorling, D., v. Schuhmann, J., Snethlage, R. & Weinbruch, S., Traffic-related immissions and their impact on historic buildings: implications from a pilot study at two German cities, Environmental Earth Sciences 69 (2013), 1135 - 1147.

[2] Institut für Steinkonservierung, Baudenkmäler unter dem Einfluss verkehrsbedingter Immissionen, Institut für Steinkonservierung e.V., Mainz, IFS-Bericht Nr. 49 (2015), 163 pp.

[3] Auras, M., Bundschuh, P., Eichhorn, J., Kirchner, D., Mach, M., Seewald, B., Scheuvens, D. and Snethlage, R., Traffic-induced emissions on stone buildings, In: Hughes J.J. and Howind T. (eds.), "Science and Art: A Future for Stone, proceedings of the 13th International Congress on the Deterioration and Conservation of Stone", University of the West of Scotland, Paisley (2016), 3-11.

4 Kucera, V., Model for multi-pollutant impact and assessment of threshold levels for cultural heritage, project period 1 January 2002 to 30 April 2005, Publishable Final report, www.corr-institute.se (2005), 52 pp.

5 Rumpel, K., Ein Verfahren zur Feststellung potentieller Immissionsraten: Oberflächenaktive Monitore SAM (Surface active monitoring), Monatsberichte aus dem Messstellennetz des Umweltbundesamtes 7 (1984), 2-15.

6 Ferm, M., Svanberg, P.-A., Cost-efficient techniques for urban- and background measurements of SO2 and NO2, Atmospheric Environment 32 (1998), 1377-1381.

7 Ferm, M., De Santis, F. Varotsos, C., Nitric acid measurements in connection with corrosion studies, Atmospheric Environment 39 (2005), 6664-6672.

8 Umweltbundesamt, Nationale Trendtabellen für die deutsche Berichterstattung atmosphärischer Emissionen 1990–2014 klassische Luftschadstoffe (Endstand 09.02.2012). Umweltbundesamt, Dessau, http://www.umweltbundesamt.de/sites/default/files/medien/376/dokumente/emissionsentwicklung_1990_-_2014_fuer_klassische_luftschadstoffe.xlsx (2016).

9 Bundschuh, P. & Auras, M., Anwendung von Dosis-Wirkungsbeziehungen auf das Münchener und Mainzer Untersuchungsgebiet, In: Wirkungen verkehrsbedingter Immissionen auf Baudenkmäler – Eine Pilotstudie zu den Innenstädten von Mainz und München. Institut für Steinkonservierung e.V., Mainz, Bericht Nr. 37 (2011) 47 – 53.

10 Bundschuh, P., Auras, M., Seewald, B., Snethlage, R., Verkehrsbedingte Immissionen und Belastung des städtischen Denkmalbestandes, In: Baudenkmäler unter dem Einfluss verkehrsbedingter Immissionen. Institut für Steinkonservierung e.V., Mainz, IFS-Bericht Nr. 49 (2015), 19-33.

11 Scheuvens, D., Dirsch, T., Moissl, A., Küpper, M., Weinbruch, S., Partikuläre Schadstoffe an Baudenkmälern. In: Baudenkmäler unter dem Einfluss verkehrsbedingter Immissionen, Institut für Steinkonservierung e.V., Mainz, IFS-Bericht Nr. 49 (2015), 79-99.

12 Bundschuh, P., Auras, M., Kirchner, D., Scheuvens, D., Seelos, K., Expositionsprogramm zur Wirkung verkehrsbedingter Immissionen auf Natursteinoberflächen, In: Baudenkmäler unter dem Einfluss verkehrsbedingter Immissionen, Institut für Steinkonservierung e.V., Mainz, IFS-Bericht Nr. 49 (2015), 53-77.

Wick action in cultural heritage

Leo Pel* and Raheleh Pishkari
Eindhoven University of Technology, The Netherlands
* l.pel@tue.nl

Abstract

Salts crystallization is one of the main degradation mechanisms of historical objects, e. g., masonry. In this study we looked at a special case often encountered in marine environment, i.e., wick action. This is a steady state situation in which one end of an object is continuously absorbing a salt solution, e. g., sea water, whereas at the same time at other side there is continuous drying. As a result there will be a continuous flux of ions towards the drying surface and the concentration at the drying surface will slowly increase, resulting eventually in crystallization. In this study we looked at wick action for a 1 m NaCl solution using a biomicritic limestone from Sardinia, which is found in many cultural heritage objects. To measure both moisture and salt content simultaneously, we have used a specially designed Nuclear Magnetic Resonance (NMR) set-up. The wick action experiment was performed for over 40 days. The results show that the concentration over 40 days slowly increases at the top until the saturation concentration is reached. It is shown that the concentration profiles can be modelled by a simple analytic solution of the advection-diffusion equation describing the ion transport.

Keywords: Wick action, drying, salt crystallization, Nuclear Magnetic Resonance

1. Introduction

Salts crystallization is one of the main degradation mechanisms of historical heritage objects, such as masonry. Moisture penetration can advect dissolved ions along with it, into a porous material. The salt ions can accumulate and crystallize in the pores and can as a result produce cracks due to crystallization pressure. To get a better insight into the salt transport mechanism we picked up on a special case when a porous material is in contact with salt solution on one side and, at the same time, exposed to drying conditions on the opposite side. This situation is often encountered in cultural heritage in marine environment.

A well-known example is the housing along the canals in the historic city of Venice. Here there is a continuous supply of fresh sea water as a source of salt which is absorbed, whereas the top of the masonry is drying, giving rise to continuous salt damage. This continuous transport of a salt solution, combined with drying, is often referred to as wick action *(see also Fig. 1)*.

During wick action salt ions will move by two mechanisms, advection and diffusion.[1,2] Advection is the process of ions moving along with the moisture flow, whereas diffusion is dependent on the concentration gradient, i. e., diffusion tries to level off any concentration gradient. As a result the net ion flux will be a competition of these two processes. Due to the continuous flow of salt ions towards the drying face there will be a continuous accumulation and as soon as

Evaporation

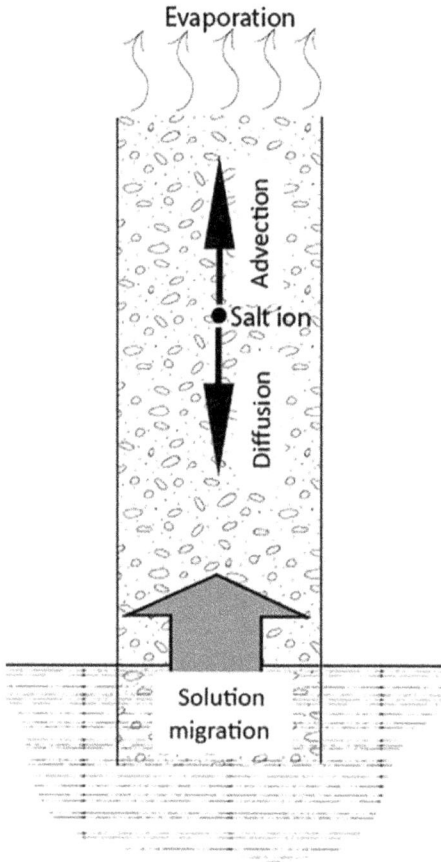

Figure 1: A schematic representation of wick action: one side of an object is absorbing a salt solution, whereas at the same time other side is drying. As a result there will be a continuous flow of ion towards the drying surface.

the maximum solubility limit has been reached, there will be crystallization, either inside the materials (subflorescence) or outside (efflorescence). In this study we have focused on wick action in the stable situation where the drying front stays at the top of the sample, i.e., the surface where the evaporation takes place. We have used Nuclear Magnetic Resonance (NMR) to measure both moisture and ion profiles non-destructively and quasi-simultaneously during wick action experiments. These experimental results have been compared to a simple analytic solution of the advection-diffusion equation describing the ion transport.

2. Theory

In order to get a better understanding of the salt concentration profiles which will develop during wick action we can have a look at combined ion and moisture transport. As long as the sample stays saturated during the wick action, the ion transport can be described by an advection-diffusion equation as given by:

$$\frac{\partial c}{\partial t} = \frac{\partial}{\partial x}\left(D\frac{\partial c}{\partial x} + uc\right) \tag{1}$$

where c [m] is ion concentration, which will be a function of both the time t (s) and position x (m), D [m²/s] is the effective diffusion coefficient of the ions in the porous material and u [m/s] is the macroscopic liquid velocity of the liquid in the porous material, i.e, the Darcy speed. This equation will be valid as long as there is no crystallization taking place else a sink term has to be added to the right-hand side of this equation.

Therefore the right hand side is describing the total ion flux, consisting of diffusion, whereas the second part describes the advection of the ions along with the liquid flow. The competition of these two can be characterized with a dimensionless number, i.e., the Peclet number. In this case, based on Eq 1. it can be defined as:

$$Pe = \frac{uL}{D} \tag{2}$$

where L is a so-called characteristic length scale which in this case can be chosen as the length of the sample. In the case Pe>1 advection will be dominant and there will be concentration gradient. Whereas in the case of Pe<1, diffusion is dominant and we expect a homogenous distribution of salt. This Pe-number was also found to be very useful to give an indication of the effect of poulticing.[3]

In the case of drying as studied here, the boundary condition for ions at the drying surface is given by a no flux boundary, i.e., q=0, and hence:

$$D\frac{\partial c}{\partial x} + uc = 0 \qquad (3)$$

As in our experiment we have a constant boundary condition we can assume that in first order the liquid flow is constant, i.e., u= constant. As also the concentration of the absorbed salt solution is constant, i.e., c= co , we can in this case solve the differential equation giving the concentration c(x,t) as a function of time and position, i.e.;

$$c(x,t) = a(t)e^{-\frac{u}{D}x} + c_o \qquad (4)$$

Where a(t) is a constant which is a function time. Hence this solution tells us that the concentration in the sample can be described by an exponential decay. In addition, it indicates that characteristic width of the salt concentration peak given by 4D/u, is determined by the ratio of the liquid velocity and the diffusivity.

3. Material and methods

For this study we have looked at a biomicritic limestone from Sardinia which has been used in many objects there. Is has a porosity of 0.34, a pore size with the maximum distribution around 1.6 μm and has only very minor magnetic impurities. For the experiments we used cylindrical samples with a diameter of 20 mm and length of 100 mm, which were drilled out of a larger block. After rinsing the sample in clean water, the sample was dried at 400°C until constant weight. The experiments were started with samples which were vacuum saturated with distilled water. In order to perform 1D experiment the specimens were isolated on the sides with an epoxy coating. The experimental setup is given in *Figure 2*.

The sample holder is a Teflon cylinder with a reservoir at the bottom and air flow inlet and outlet at the top to provide drying at the top of the sample. The bottom of the sample is in contact with a 1m NaCl solution. In order to maintain a constant level the pump is controlled by using an electrical sensor (max level fluctuations are in the order of 3 mm). To induce drying, the top of the sample is

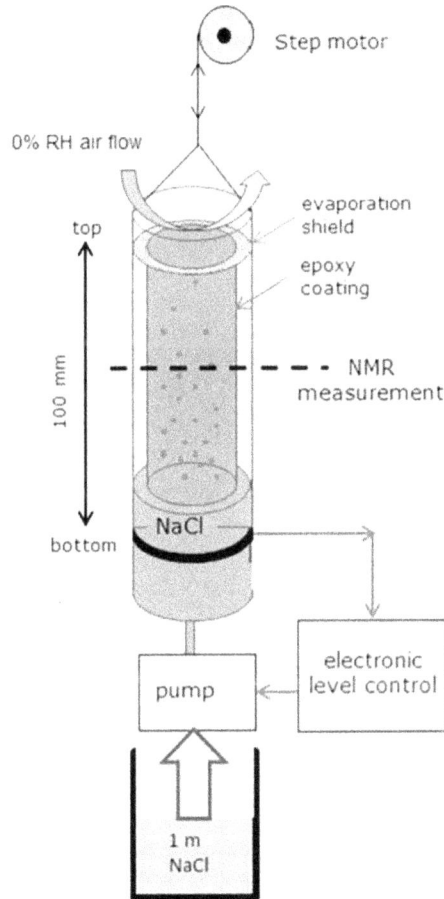

Figure 2: A schematic representation of the setup for measuring the wick action using NMR. With an electronic level control and a pump the reservoir level is kept constant.

exposed to a constant air flow of 10 L/m at approximately 0% relative humidity. An plastic evaporation shield has been added as to separate the drying and absorption part, i.e., to limit the influence of the drying on the bath concentration. Both the moisture and Na-content are measured using Nuclear Magnetic Resonance (NMR).[4,5] Using NMR we are able to non-destructive and quantitatively measure the moisture and Na-content at a fixed position, with a 1D-resolution in the order of 2 mm for hydrogen and 8 mm for Na. Hence every point represents an average over 2 to 8 mm. A stepper motor was used to move the sample holder vertically inside the NMR to be able to measure the moisture and Na profiles over the complete sample. Measuring both a

moisture and Na-profile takes in the order of 3 hours. The total experiment lasted up to 40 days.

4. Experimental results

The NMR signal was collected from cross sections of the sample in 1.27 mm steps along the axis of the cylindrical sample. Both the moisture and Na-concentration profiles were measured every 3 hours for 40 days. The resulting Na concentration profiles for every 4 days are given in *figure 3*.

At position x=100 the sample is in contact with the reservoir of 1 molal NaCl solution, whereas at x=0 the sample is drying. As can be seen the concentration

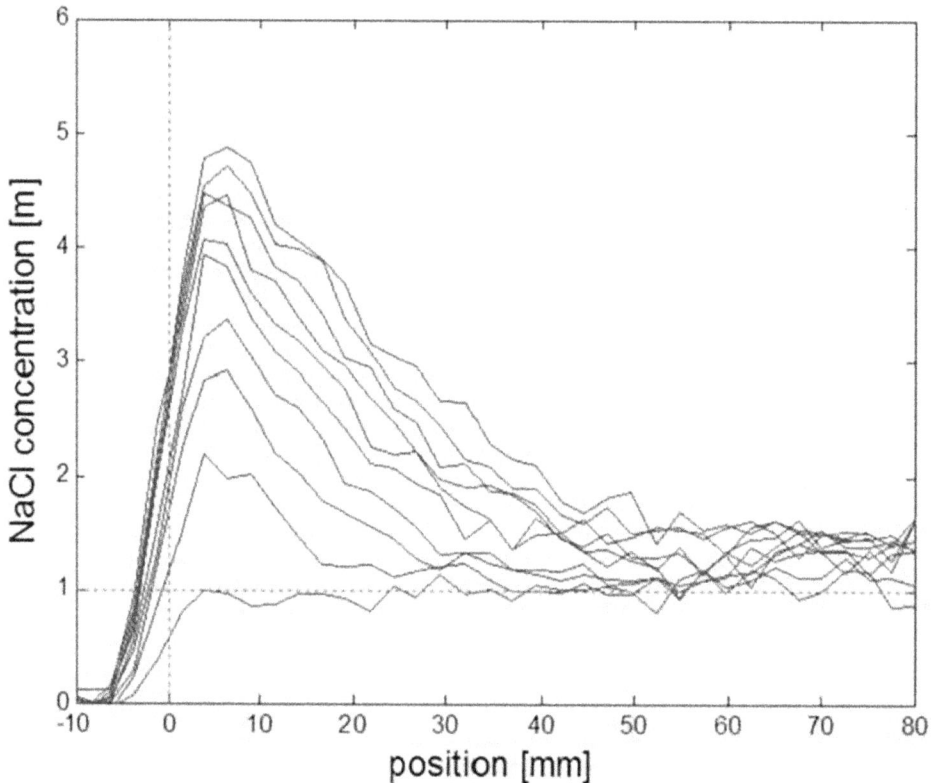

Figure 3: The measured Na-concentration profiles using NMR during wick action of a sample of 100 mm in length and a reservoir of 1m at x=100 mm. The profiles are given for every 4 days for a total time of 40 days.

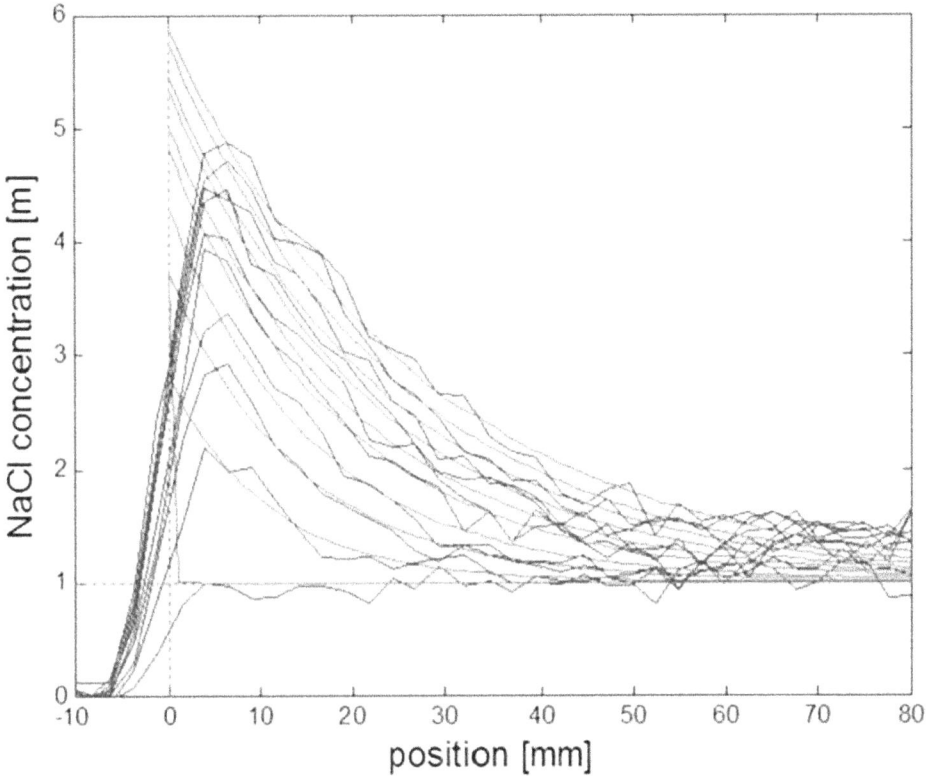

Figure 4: The measured Na-concentration profiles during wick action of a sample of 100 mm in length and a reservoir of 1m. The profiles are given for every 4 days for a total time of 40 days. The smooth curves represent fits of the model to each individual profile.

at the top is slowly increasing with time indicating Pe>1, whereas at the bottom of the sample the concentration remains almost constant reflecting the constant concentration of the reservoir.

However at the top of the sample the concentration does not raise to 6.1 molal as to be expected. Here one has to take into account the 1D resolution of 8 mm, i.e., each point represents the average over 8 mm. Indeed looking in more detail one can see the concentration, which is measured every 1.27 mm, is slowly rising from almost zero 0 molal at -4 mm representing a point outside of the sample, to 5 molal at 4 mm inside of the sample. Hence the 1D resolution is smoothing off the real concentration profile within the sample, and the maximum concentrati-

on reached within the sample, i.e., the top, will be higher.

In order to correct for this problem, we have fitted the model as derived (see Eq. 4) to each individual profile. The results are given in figure 4. As can be seen each profile can be approximated well by an exponential function. We can also see that the interpolation with x=0 indicates that the maximum salt concentration of 6 molal is reached after 40 days, which could also been seen from crystals forming at the top of the sample.

5. Conclusions

Wick action in combination with evaporation at the top of a sample can be described by a simple analytic model, which only takes into account the liquid speed and the diffusivity. The measured concentration profiles match well with a simplified model. Also NMR has been shown to be an effective method to measure the concentration profiles during wick action over a long time and non-destructively.

References

[1] Y.T Puyate and.C.J Lawrence, Effect of solute parameters on wick action in concrete. Chemical Engineering Science (54) (1999) 4257-4265.

[2] Y.T Puyate and.C.J Lawrence, Steady state solutions for chloride distribution due to wick action in concrete, Chemical Engineering Science (55) (2000) 3329-3334.

[3] L. Pel, A. Sawdy and V. Voronina, Physical principles and efficiency of salt extraction by poulticing, Journal of Cultural Heritage (11) (2010) 59–67.

[4] L. Pel, P.A.J. Donkers, K. Kopinga and J.J. Noijen, 1H, 23Na and 35Cl imaging in cementitious materials with NMR, Appl Magn Reson (47) (2016) 265–276.

[5] L. Pel and H.P. Huinink, Building Materials Studied by MRI, Encyclopedia of Magnetic Resonance, by John Wiley & Sons, Ltd 2012 (DOI: 10.1002/9780470034590. emrstm1294).

Measurement techniques and experimental studies

A preliminary study on dynamic measurement of salt crystallization and deliquescence on a porous material surface using optical microscope

Masaru Abuku[1]*, D. Ogura[2] and S. Hokoi[2]
[1] Faculty of Architecture, Kindai University, Japan
[2] Graduate School of Engineering, Kyoto University, Japan
* abuku@arch.kindai.ac.jp

Abstract

This paper reports on our preliminary attempt to perform dynamic laboratory measurements of salt crystallization and subsequent deliquescence on a surface of a porous material specimen (autoclaved aerated concrete as a model material) by using a high-resolution digital optical microscope that allows generating three-dimensional images. A specimen that is initially filled with a NaCl solution is dried in a room at ~26°C and ~40% RH for 14 days and is then wetted in a desiccator at ~95% RH during the next 14 days. A temporal change of the spatial distribution of the volume occupied by salt crystals on the specimen surface is determined based on the contrast between two microscopic images successively taken with a certain interval of time. The mass of the specimen is also measured simultaneously to determine the rate of water evaporation/absorption. The results show that salt crystallization during drying is not spatially uniform throughout the specimen surface, which likely depends on the spatial distribution of the pore structure in the specimen, and salt crystal deliquesces faster near the edges of the specimen than in the centre, which indicates a stronger dependency of salt deliquescence on the geometry of the specimen than on the inner pore structure of the specimen.

Keywords: light microscopy, efflorescence, sodium chloride, absorption/desorption

1. Introduction

Cycles of crystallization and deliquescence of salt can be often observed on a surface of building and heritage materials such as stones, bricks, mortar, etc. In general, salt crystals grow in cold seasons and disappear in warm seasons, often as a result of hygrothermal responses of the materials to changes of ambient air temperature and humidity, rain, solar radiation and so on. To physically understand such phenomena and propose a sufficient countermeasure, it could be useful to quantitatively determine the rates of water evaporation and the subsequent salt crystallization and those of water absorption and the subsequent salt deliquescence on the surface of the material.

Water and salt transport in porous building materials has been steadily and rather well studied over the last years. This phenomenon can be characterized by two main material properties: the sorption isotherm and the liquid water conductivity, both of which are dependent on the salt concentration, water content and temperature. Apart from water and salt transport, crystallization of salt in pores of a porous material or at/near the surface of the wall still needs fundamental studies for quantification. Regardless of difficulty of measurement, recent non-destructive measurement techniques enabled to visualize the volume of salt in pores of the material.[1-3] Another approach to quantify the amount of salt in a porous material is to measure the pore size distribution of the material, i.e. to show that although the total pore volume is basically reduced due to the presence of salt crystals, the specific pore volume for some pore sizes can increase.[4] This means that salt crystals can provide a new pore structure that exhibits different water and salt transport properties in the material.[5-6]

Compared to crystallization in pores, less attention has been paid to crystallization at surfaces of porous materials in the past[7-15], because in-pore crystallization is more harmful for damage of porous materials. However, recently, the work of Veran-Tissoires[13-15] using 1 mm glass beads gave an important motivation to understand the phenomena at the material surface, indicating that salt structures forming efflorescence are porous and transfer salt solution by capillarity, influencing the water evaporation rate at porous material surfaces, which is called the screening effect. Rad and Shokri[10] also studied salt precipitation on sand grains with an average particle size of 0.48 mm and the relation of evaporation with salt precipitation. Rad et al[11] further conducted three-dimensional imaging of using x-ray microtomography, to understand more in detail the relation of the pore sizes and the initial salt concentration with the dynamics and patterns of salt precipitation during drying. Our current study focuses on salt on a material surface that crystallizes during drying and furthermore deliquesces during subsequent re-wetting. It reports on our preliminary attempt to perform dynamic laboratory measurements of salt crystallization and subsequent deliquescence on a surface of a porous material specimen (autoclaved aerated concrete as a model material) by using a high-resolution digital optical microscope that allows generating surface three-dimensional images. A temporal change of the height of salt crystals on the specimen surface is determined based on contrast between two microscopic images successively taken with a certain interval of time.

2. Methodology

Dynamic laboratory measurements of salt crystallization and subsequent deliquescence on a surface of a porous material specimen (Autoclaved Aerated Concrete (AAC) as a model material) are carried out with use of a high-resolution digital optical microscope that allows generating three-dimensional images. The spatial resolution at the top surface of the specimen is horizontally 10 µm x 10 µm; vertically the stage moves with a resolution of 0.1 µm, to calculate the height of the object (the specimen surface or precipitated salt) based on focus. The dry bulk density of the specimen is 507 kg/m^3. The moisture transport properties of the same type of AAC as this specimen is given in e.g.[16] The mass of the specimen as well as the temperature and humidity of the ambient air is also measured simultaneously. A temporal change of the height h of salt crystals on the specimen surface is determined based on contrast between two microscopic images successively taken with a certain interval of time.

The following steps are taken:

(1) A cubic specimen of 27 cm^3 is dried completely to determine the porosity and dry density of the specimen. Then, the surfaces of the specimen except the top and bottom surfaces are made vapor tight.

(2) The specimen is immersed in pure water and then a drying test is conducted to know how fast the specimen is dried in the laboratory. The temporal change of the mass of the specimen is measured.

(3) The specimen is again completely dried and then immersed in saturated NaCl aqueous solution with a salt concentration of 0.1 kg/kg. Note that in this paper, the salt concentration is

defined by the mass of salt divided by the mass of the solution. Afterwards, the bottom of the specimen is made vapor tight; only the top of the specimen is open to vapor transfer.

(4) The specimen is stored at a constant temperature (~26°C) and humidity (~40% RH). The height h of salt crystals on the specimen surface is recorded with a certain interval of time using the microscope that can allow to generate three-dimensional images. The weight of the specimen is also measured to determine the evaporation rate. The drying process is continued for approximately 14 days.

(5) The specimen is stored at almost the same temperature but at a humidity of ~95% RH. Salt crystals on the surface and the specimen's weight are recorded, by the latter of which the rate of water vapor absorption is determined. The wetting process is continued for approximately 14 days.

In the following sections, results are discussed based on the properties of the specimen and salt at the given temperature and humidity conditions.

3. Results and discussion

Figure 1 shows photographs of the center (2.4 cm x 2.4 cm) of the top surface of the specimen taken at t = 0.71, 1.79, 2.79 and 6.79 days during the drying stage. To omit the effect of the edges of the top surface on three-dimensional image analysis, only a square of 2.4 cm x 2.4 cm is taken though the top surface area of the specimen is 3 cm x 3 cm. Based on the difference of two three-dimensional images successively taken with a certain interval of time, the increment Δh of the height h of salt on the specimen surface

at the drying stage is calculated. Δh and h at t = 1.79, 2.79 and 6.79 days are illustrated in *Figure 2*. Note that h is equal to 0 at t = 1.79 days and for example, h at t = 2.79 days is the sum of h at t = 1.79 days and Δh at t = 2.79 days.

The results show that during drying, there is a large spatial distribution in the crystallization rate in a shorter time scale (~one day) as also measured by Rad et al[11]; but at the end of drying, salt is rather uniformly distributed on the surface. Although salt crystals first grow at some specific areas, it seems that crystal growth stops after some time. However, next salt crystal growth occurs at other areas, where successive water evaporation occurs due to a lower liquid water conductivity. This could be explained by the intrinsic spatial distribution of the pore structure of the specimen[11], which determines the in-pore liquid water transport and the evaporation at the surface.

Looking at e.g. *Figure 2b*, a very larger increase of h is observed at some large open pores on the specimen surface seen in e.g. *Figure 1a*. This can be attributed to the difficulty of determining the height of the surface of the material in the pores.

Photographs of the center of the top surface of the specimen taken at t = 0.47, 0.86, 1.24 and 1.86 days during the wetting stage are given in *Figure 3*. The decrement Δh of the height h of salt on the specimen surface at the wetting stage and h at t = 0.86, 1.24 and 1.86 days are illustrated in *Figure 4*. Note that h is equal to 0 at t = 1.86 days.

Although the spatial distribution of salt crystals that are deliquescing is more uniform compared to that of crystallization and independent of the in-pore structure of the specimen, it seems clear that deliquescence occurs faster near the edges of the specimen than the center of the surface due to a geometry effect on the heat and moisture transfer. Note that

(a) 0.71 days   *(b) 1.79 days*   *(c) 2.79 days*   *(d) 6.79 days*

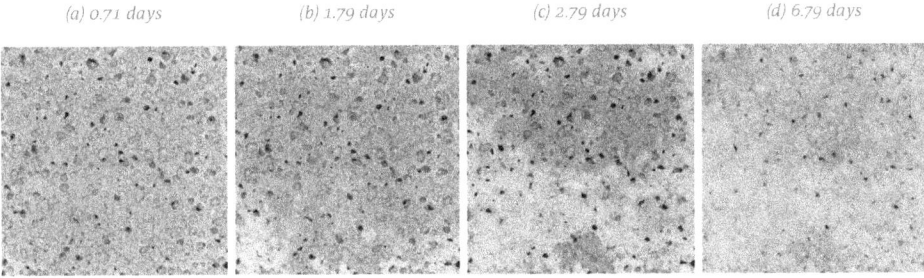

Figure 1: Photographs of the center (2.4 cm x 2.4 cm) of the top surface of the specimen taken at the drying stage

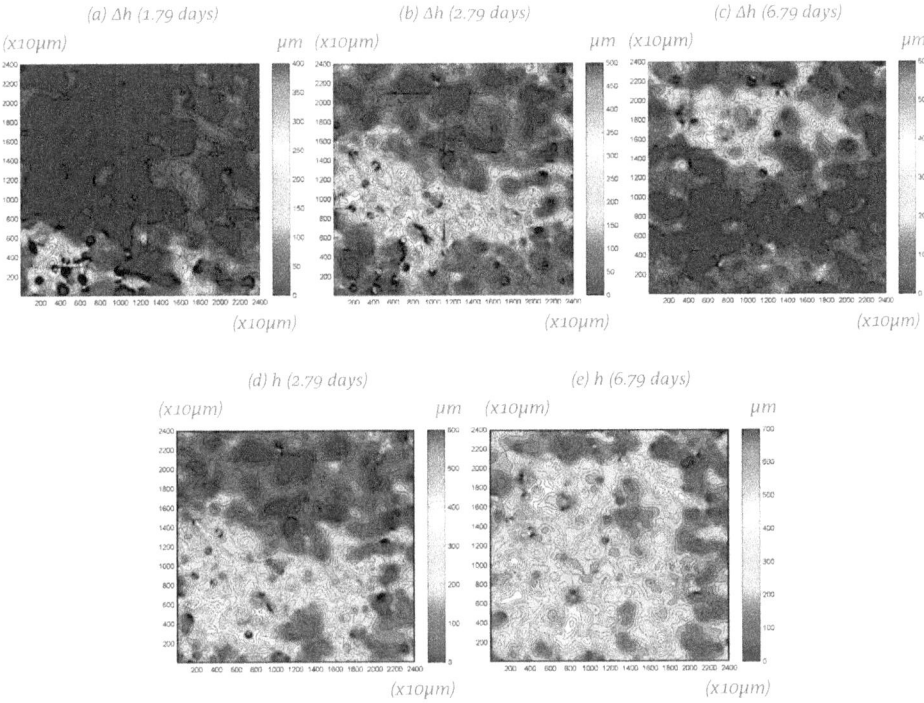

(a) Δh (1.79 days)   *(b) Δh (2.79 days)*   *(c) Δh (6.79 days)*

(d) h (2.79 days)   *(e) h (6.79 days)*

Figure 2: The spatial distribution of (a, b, c) the height change Δh of salt crystallizing on the top surface (2.4 cm x 2.4 cm) of the specimen for three consecutive drying periods and (d, e) the height h at the end of the last two periods (2.79 and 6.79 days). (a) Δh for 26 hours between 0.71 and 1.79 days; (b) Δh for 24 hours between 1.79 and 2.79 days; and (c) Δh for 96 hours between 2.79 and 6.79 days.

when the specimen absorbs moisture, the temperature of the surface gets higher due to heat of condensation. Because the change rate of the temperature is larger near the edges of the specimen, the solubility near the edges may have become more increased due to the temperature increase. However this effect is considered to be extremely small becau-se of the low temperature dependency of the solubility of NaCl.

A clear difference in time is observed between crystallization and deliquescence processes. This can be explained as follows. The specimen was completely wetted with the salt solution at the beginning of drying and salt solution transfer in the specimen took some time.

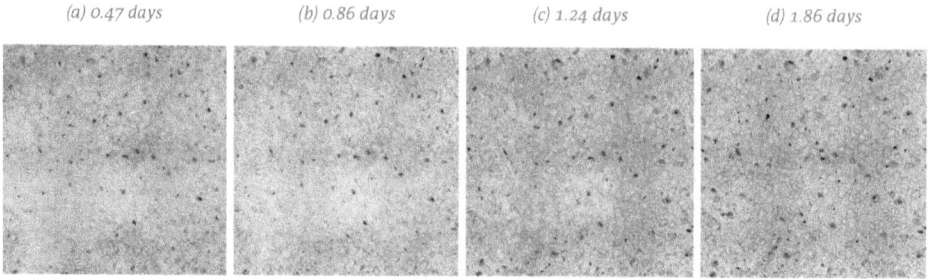

(a) 0.47 days   (b) 0.86 days   (c) 1.24 days   (d) 1.86 days

Figure 3: Photographs of the center (2.4 cm x 2.4 cm) of the top surface of the specimen taken at the wetting stage

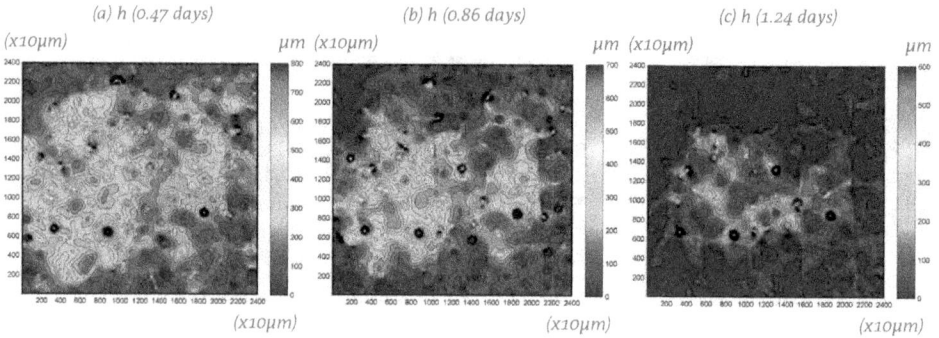

(a) h (0.47 days)   (b) h (0.86 days)   (c) h (1.24 days)

Figure 4: The spatial distribution of the height h of salt deliquescing on the top surface (2.4 cm x 2.4 cm) of the specimen at t = 0.47, 0.86 and 1.24 days of wetting periods. h is defined to be 0 at 1.86 days.

This means that the drying rate was dependent on the salt solution permeability of the specimen. As a result, crystallization continued for at least 7 days of the drying process. *Figure 5* shows the temporal change of the average moisture content w in the specimen that contained salt water, calculated from the dry mass of the specimen and the mass of water in the specimen, and the content of precipitated salt estimated based on w and the initial salt content; *Figure 6* does the same during wetting. In *Figure 5*, the data for the specimen that contained pure water are also plotted. Although a rapid change of the evaporation rate for pure water is observed at approximately 2 days, this would be mainly due to the typical transition from a capillary -driven to a vapor-driven drying and partly due to a large change of the humidity

condition (>10% RH) which was not well controlled during the experiment. The evaporation rate from the specimen with the salt solution is much lower than that from the specimen with pure water.

If wc > 0, the water content is not sufficient to dissolve all the salt crystals both in the specimen and at the surface of the specimen. In *Figure 6*, wc > 0 after 2 days indicates that salt crystals still surely remain during absorption, which can lower the vapor absorption rate.

In our current work, we only conducted measurements under limited conditions, but future work should address more different conditions to enable to develop a model that can predict cycles of crystallization and deliquescence of salt on and near the specimen surface as a function of the type of a porous material and salt, the initial salt concentration, the ambient temperature and relative humidity, and so on.

4. Conclusions

In this paper, we reported our preliminary dynamic measurements of salt crystallization and subsequent deliquescence on a surface of a porous material specimen (autoclaved aerated concrete as a model material) with use of a high-re-

solution digital optical microscope that allows generating three-dimensional images. The obtained results are analysed to quantify the spatial distributions of the rates of salt crystallization and subsequent deliquescence on the surface and their spatially averaged values. The results also showed that salt crystallization during drying is not spatially uniform on the specimen surface and depends on the spatial distribution of the pore structure in the specimen as demonstrated by Rad et al[11], and salt crystals deliquesce faster near the edges of the specimen than in the centre, which indicates a stronger dependency of salt deliquescence on the geometry of the specimen than the inner pore structure of the specimen. The obtained data and future measurements with different conditions are to be used for model development and validation.

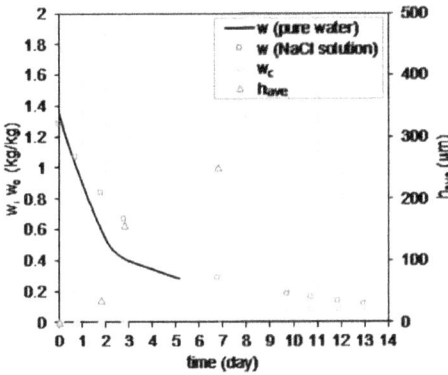

Figure 5: The average moisture content w, the estimated precipitated salt content wc and the average height have of salt on the specimen surface for the drying process of 13 days. The solid line shows the measurement result of w for pure water evaporation from the same specimen for 5 days. have is defined to be 0 at 0 day.

Acknowledgements

This work was supported by JSPS KAKENHI Grant Number 26709043. The authors would like to thank the reviewer for the valuable comments.

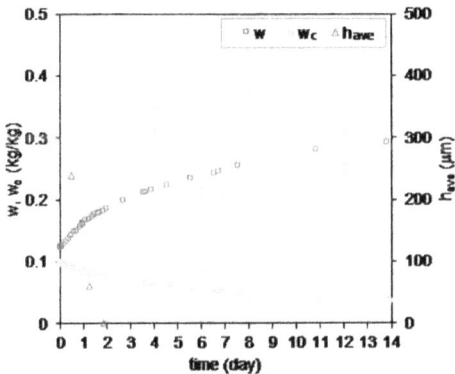

Figure 6: The average moisture content w, the estimated precipitated salt content wc and the average height have of salt on the specimen surface for the wetting process of 8 days. have is defined to be 0 at 1.86 days.

References

[1] H. Derluyn, M. Griffa, D. Mannes, I. Jerjen, J. Dewanckele, P. Vontobel, A. Sheppard, D. Derome, V. Cnudde, E. Lehmann, J. Carmeliet, Characterizing saline uptake and salt distributions in porous limestone with neutron radiography and X-ray microtpmgraphy, Journal of Building Physics 36 (2013) 353-374.

[2] I. Daher, Salt Transport Experiments in Fractured Media, PhD thesis, Imperial College London, 2016.

[3] C. J. Graham, A petrographic investigation into the durability of common replacement sandstones to the crystallisation of de-icing salts, PhD thesis, University of Glasgow, 2016.

[4] N. Aly, M. Gomez-Heras, A. Hamed, M. Alvarez de Buergo, F. Soliman, Porosity changes after different temperature regimes for a salt weathering simulation test on Mokkattam limestone (Egypt), In: Proceedings of the 3rd International Conference on Salt Weathering of Buildings and Stone Sculptures (SWBSS2014), Brussels, Belgium, October 14-16, 2014, 247-257.

[5] S. Gupta, H. P. Huinink, M. Prat, L. Pel, K. Kopinga, Paradoxical drying of a fired-clay brick due to salt crystallization, Chemical Engineering Science 109 (2014) 204-211.

[6] L. Pel, S. Gupta, Paradoxical drying due to salt crystallization; the effect of ferrocyanide, In: Proceedings of the 3rd International Conference on Salt Weathering of Buildings and Stone Sculptures (SWBSS2014), Brussels, Belgium, October 14-16, 2014, 77-87.

[7] S. Dai, H. Shin, J. C. Santamarina, Formation and development of salt crusts on soil surfaces, Acta Geotechnica 11 (2016) 1103-1109.

[8] M. Dueñas Velasco, P. Duru, M. Marcoux, M. Prat, Efflorescence fairy ring and salt centripetal colonization at the surface of a drying porous medium containing a salt solution. Impact on drying curve., In: Proceedings of the 4th European Drying Conference (EuroDrying'2013), Paris, France, October 2-4, 2013, 1-9.

[9] F. R. Janmahomed, Salt crystallization at the surface of consolidated porous media as determined by microCT imaging, Student report, Delft University of Technology, 2012.

[10] M. N. Rad, N. Shokri, Nonlinear effects of salt concentrations on evaporation from porous media, Geophysical Research Letters 39 (2012) L04403.

[11] M. N. Rad, N. Shokri, M. Sahimi, Pore-scale dynamics of salt precipitation in drying porous media, Physical Review E 88 (2013) 032404.

[12] M. N. Rad, Pore-scale investigation of salt precipitation during evaporation from porous media, PhD thesis, University of Manchester, 2014.

[13] S. Veran-Tissoires, M. Marcoux, M. Prat, Discrete salt crystallization at the surface of a porous medium, Physical Review Letters 108 (2012) 054502.

[14] S. Veran-Tissoires, M. Marcoux, M. Prat, Salt crystallization at the surface of a heterogeneous porous medium, Europhysics Letters 98 (2012) 34005.

[15] S. Veran-Tissoires, M. Prat, Evaporation of a sodium chloride solution from a saturated porous medium with efflorescence formation, Journal of Fluid Mechanics 749 (2014) 701-749.

[16] D. Ogura, S. Hokoi, T. Shimizu, H. Noguchi. Influence of hysteresis in sorption isotherm and moisture conductivity on condensation and evaporation processes, Journal of Environmental Engineering (Transactions of AIJ) 643 (2009) 1065-1074.

Diagnostics and monitoring of moisture and salt in porous materials by evanescent field dielectrometry

Cristiano Riminesi*[1] and R. Olmi[2]
[1] Institute for the Conservation and Valorization of Cultural Heritage (ICVBC), National Research Council, Firenze, Italy
[2] Institute of Applied Physics (IFAC), National Research Council, Firenze, Italy
*cristiano.riminesi@cnr.it

Abstract

Moisture and salts are the main causes of decay of porous materials, like wall paintings, stones, plasters and cement-based artefacts. Water is the ‚driving force' of decay, such as the detachment of the painted layer, the whitening of surfaces due to the crystallization of salts (efflorescence), and the weakening of the cementing binder. Early diagnostics of water content and detection of the presence of soluble salts inside the material is a key issue for understanding the degradation processes in such kind of materials and for improving their schedule maintenance. In this contribution a non-invasive microwave system based on evanescent field dielectrometry is described. The method was tested in the laboratory on moistened plaster samples, some of them containing salts at different concentrations. Measurements on water-saturated and oven-dry samples provide the basis for calibrating the instrument for on-site measurement of masonry structures, wall paintings and concrete historical buildings too. The obtained results prove the usefulness of the method as a tool for diagnostics and for monitoring the effectiveness and durability of restoring interventions.

Keywords: sub-surface investigation, moisture and salt content, dielectrometry, resonant technique, SUSI© system, plaster, stone, concrete.

1. Introduction and research aims

In this contribution the authors describe the Evanescent-Field Dielectrometry (EFD) system called SUSI©* for determining the water content and detecting the presence of soluble salts inside porous materials, and present its application. The water and salt content are key issues for the maintenance of works of art, such as wall paintings, stone artefacts, masonry and cement-based artefacts.

The SUSI© system is an electromagnetic (EM) diagnostics method based on the existence of the relation between physical characteristics, in particular water content, and the complex permittivity of hygroscopic porous materials.[1-6]

EM diagnostics methods can be usefully employed for such tasks, when applied with non-destructive modalities. Among the EM tools, dielectric spectroscopy appears as an eligible technique, i.e. the study of the spectral response of the dielectric permittivity of porous materials can help in determining their state of conservation. In particular, thanks to the dielectric contrast between water and dry material, the presence of moisture is easily detectable. Moreover, ionic conductivity – related to the presence of salts in solution – is measured as well.

A system for dielectric spectroscopy is generally too complex and too expensive to be used for real-time monitoring. Narrow-bandwidth, resonant dielectrometry is a viable solution, because it allows to realize portable, low-weight instruments independently measuring the moisture and salt content. Humidity content mea-

* Italian acronym for Sensore per la misura di Umidità e Salinità Integrato. integrated sensor for measuring humidity and salinity

surements based on EFD techniques have been developed for the diagnostics of wall paintings, masonry and cement-based materials.[7-11]

In particular, water inside building materials is known to be the main source of damage in masonry for the following reasons:

- water diffusion inside masonry and its evaporation from the surface can deteriorate the finish of the surface, and in the presence of a wall painting may degrade the painted layer or the preparation layer – detachment and loss of cohesion;

- the water diffusion can induce the growing of microorganisms on the exposed surfaces that induce biodeterioration;

- or, if the support contains salts, a high risk of whitening and salts crystallization could appear as water diffusion affects the surface.

A sub-surface diagnostic can reduce and prevent such kinds of risks. The same result could be obtained by sampling the support in depth, but this practice can't be applied on surface or material with artistic value (wall paintings, monumental stones, etc.). For example, if we want to assess the effectiveness and durability of a desalination treatment on a wall painting, we have to monitor the residual quantity of salts in the substrate. The availability of a non-invasive and portable tool for sub-surface investigation of moisture content and for detecting the presence of salts, like the SUSI system, can be of great help in the preservation and quality control of the treatment.

The SUSI system has been successfully used in the past for the following applications:

(1) – preliminary screening of state of conservation of the support - maps of moisture content and salts content can be achieved for focusing the sampling[9];

(2) – monitoring the effectiveness and durability of restoration or maintenance interventions[10];

(3) – giving insight about the absorption dynamics of water-based products in areas to be treated during the restoration phase.[8]

2. Materials and methods

The SUSI is based on EFD technique that operates in the microwave range. The diagnostics parameters achieved by this instrumentation are the moisture content (MC) and the Salinity Index (SI) related to the presence and amount of salts. A photo of the system is shown in *Figure 1*. It consists of a two-port resonant probe developed for measurements on solid materials, connected to a scalar network analyser (SNA) for measuring the sensor response (the S21 scattering parameter).[7] The system is completed by a numerical code, running on a PC.

The probe (on the bottom left side) is an open resonant cavity able to check the material without damaging it.[7] Both, the

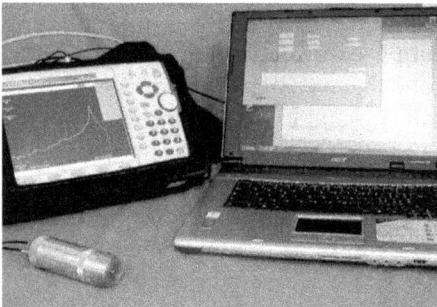

Figure 1: The SUSI© system (US Patent Specification 7,560,937 B2)

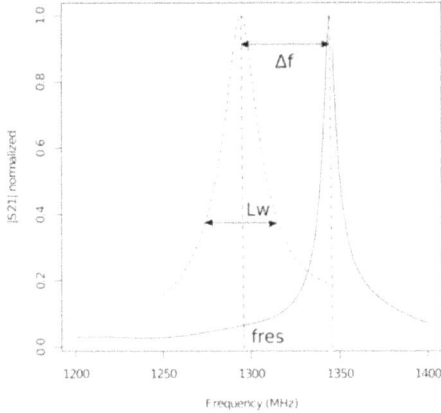

Figure 2: Principle of the SUSI[c] system

MC and SI related to the material in contact with the probe are calculated in real time by means of the numerical code.

The measurement principle is shown in *Figure 2*. The normalized S21 measured on the material (dashed line) allows to calculate the shift of the resonance frequency (Δf) with respect to the response in air (solid line), and the width L_w at 3 dB or the quality factor Q of the bell-shaped curve. The resonance frequency ranges between 0.9-1.5 GHz for stone materials. The probed volume directly depends on the size and geometry of the sensor; in this case the investigated volume is approximately a semi-spherical volume of 2 cm radius within the material.

The diagnostic parameters (MC and SI) are directly computed from the measured quantities Δf and Q, by resorting to a calibration procedure.[8] The dielectric permittivity, if needed, can be obtained offline by an inversion procedure involving the development in orthogonal modes at the coaxial opening.[7]

The MC parameter has been demonstrated to be linearly related to Δf [8]:

$$MC = \alpha\Delta f + \beta \qquad (1)$$

Therefore, the „calibration" parameters α and β relative to a material can simply be obtained by two measurements, on a dry sample and on a water-saturated one.

The salinity index SI is related to the quality factor Q of the resonant probe. Q can be simplified as the sum of the reciprocals of an „unloaded" term, depending on the geometry and on the electric/dielectric properties of the materials constituting the microstrip cavity, and of a „loaded" term depending only on the dielectric properties of the material facing the sensor. Operatively, SI is computed by using the following expression:

$$SI = \frac{f_r f_o}{2\Delta f^2}\Delta\left(\frac{1}{Q}\right) \qquad (2)$$

where fr is the resonance frequency measured on the material, fo is that in air, and $\Delta(Q^{-1})$ is the variation of the reciprocal of Q among air and material conditions.

SI has a relatively simple link with the loss tangent (tan δ) of the material. Following a definition for SI used in soil science[12], in terms of the dependence of material electrical conductivity (σ_c) on its dielectric constant (ε'_c)

$$SI = \frac{\partial\sigma_c}{\partial\varepsilon'_c} \qquad (3)$$

and using the relation between conductivity and dielectric losses, the following expression for SI, in terms of the saturation degree S of the porous media, for a given porosity φ (the water content is φS), is obtained:

$$SI = K\frac{\partial\varepsilon''_c}{\partial s}\left(\frac{\partial\varepsilon'_c}{\partial s}\right) = K\tan\delta + K\varepsilon'_c\frac{\partial\tan\delta/\partial s}{\partial\varepsilon'_c/\partial s} \qquad (4)$$

with K an arbitrary constant. Eq. (4) states that SI includes a linear term in the tan δ and a function of tan δ, ε'_c and S.

The relation between (4) and (2) can be obtained by specifying a dielectric model for the material under investigation. In terms of a simple three-phase model[13], using the so called Complex Refractive Index Method (CRIM), the composite material is represented as a mineral/water/air mixture with a complex permittivity given by:

$$\varepsilon'_c = \left[(1-\phi)\varepsilon_m^{1/2} + (1-S)\phi\varepsilon_a^{1/2} + \phi S\,\varepsilon_w^{1/2}\right] \quad (5)$$

where ε_m is the permittivity of the solid matrix, ε_a and ε_w are those of air and water, respectively. Although developed for one-dimensional layered structures, the CRIM model can be demonstrated to satisfy the Hashin & Shtrikman bounds[14] and, therefore, to be a coherent model also for more complex structures.

Figure 3 shows the dependence of SI computed by (4), with ε'_c given by (5), on the saturation degree and on the conductivity of a saline solution filling the solid matrix, for a frequency of 1 GHz and for a porosity φ = 20%. The SI is arbitrarily normalized, choosing a value for the K constant in (4), such as to assume a full-saturation value of 10 for a conductivity of 10 S/m. We observe that the SI slightly depends on the saturation degree, i.e. on

the water content, but SI values relative to different material conductivities are very well separated.

The measured SI, computed by eq. (2), can be easily demonstrated to be related to the dielectric properties as in (4), with a suitable choice of the K constant.

3. Results

3.1. Calibration procedure

The calibration procedure is adopted to relate the measured MC with a water content calculated by gravimetric approach. In this regard, on each type of porous material we can apply the following procedure:

1. Samples in ambient conditions are weighted and measured by SUSI system;

2. Samples are dried in oven at 65°C for three days, until to steady-state condition, then weighted and measured by SUSI system;

3. Samples are water saturated in a vacuum cell with inner pressure of about 2-3 kPa, in order to obtain maximum filling of pores accessible to water, then weighted and measured by SUSI system.

The SUSI measurements are taken when the sample weight reaches steady-state conditions. At least two cycles of drying/wetting should be performed on each sample. Steps 2 and 3 are functional to obtaining the calibration parameters, α and β of eq. (1), for each class of materials.

The calibration procedure for cement mortar samples is described here as an example. *Table 1* summarizes the cement mortar samples made in laboratory and used in this study.

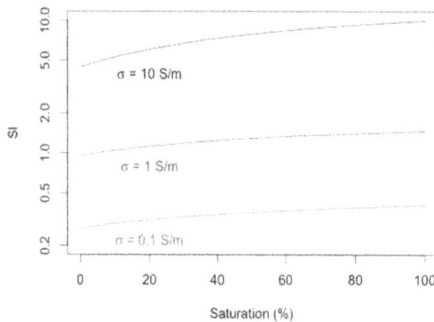

Figure 3: SI versus the saturation degree and the water conductivity

Name	Description	Photo
M1	Size: 4x4x16 cm	
M2	Composition: Portland CEM I cement, binder:sand (weight ratio) : 1:3	
P1	Size: 4x4x16 cm Composition: Hydraulic lime, binder:sand (weight ratio) : 1:3 Size: 4x4x16 cm Composition: Natural prompt cement, no sand	

Table 1: Cement mortar samples made in laboratory

The calibration procedure consists in determining the parameters α and β of eq. (1), relating the SUSI-measured un-calibrated MC (uMC) with the gravimetric MC (MCg) obtained on a dry basis.

Figure 4 shows the calibration curve for sample M1 *(Table 1)*. We observe an uneven spreading of uMC values among samples at low water content. This depends on the characteristics of the surface of the material, less density on the surface than in the bulk, due to a no-homogeneity associated with the sample preparation. This affects the water content in different way in the surface respect to the bulk.

For any given type of porous materials, the parameters α and β can be implemented in the calibration section of the software managing the instrument.

The SI measurement does not require any calibration. Actually, the salinity index is assumed only as a semi-quantitative measure of the salt content, as we are not able to distinguish among salt species only on the basis of the dielectric measurement.

3.2. Application for diagnostics and monitoring

Two applications of the SUSI system for diagnostics and monitoring are briefly described in this sub-section:

- Mapping of the MC and SI on the wall painting of St. Clement at mass and the legend of Sisinnius in the St. Clement Basilica, Rome (Italy)[9];

- The assessment of the effectiveness of an extractive poultice of salts from the wall paintings in the Allori's loggia in the Pitti Palace in Firenze (Italy).[10]

About the first application, the St. Clement Basilica is built on three different levels. The middle level is located below the road level at about 12 m. At this level are located several precious ancient wall paintings. These wall paintings are subject to hard environmental conditions: high relative humidity of the ambient air (about 90% – constant during the year - with a temperature ranging from 12°C to

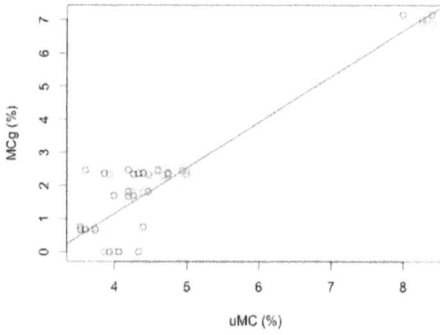

Figure 4: Gravimetric MC (MCg) versus uMC for sample M1

Figure 6: Maps of the distribution of (a) moisture content and (b) of the salinity index

Figure 5: St. Clement at mass and the legend of Sisinnius wall painting. The solid red framework indicates the area investigated by SUSI system and circles represent the points where measurements were performed

22°C), and capillary rise of water from the ground below.[15] On the wall paintings, the typical degradation processes induced by rising damp were present: whitening of the painted layer, salt crystallization, biological microorganism attacks (algae, bacteria, etc.). In particular, on the St. Clement at mass and the legend of Sisinnius, wide areas of salt efflorescence and thick encrustations were present.

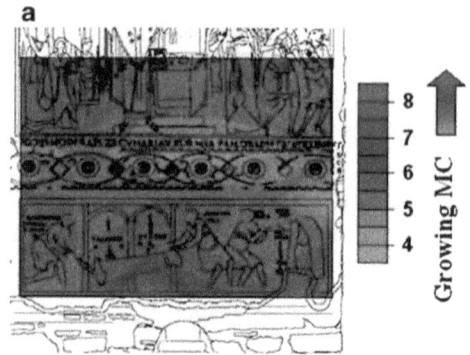

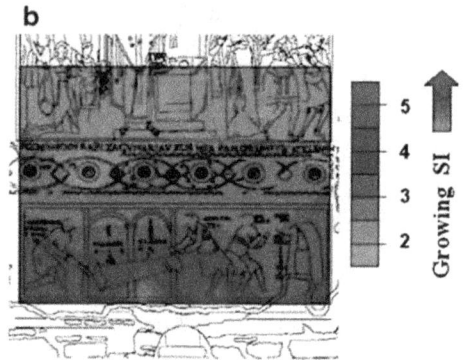

The measurements were performed on the points indicated on *Figure 5*, keeping the sensor in contact with the surface for less than 30 s (the time necessary for the data acquisition). The maps, in terms of MC and SI, are obtained by interpolating the collected data.

SUSI measurements showed that the wall painting exhibited a decreasing gradient of MC from the bottom to the top (until 1.50 m in height) of the investigated area *(Figure 6a)*, ranging from 3.5% (red) to 8% (blue). The SI was higher in the first meter from the floor *(Figure 6b)*, ranging between a minimum of 1.5 (negligible salts) in the upper part of the investigated area to a maximum of 5.5 (quite high salt concentration) in the lower part.

As regards the latest application, the results of tests on the wall paintings at the Allori's loggia in the Pitti Palace in

Figure 7: The wall painting of the Allori's small loggia in the Pitti Palace.

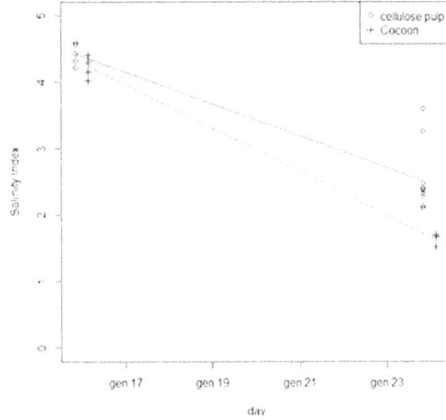

Figure 8: Plots of Salinity Index for two kinds of extractive poultices

Firenze are summarized. *Figure 7* shows the portion of the wall painting where the poultices were applied. On the area labelled with A, a cellulose pulp (bc1000 and bc 200) was applied; on area B the mixing of cellulose pulp, sepiolite clay and sand in different ratios (6 parts of arbocel bw 40/6 parts of sepiolite clay/6 parts of sand was applied); and, on area C the Cocoon® by Westox was applied. The effectiveness of each poultice was evaluated by SUSI measurements applying the following protocols:

a. preliminary measurement by SUSI system before the application of extractive poultices on the selected area on a regular grid;

b. new measurements after from the removal of extractive poultices.

Figure 8 shows the results regarding two kinds of poultice: cellulose pulp and Cocoon. The Cocoon poultice demonstrates a better effectiveness after 7 days with respect to the cellulose pulp, while the poultice made by mixing cellulose pulp and sand didn't provide good results.

4. Conclusions

The SUSI system based on EFD technique can be employed for in situ diagnostics of buildings materials, in particular for diagnosis and monitoring of the presence of water salts up to 2 cm in depth (for the probe used in this work). The resonant approach guarantees an excellent sensitivity on the measurement of the moisture content and also allows to simultaneously and independently detect the presence of soluble salts. A rough quantification of the salt concentration is made based on a Salinity Index, empirically introduced and theoretically justified based on a simple 3-phase model of material. The measurement system, calibrated on samples with proper characteristics (plaster, stone, concrete, etc.), is a promising technique to be applied in diagnosis and monitoring of materials used in historical buildings and artworks.

5. References

[1] A. Kraszewsk, Microwave Aquametry (New York: IEEE), 3-34, 1996.

[2] R. Rouveure, et al., "A microwave sensor for agricultural implements", IGARSS '02, IEEE T Geosci Remote, 5, 3020–2, 2002.

[3] R.A. Yogi, et al., "Microwave microstrip ring resonator as a paper moisture sensor: study with different grammage", Meas. Sci. Technol., 13, 1558–62, 2002.

[4] B.L. Shrestha, et al., "Prediction of moisture content of alfalfa using density-independent functions of microwave dielectric properties", Meas. Sci. Technol., 16, 1179–85, S 2005.

[5] W. Meyer and W.M. Schilz, "Feasibility study of density-independent moisture measurement with microwaves", IEEE Trans. Microw. Theory Tech., 29, 732–9, 1981.

[6] S. Trabelsi and S.O. Nelson, "Density-independent functions for on-line microwave moisture meters: a general discussion", Meas. Sci. Technol., 9, 570–8, 1998.

[7] R. Olmi, et al., "Diagnostics and monitoring of frescoes using EFD", Meas. Sci. Technol., 17, 2281-88, 2006.

[8] R. Olmi and C. Riminesi, "Study of water mass transfer dynamics in frescoes by dielectric spectroscopy", Il Nuovo Cimento, SIF, 2008.

[9] V. Di Tullio, et al., "Non-destructive mapping of dampness and salts in degraded wall paintings in hypogeous buildings: the case of St. Clement at mass fresco in St. Clement Basilica, Rome", Anal Bioanal Chem, 396, 1885–96, 2010.

[10] R. Manganelli Del Fà, et al., "Non-destructive testing to perform service of the evaluation of conservation works", Proc. of the 6th Inter. Conf. on Emerging Technologies in Non-destructive Testing (ETNDT 2016), 457-463, 2016.

[11] C. Riminesi, et al., Moisture and salt monitoring in concrete by evanescent field dielectrometry, Meas Sci Technol, 28(1), 2017.

[12] M. A. Malicki, et al., „Evaluating soil salinity status from bulk electrical conductivity and permittivity", Eur. J. Soil Sci., 50, 505-514, 1999.

[13] K. Roth, et al., "Calibration of time domain reflectometry for water content measurement using a composite dielectric approach", Water Resour. Res., 26, 2267– 2273, 1990.

[14] Z. Hashin and S. Shtrikman, „A variational approach to the elastic behavior of multiphase minerals", J. Mech. Phys. Solids, 11(2), 127-140, 1963.

[15] M. Coladonato, et al., "Environmental study for the safety of frescos in hypogeous site: the relationship between the water quantity and the soluble salts into the masonry", Proc. 6th Internat. Conf. Non-destructive Testing. Microanal. Methods and Environ. Eval. Study and Conserv. Works of Art, Rome, 1996.

Determination of the water uptake and drying behavior of masonry using a non-destructive method

Amelie Stahlbuhk*, Michael Niermann and Michael Steiger
[1] Department of Inorganic and Applied Chemistry, University of Hamburg, Germany
*stahlbuhk@chemie.uni-hamburg.de

Abstract

Historic masonry of different types often suffers from various enrichments in surface and near-surface areas. Salts, atmospheric pollutants and other materials from external sources are possible causes for these enrichments which can directly influence the water uptake and drying of the masonry. Thus, a compaction of the superficial porous network can be expected in the case of crust formation, an enhanced water uptake in the case of hygroscopic salts. Both of these lead to a change of the intrinsic and unaffected water balance. Besides damaging effects, possibly provoked by crusts, and elevated salt contents, it is of interest to investigate how affected walls behave at a given relative humidity, e.g. threshold values for enhanced water uptake or release influenced by salts. The present study reports on a non-destructive chamber method used to investigate the impact of encrustations and salts on the masonry's behavior concerning drying by determining moisture flow. Advantages of the chamber are its easy and non-destructive application on the masonry using a sealing material that is solely pressed on the wall without leaving residues and the use of harmless water vapor. An investigation of wall paintings in the cloister of the St. Peter Cathedral in Schleswig which in parts suffer from massive yellowish encrustations is presented to demonstrate the promising results obtained with the chamber method.

Keywords: non-destructive method, water uptake and drying, encrustations

1. Introduction

Masonry of different kind is often affected by various types of enrichments at surfaces and near-surface regions. Such masonry includes e.g. bare brickwork, plastered walls, wall paintings and other monuments. Porous building materials are used for the construction of those objects since antiquity. Due to their surface and their porous network they are directly exposed to the atmosphere. Having an additional connection to the ground water, atmospheric and anthropogenic influences can act on them. Enrichments can be present in form of salts, atmospheric pollutants and other materials from external sources. Salts are mostly introduced via capillary rise of ground water or anthropogenic effects like usage of de-icing salt, agriculture, animal excrements and cleaning or conservative actions.[1] Also anthropogenic emissions, e.g. caused by the combustion of oil and coal, lead to the deposition of pollutants on the porous objects which are a source for salts as well, especially for the formation of gypsum.[2] In addition, cleanings and conservative procedures can introduce other materials like resins or fixatives whose ageing products can also enrich over the course of years.[3,4] Sulfur dioxide pollution was a problem in industrial areas until the 1990s. With the distribution of exhaust gas purification for power plants and a change in lifestyle towards a declining use of coal in domestic heating the problem was reduced, at least in Europe.[5,6] Regarding the formation of crusts, gypsum as a result of SO_2 impact, might be the most popular one.[7,8]

The processes of water uptake and release in a porous medium in contact with its surroundings are described by the sorption equilibrium. As a function of the materials' pore size distribution, its water content and the ambient temperature and relative humidity (RH) water vapor may be adsorbed from or released to the atmosphere. Salts in the pores of building materials not only have a high damage potential due to the generation of crystallization pressure.[9] Salts also influence the water uptake and drying behavior of such porous matter. Each salt has a characteristic, temperature dependent value of the relative humidity – the deliquescence humidity – at which it starts to pick up water from the atmosphere to form a saturated solution. Regarding salts in building materials, the hygroscopic behavior of the salt and that of the material both influence water uptake and release. Very hygroscopic salts, like nitrocalcite, $Ca(NO_3)_2 \cdot 4H_2O$, with a high solubility take up high amounts of water also at moderate relative humidities[2], theoretically, leading to a saturation of the pore space with salt solution and consequently to a much higher water content within the masonry than in the salt free material at the same relative humidity. On the other hand salts with an extremely low solubility like gypsum, $CaSO_4 \cdot 2H_2O$, pick up water vapor only at extremely high humidities (>99.9% RH). Furthermore the behavior concerning water uptake and release is more complicated in the case of salt mixtures like they are commonly present in real objects.

Next to the enrichment of salts also other materials of various origins accumulated on the material surface may have a direct influence on the water upta-

ke and drying behavior of masonry, thus, they may change the natural water balance of the considered object. While hygroscopic salts enhance the water uptake, the formation of superficial crusts of slightly soluble salts or non-saline deposits result in a compaction of superficial pores leading to a hindrance of water sorption. Another effect that leads to an impairment of the natural water balance of porous materials is hydrophobing. In this case the pores are not blocked but hydrophobic agents coat the pore walls and the material surface, thereby affecting water uptake and drying.

In this work we report on the application of a chamber method[10] that was originally developed to investigate the deposition of air pollutants such as SO_2 onto material surface.[11,12] In contrast to the deposition of atmospheric pollutants, the water sorption is a reversible process. Nonetheless, the chamber method is also suitable for the observation of moisture flows across the material surface in both directions, i.e. hygroscopic water uptake and evaporation during drying. Questions concerning the behavior during drying of masonry containing salts or other foreign substances and threshold values for augmented water uptake and release due to salts may be examined with this non-destructive method.

2. Experimental

2.1. Chamber method

For the measurement of moisture flow a chamber as shown in *Figure 1* was used. The cylindrical chamber (1) made of POM (polyoxymethylene) and PMMA (polymethylmethacrylate) with a diameter of 100 mm is mounted on a POM-base plate with dimensions 300x150 mm². For the measurements, the plate with a cylindrical aperture on the back side is gently pressed against the wall under

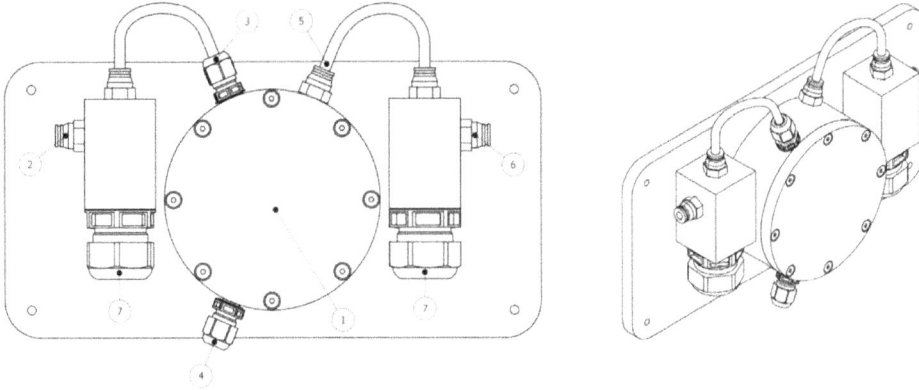

Figure 1: Schematic diagrams of the chamber used for the measurements of moisture flow, showing a front view (left) and a 3D representation (right). Numbers represent (1) cylindrical chamber, (2), (3) inlet and (5), (6) outlet ports for the air flow, (4) external power supply for the ventilating fan and (7) positions for the RH/T sensors at the entrance and the exit of the chamber.

investigation using sealing strips (Tesamoll P-profile) which can be removed without residue after the measurements. An adjustable air flow with controlled relative humidity passes the chamber through the inlet (2, 3) and outlet ports (5, 6), respectively. The chamber is ventilated using a fan that is installed on the lid and connected to the external power supply via inlet (4). Ventilation is necessary to achieve ideal mixing in the chamber, thus, to avoid influences of Brownian diffusion on the overall moisture exchange rates. In a well-mixed chamber, the relative humidity at the exit equals the RH inside the chamber and transport of water vapor to the surface or evaporation from the surface are the rate limiting processes. The base plate also carries two RH/T sensors (Hygroclip, Rotronic Messgeräte GmbH) that are used to monitor continuously the relative humidity and temperature both at the entrance and the exit of the chamber (7). The two sensors (±0.8 % RH and ±0.2 K) are conected to a Hygrolab C1 station (Rotronic Messgeräte GmbH).

In the experiments that were carried out on site, the drying behavior of the masonry was investigated. The airflow was generated by a membrane pump

such that typical volume fluxes were 3.9 L·min^{-1}. Incoming air was dried in a silica gel drying tube positioned between membrane pump and chamber entrance. For other applications, the drying tower may be replaced with a humidification system using saturated salt solutions to control the relative humidity at the chamber entrance. The flow rate was measured using a gas meter connected to the chamber exit (6).

During the measurements the only data captured are relative humidity and temperature at the entrance and the exit of the chamber. Known quantities are the volumetric flow rate Q (in m^3·h^{-1}) of the air flow and the surface area A (in m^2) of the wall that is exposed to the cylindrical chamber during the experiment. The relative humidity φ is defined as

$$\varphi = p_w/p_{w,0} \tag{1}$$

where p_w and $p_{w,0}$ are the water vapor pressure and the saturation water vapor pressure, respectively. The latter was calculated using the Wagner Pruss equation [13]. The absolute humidity f is given by:

$$f = p_w M_w/(RT) \tag{2}$$

where M_w is the molecular mass of water (0.0180153 kg·mol⁻¹), R is the gas constant (R = 8.3144 J·mol⁻¹·K⁻¹) and T is the absolute temperature.

The difference of the absolute humidities at the exit (f_{out}) and at the entrance (f_{in}), i.e. $\Delta f = f_{out} - f_{in}$, is the result of either evaporation ($\Delta f > 0$) or water uptake from the air ($\Delta f < 0$). The moisture flux F in units of g·m⁻²·h⁻¹ is then given by:

$$F = \Delta f \, Q/A \qquad (3)$$

For visual presentations the mass of water, evaporated during drying, (or the mass of water, adsorbed from the atmosphere in case of water uptake, respectively) per unit area (in g·m⁻²) will be used which is obtained by multiplying the moisture flow rate with the corresponding time (in h). Hence, the transported quantity of water vapor in g·m⁻² is depicted at any point of the measurement.

2.2. Cloister of St. Peter Cathedral

In this study, wall paintings in the cloister of the St. Peter Cathedral in Schleswig, Northern Germany, were investigated. The so-called "Schwahl" is furnished with wall paintings from the 14th century showing monochrome scenes of Christ's life in 22 bays and polychrome paintings of apostles, mythical creatures and leaf tendrils in the vaults and on the courtyard-sided walls. The restauration history of this object started quite early and involves several interventions including revisions, whitewashes and reparations which are not completely recorded in the years before the late 19th century [14]. Regarding the state of conservation the walls of the object showed high contents of nitrates and chlorides in most parts, gypsum enrichments on the surface and damage in the form of scaling and sometimes loss of pigmented areas. Due to a very humid climate the salts remain dissolved most of the time which is also the reason for the high moisture content of the masonry. Even though success was noted in the last years concerning salt reduction and gypsum conversion, the removal of yellowish encrustations on parts of the bays was not effective so far. These inhomogeneous encrustations most likely comprise organic materials which are probably alteration products of organic substances introduced in former restoration campaigns. With their massive appearance it was expected that they would not only affect the object in an aesthetic manner but also the natural water balance of the walls by clogging superficial pores.

The chamber method was used on different parts of the most affected bay no. 5 which is depicted in *Figure 2*. The area under the frieze is not pigmented and comprises a fresh and even lime plaster which does not show any visible encrustations (area 1). Within the frieze which is decorated with paintings of ani-

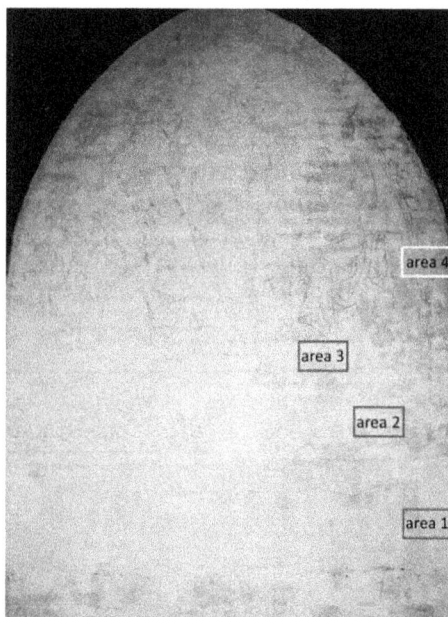

Figure 2: Bay no. 5, cloister of St. Peter Cathedral in Schleswig. Measurement areas 1 to 4 are depicted.

mals (area 2) a patchy yellowish encrustation was clearly visible. The same situation was noted in area 3, while in area 4 on the upper right side of the wall painting a massive and continuous crust was observed. To all these areas the chamber was applied to measure the moisture flow during drying, implemented by passing a dry airflow through the chamber. Two or three replicate measurements were conducted on all areas except area 3.

3. Results and discussion

The results of the multiple measurements at the different positions on the wall are represented in *Figure 3*. In all measurements an increase of the mass flow with time is observed which demonstrates the ongoing release of water vapor from the wall to the atmosphere. The curves representing the measurements in the freshly limed area 1 show the highest mass flow. Compared to

that, the results from areas 2 and 3 with patchy encrustations show a significantly reduced water release. For test area 4 the moisture flow is even smaller and the curves reach a plateau after less than one minute of drying which means that there is no further release of water from the wall.

The curves representing replicate measurements in the same areas show some variability which might reflect minor shifts of the chamber position on the wall between two measurements. This includes e.g. a higher or lower extent of encrustations. Different initial moisture contents of the exposed parts of the wall might also be an important cause of the observed differences. Every measurement causes a reduction of the moisture content at a test position due to partial drying. If there is not sufficient time for re-humidification between two consecutive measurements, the initial moisture content is lower in the next run. Nevertheless, the curves belonging to the same

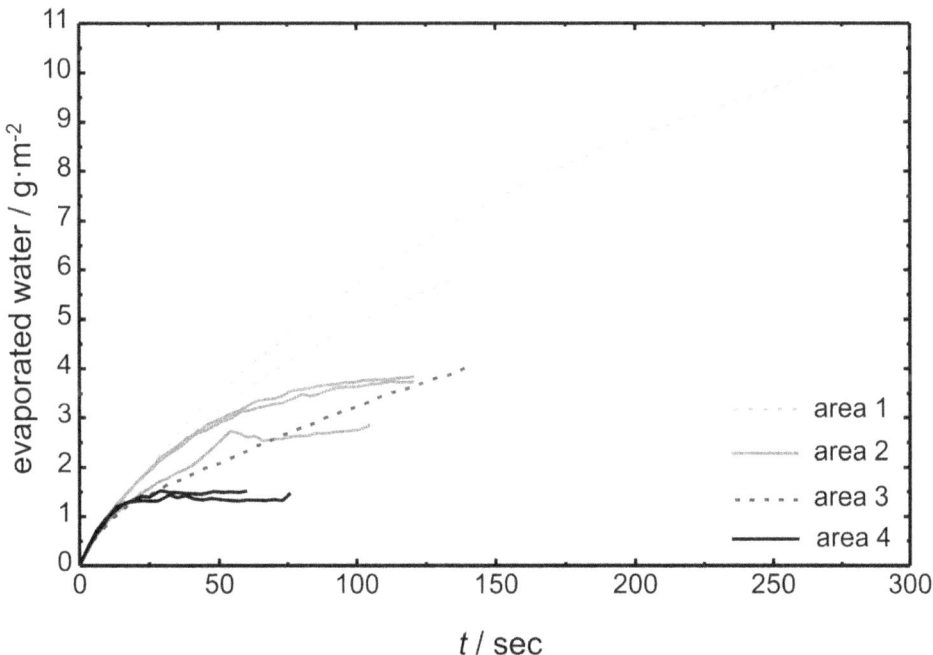

Figure 3: Evaporated water versus time curves for the four areas in bay no. 5 (see Figure 2).

measurement position or to areas with a similar state of the surface can be clearly grouped together and show similar drying behavior. The relative drying rates observed for these groups seem to be plausible as it was found that an area without any visible encrustation shows significantly higher drying rates than areas covered with patchy or even extensive crusts.

Before the results will be discussed, the two distinct drying stages of porous materials may be discussed. During the first stage climatic conditions, air flow rate and surface texture are the limiting factors. Capillary moisture transport to the drying front at the surface provides more moisture than the amount that can evaporate, as climatic and transition conditions limit evaporation in this stage. Consequently, these conditions determine the slope of the curve *(see Figure 3)* while the material inherent properties, e.g. the moisture transport properties, influence the duration of the first stage. With decreasing moisture content within the material the liquid flow to the surface declines. As less water is transported to the surface, the drying rate decreases and the drying front shifts into the material.[15, 16, 17] Regarding the specific case of salt contaminated materials, as the pore solution is transported to the surface due to capillary forces, salts crystallize and accumulate on the surface or in near surface regions. Hence, they provoke pore clogging, hindering capillary transport to the surface and slowing down the drying rate.[18]

The curves shown in *Figure 3* are quite similar during the first seconds with a nearly linear initial increase, i.e. a constant drying rate, characteristic of the first drying stage. The limiting conditions at this stage were quite similar in all experiments thus, it is not surprising that equal initial drying rates are observed. In contrast, the material characteristics might differ between the positions

of measurement leading to variations of the duration of the first drying stage. The decreasing slope of the curves of areas 2 and 4 represents the decreased drying rate at the beginning of and during the second drying stage. In case of area 1, a change of the slope is not clearly apparent, consequently assuming that the first drying stage was not terminated during the duration of the experiment seems reasonable. Without hindrance in form of a crust like in area 1, water vapor is transported uninhibitedly to the surface and the drying front can recede into the material in the second stage. If crusts represent an obstacle (see area 4), capillary transport to the surface is hindered as it was already mentioned in the preceding section. From microscopic images of samples from bay no. 5 it is known that the yellowish crusts cover parts of the render and block the pores to a large extend as they comprise a dense layer. Thus, it can be assumed that pore clogging is not a consequence of salt crystallization close to the surface induced by drying but an effect caused by (most likely organic) deposits on the wall paintings which were already present before the experiments. In case of area 4, the course of the mass loss curve reaches a plateau after only 20 seconds, suggesting that only the condensed water on the surface of the crust evaporated during drying.

These observations demonstrate that the simple setup with the flow chamber yields useful and promising results concerning the water transport across the surfaces of encrusted wall paintings. Besides, this hindered water vapor transport caused by the encrustations has another consequence for the affected objects which could pose a risk. Dissolved salts within the porous network tend to crystallize on the surface when the relative humidity decreases to form subflorescences or efflorescences. If parts of the surface are covered with impermeable crusts, salts accumulate in uncovered areas adjacent to the crusts causing a risk. Also flaking of the crust is a possible scenario e.g. in the case of heavy accumulations and small cracks within the dense layer which allows drying of the underlying wall and crystallization of salts. The chamber method is a useful tool to detect and validate such critical circumstances on the object. The main advantage for the use on sensitive objects is the non-destructive character enabled by a sealing strip which is normally applied for the sealing of windows and doors. Limitations result for very uneven or bent walls where it is not possible to seal the room between chamber aperture and wall for level differences greater than the height of the strip. However, bare brickwork and other non-decorated walls might be investigated by using another sealant (Terostat IX, Loctite, Henkel) which adheres to the wall but can be removed without residues as well.

4. Conclusions

A non-destructive chamber method was successfully used to study the influence of encrustations on wall paintings on the drying behavior of the wall. Originally, the chamber method was used to determine the deposition of atmospheric pollutants to material surfaces. In the present study, a simplified version of the chamber was applied in an investigation of water release from wall paintings. It was shown that complete coverage with encrustations hinders the drying process of the render and the underlying wall by clogging superficial pores and blocking the passage for the intrinsic water vapor exchange between masonry and atmosphere. In addition, it was found that the extent of encrustation controls the drying behavior of the surfaces. Patchy and irregular crusts show mass flow curves between unaffected areas and fully covered areas. With these accessible ob-

servations potentially critical situations for the object considered as a consequence of hindered water vapor transport and accumulation of salts can be derived. These promising results induce other interesting applications of the chamber method on real objects like, for example an investigation of the hygroscopic water uptake of walls contaminated with salts, or, the monitoring of the dynamics of the drying and water uptake behavior in response to daily or seasonal cycles of the climatic conditions.

Acknowledgments

Financial support of this research by the Deutsche Bundesstiftung Umwelt is gratefully acknowledged.

References

[1] A. Arnold, K. Zehnder, Monitoring wall paintings affected by soluble salts. In: S. Cather (Ed.), The conservation of wall paintings. Proceedings of a symposium organized by the Courtauld Institute of Art and the Getty Conservation Institute, London, July 13-16, 1987. Getty Conservation Institute, Los Angeles, 1996, 103–135.

[2] M. Steiger, A. E. Charola, K. Streflinger, Weathering and deterioration. In: S. Siegesmund, R. Snethlage (Eds.), Stone in architecture. Springer, Berlin, Heidelberg, 2014, 225–315.

[3] P. Mora, L. Mora, P. Philippot (Eds.), Butterworths series in conservation and museology, Butterworths, London, 1984.

4 L. Borgioli, P. Cremonesi, Le resine sintetiche usate nel trattamento di opere policrome, Il prato, Saonara (Padova), 2005.

5 P. Brimblecombe, Air pollution and architecture: Past, present and future, J. Architect. Conserv. (6) (2000), 30–46.

6 S. J. Smith, J. van Aardenne, Z. Klimont, R. J. Andres, A. Volke, S. Delgado Arias, Anthropogenic sulfur dioxide emissions: 1850-2005, Atmos. Chem. Phys. (11) (2011), 1101–1116.

7 M. Steiger, Salts and crusts. In: P. Brimblecombe (Ed.), The effects of air pollution on the built environment. Imperial College Press, London, 2003, 133–182.

8 M. Steiger, Air pollution damage to stone. In: P. Brimblecombe (Ed.), Urban pollution and changes to materials and building surfaces. Imperial College Press, London, 2016, 65–101.

9 M. Steiger, Crystal growth in porous materials – I: The crystallization pressure of large crystals, J. Cryst. Growth (282) (2005), 455–469.

10 A. C. Hill, A special purpose plant environmental chamber for air pollution studies, J. Air Pollut. Control Assoc. (17) (1967), 743–748.

11 A. Behlen, C. Wittenburg, M. Steiger, W. Dannecker, Dry deposition of NO, NO2, HNO3, and PAN on historic building stones. In: J. Riederer (Ed.), International congress on deterioration and conservation of stone – Proceedings. Möller Druck und Verlag GmbH, Berlin, 1996, 377–386.

12 U. Wiese, A. Behlen, M. Steiger, The influence of relative humidiy on the SO_2 deposition velocity to building stones: A chamber study at very low SO_2 concentration, Environ. Earth Sci. (69) (2013), 1125–1134.

13 W. Wagner, A. Pruss, International equations for the saturation properties of ordinary water substance. Revised according to the international temperature scale of 1990, J. Phys. Chem. Ref. Data (22) (1993), 783–797.

14 B. Löffler-Dreyer, Von Truthähnen und Füchsen – zur Restaurier-ungsgeschichte der Wand- und Gewölbemalereien im Schwahl des Schleswiger Doms, DenkMal! Zeitschrift für Denkmalpflege in Schleswig-Holstein (21) (2014), 98–105.

15 C. Hall, W.D. Hoff, M.R. Nixon, Water movement in porous building materials – VI. Evaporation and drying in brick and block materials, Build. Environ. (19) (1984), 13–20.

16 G.A. Scheffler, R. Plagge, Ein Trocknungskoeffizient für Baustoffe, Bauphysik (31) (2009), 125-138.

17 T. Diaz Gonçalves, V. Brito, L. Pel, Water vapor emission from rigid mesoporous materials during the constant drying rate period, Drying Technol. (30) (2012), 462–474.

18 R.M. Espinosa-Marzal, G.W. Scherer, Impact of in-pore crystallization on transport properties, Environ. Earth Sci. (69) (2013), 2657-2669.

Measurement of salt solution uptake in fired clay brick and identification of solution diffusivity

Etsuko Mizutani[1]*, D. Ogura[1], T. Ishizaki[2], M. Abuku[3] and J. Sasaki[2]
[1] Graduate School of Engineering, Kyoto University, Kyoto, Japan
[2] Tohoku University of Art and Design, Yamagata, Japan
[3] Faculty of Architecture, Kindai University, Higashi-Osaka, Japan
* be.etu@archi.kyoto-u.ac.jp

Abstract

Salt solution uptake rates in fired clay brick were measured with γ-ray to investigate the influence of salt on solution diffusivity for appropriate numerical analysis of salt solution transfer and crystallization in porous materials. NaCl and Na_2SO_4 are the main salts confirmed in Hagia Sophia (Aya Sophia Museum), Istanbul, Turkey, and are used in this experiment. We identified solution diffusivity as a function of solution content by a numerical analysis of solution transfer. The main results are as follows. Solution uptake tests demonstrated that saturated solutions take approximately six times longer to attain a steady state than the time needed for pure water to obtain steady state. This was thought to be caused by the viscosity of the dissolved salts and the presence of crystallised salt in the material. We identified solution diffusivity by multiplying moisture diffusivity for pure water by a constant to reproduce the measured solution content in the cases where large amounts of efflorescence was observed at the material surface. In the case of a saturated solution of Na_2SO_4, which is considered the salt that most frequently precipitates into material, it was necessary to correct the solution diffusivity equation by considering the effect of in-pore salt precipitation.

Keywords: Solution diffusivity, γ-ray attenuation, fired clay brick, sodium salts

1. Introduction

Salt weathering is one of the main causes of the deterioration of brick masonry walls, which can be found at the Hagia Sophia (Aya Sophia Museum) in Istanbul *(see photo 1)*. There are many sources of salt in the environment, including ground water, atmospheric deposition, and originally contaminated salts in the materials. In Hagia Sophia, atmospheric deposition and originally contaminated salts in the materials are assumed to be the main sources of salt that cause weathering of the structures.[1] In the Hagia Sophia case, there exist complex humid–dry and wet–dry cycles that complicate the deterioration process that comprises adhesion, dissolution, and infiltration processes that transfer salts from the sources in the environment. Therefore, the process is much more complex than when ground water is the main salt source. The objective of this study was to develop an analytical model of the simultaneous transfer of heat, moisture, and salt and the subsequent crystallisation and dissolution of the salt for reproducing the deterioration mechanism caused by these complex salt deterioration processes.

Salt, which exists in the dissolved phase or solid phase (crystal) in porous material, causes significant changes to the material properties of the surface it contacts. This is due to a rise in viscosity of the solution and the descent of vapour pressure of the salt solution and changing pore structure in the solid phase.[2, 3] Therefore, material properties conside-

quantitatively measured by a γ-ray attenuation apparatus. In addition, we characterised solution diffusivity as a function of volumetric solution content by a numerical analysis of the salt solution transfer process. This was done to investigate the effects of dissolved salt and salt crystallisation.

2. Methodology

2.1. Outline of solution uptake test

A salt solution uptake test of NaCl and Na_2SO_4 was conducted to investigate the influence of dissolved salts and crystallised salts on solution transfer. Fired clay bricks with dimensions of 3x4x10cm were used. An epoxy resin and aluminium film were applied to the surfaces of the bricks, with the exception of the top and bottom face (3x4cm), to provide water and vapour proof seals and accomplish one-dimensional solution transfer. Specimen bricks were hung above a pure water/salt solution water bath. The bottom surface was exposed to the bath. The top surface was in contact with room air only. Initially, pure water uptake tests were conducted using specimen bricks in a dry condition. Subsequently, salt solution uptake tests were conducted with the same specimen that had been dried again. This process allowed for a comparison of the uptake of water vs salt solution. Experimental conditions during the experiments, such as the saturation ratio of the solution, the ambient temperature, and the room humidity are shown in *Table 1*.

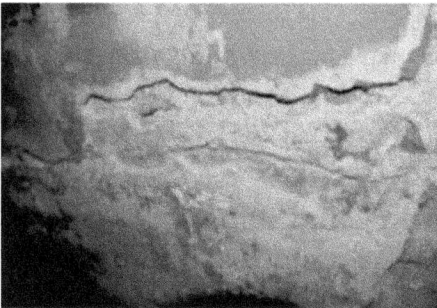

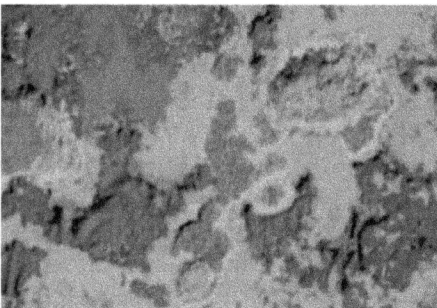

Photo 1: Deterioration of the inner walls due to salt crystallisation at Hagia Sophia, Istanbul, Turkey

ring these effects of salt are needed to predict the real phenomenon appropriately.

In this research, we investigated the solution diffusivity, which is an important aspect that influences the transfer rate of dissolved salts. We conducted pure water and salt solution uptake tests on fired clay brick material. Na_2SO_4 and NaCl are the main salts found in the walls of Hagia Sophia, and they were used in this experiment. The spatial distribution of the solution during the experiment was

2.2. Measurement set-up of volumetric salt solution content

The time evolution and spatial distribution of the volumetric content of the pure water/salt solution in the material

Number of specimen	Test 13	Test 14	Test 17
Salt	NaCl	Na_2SO_4	
Saturation ratio at 20°C	100%	100%	20%
Temperature (°C) Relative humidity (%)	20.5°C, 40–48%	20.5°C, 40–48%	20.5°C, 43–44%

Table 1: Experimental conditions

during the uptake process was experimentally measured using γ-ray attenuation. Volumetric solution content ψ_{sol} [m³/m³] was evaluated using γ-ray transmission of the material during the experiment in equation 1:

$$\psi_{sol} = -\frac{1}{\rho_{sol}\mu l} \ln \frac{l}{l_o} \qquad (1)$$

where ρ_{sol} is the density of solution [kg/m3], μ is the mass attenuation coefficient [cm²/g], l is the thickness of specimen [cm], and I_o and I are the intensity of γ-ray radiation of material at dry state and at the time of experiment [cps], respectively. μ is the specific value of a radioisotope, and the value of the water of americium is 0.2059. Prior to the experiments, the mass attenuation coefficient, μ of the salt solution was identified from the measured values of the γ-ray transmission of an acrylic water container filled with a salt solution. *Figures 1 and Figure 2* show the schematic representation of the measurement set-up and the mass attenuation coefficient of the salt solutions, respectively. The identified mass attenuation coefficients of the salt solutions, which depend on salt concentration and the type of salt used, were utilised to calculate the volumetric salt solution content.

2.3. Fundamental equations for dissolved salt and moisture transfer

The fundamental equations for dissolved salt and moisture transfer which are

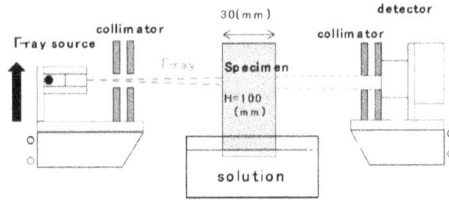

Figure 1: Schematic representation of the measurement set-up

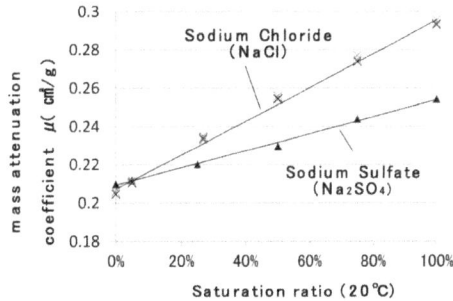

Figure 2: Mass attenuation coefficients of the salt solutions

used to identify the solution diffusivity are shown. The Boltzmann transformation method is also an effective method for the determination of solution diffusivity when gravity can be neglected.[4, 5] In this research, we conducted a numerical analysis of the salt solution and moisture transfer that included the effect of gravity.[6, 7] We identified the solution diffusivity Dsol in Eqn. (2) using time change of measured volumetric solution content distribution. For this research, we assumed that the salt crystallisation and salt diffusion due to the salt concentration gradient did not occur during the absorption of the solution. In addition, we assumed that dissolved salt moved by advection only. Under these assumptions, the mass balance of liquid phase which is consists of liquid phase water and dissolved salt ions, is expressed by equation (2):

$$\frac{\partial \rho_{sol} \psi_{sol}}{\partial t} = \nabla \cdot D_{sol} \left(\nabla \psi_{sol} - \frac{\partial \psi_{sol}}{\partial \mu_o} g \right) \qquad (2)$$

where $D\psi_{sol}$ is the solution diffusivity in function of solution content [kg/ms],

μ_o is the water chemical potential [J/kg], and g is the gravitational acceleration [m/s²]. In this paper, the literature values of solution density for NaCl solution and Na_2SO_4 solution at 20 °C are use.[8] We identified the solution diffusivity D_{sol}, used in equation (2) as $\nabla \psi_{sol}$. The sorption isotherm for the material was identified from the literature.[9]

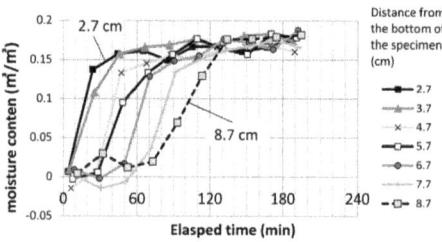

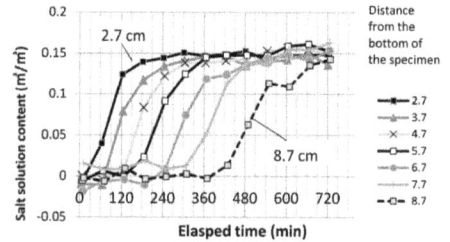

Figure 3: Change in the spatial distribution of the volumetric solution content used in TEST 13 Left: pure water, Right: NaCl saturation)

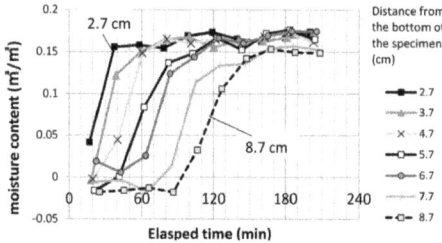

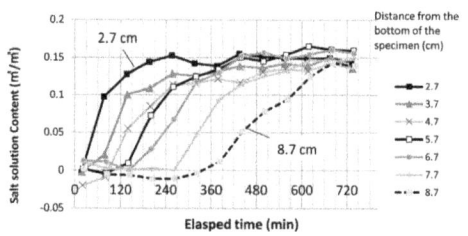

Figure 4: Change in the spatial distribution of the volumetric solution content used in TEST 14 (Left: pure water, Right: Na_2SO_4 saturation)

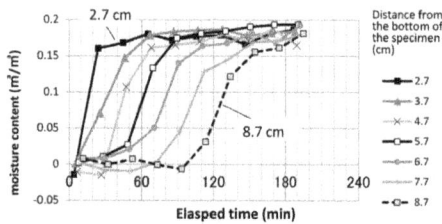

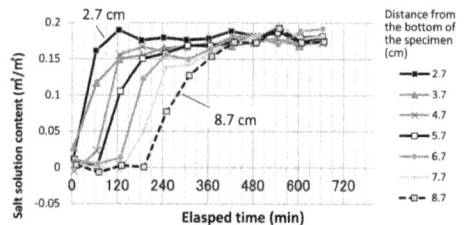

Figure 5: Change in the spatial distribution of the volumetric solution content used in TEST 17 (Left: pure water, Right: Na_2SO_4 20%)

3. Results and discussion

3.1. The time evolution of water / salt solution content distribution

Figures 3 and 4 show the change in the spatial distribution of volumetric solution content during TEST 13 (Left: pure water, Right: NaCl saturated solution) and TEST 14 (Left: pure water, Right: Na_2SO_4 saturated solution), respectively. Since we assumed that the change in the concentration of salt solution would be very little during the solution uptake tests, constant values of the mass attenuation coefficient of the solution for each experimental condition were used to calculate the volumetric solution content. The moisture uptake rate of the specimens was variable. The specimens reached a near steady state in approximately two to three hours after the beginning the absorption of the pure water. The uptake rate of the saturated solutions of both NaCl and Na_2SO_4 were significantly smaller than that of pure water, as it took approximately 12 hours for saturated solutions to reach a steady state. In addition, the uptake rate of the saturated solutions at upper locations on the specimens (7.7 cm and 8.7 cm from the bottom) became gradually slower. This tendency was assumed to be due to gravity or salt crystallisation within the material during the uptake process. This was significant in case of Na_2SO_4. *Figure 5* illustrates the change in the spatial distribution of the solution in the specimen that absorbed the Na_2SO_4 solution with a saturation ratio of 20%. The salt solution uptake rate was twice the rate of pure water, with a decreasing uptake rate at the upper locations of the specimen not seen. *Photo 2* shows a comparison of salt crystallisation at the upper surface of the specimen after enough time had passed to allow crystal growth dissipation. In the case of the saturated solution of NaCl, salt precipitation on the surface began during the

solution uptake process and continued approximately 5 days after the solution uptake experiment. On the other hand, in the case of the Na_2SO_4 saturated solution, efflorescence was not confirmed during nor after the uptake test. Na_2SO_4 easily crystallised within the specimen because Na_2SO_4 can change phase at a higher in relative humidity than NaCl. In the case of the Na_2SO_4 solution with

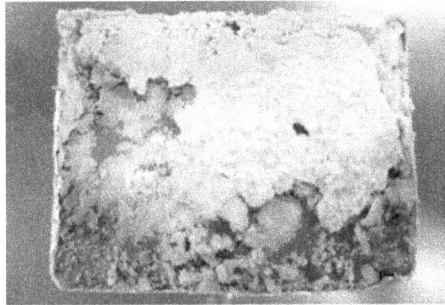

TETS 13 NaCl sat.

TEST 14 Na_2SO_4 sat.

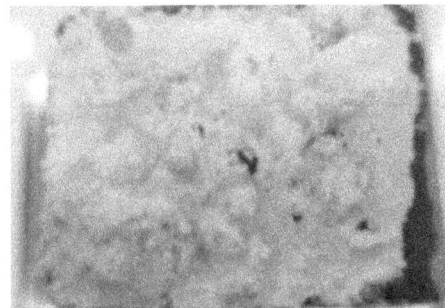

TEST 17 Na_2SO_4 20%

Photo 2: Salt crystallisation at the upper surface of the specimen

the saturation ratio of 20%, salt crystals emerged at the top surface. When we conducted the same experiment at a lower relative humidity (about 25%), the process of crystallisation of the 20% Na_2SO_4 solution was similar to that observed in the Na_2SO_4 saturated solution, TEST 14 *(see Photo 2)*. Therefore, a small difference in the concentration of the solution, ambient temperature, or ambient humidity may affect the crystallisation behaviour of Na_2SO_4, which may increase the solution uptake rate.

3.2. Identified solution diffusivity

In this research, solution diffusivity was approximated by the exponential function $D\psi_{sol}=C_1e^{C_2\psi sol}$, and the fixed values C_1 and C_2 were determined by matching the solution uptake tests. *Figure 6* shows the comparison between the measured values and the calculated values used to identify the solution diffusivity in the cases of pure water and a saturated solution of NaCl. The calculated values agree with experimental values in both cases. Therefore, we concluded that the solution diffusivity can be found by multiplying the moisture diffusivity of pure water by a constant. *Figure 7* illustrates the comparison between the measured values and the calculated values of solution content in the case of the Na_2SO_4 solution. In the case of a 20% saturated solution, the calculated values of solution diffusivity can be found by multiplying the moisture diffusivity of pure water by a constant. Generally, the results correspond with the measured values *(see Figure 7, Right)*. Alternatively, in the case of the saturated Na_2SO_4 solution, it was difficult to replicate the measured values by simply changing the solution diffusivity to that of pure water *(see Figure 7, Left)*. This discrepancy is thought to be due to the difference of crystalli-

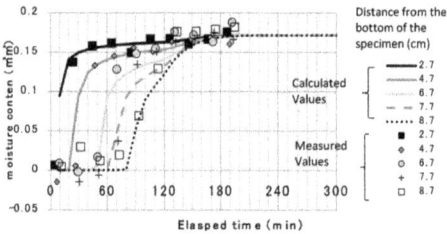

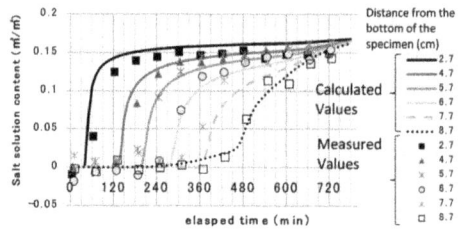

Figure 6: Comparison between the measured and calculated values of the volumetric solution content used in TEST 13 (Left: pure water, Right: NaCl sat)

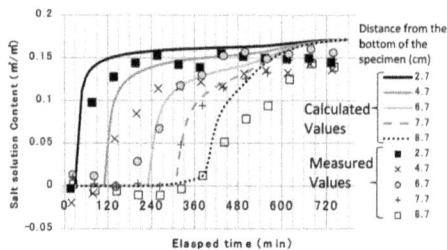

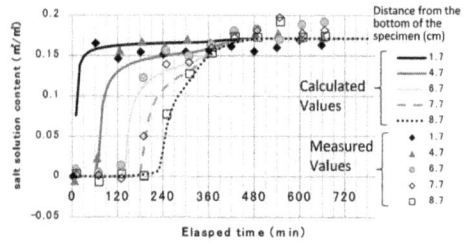

Figure 7: Comparison between the measured and calculated values of the volumetric solution content used in TEST 14 and 17 (Left: Na_2SO_4 sat, Right: Na_2SO_3 20%)

sation behaviour within the specimen. Therefore, the solution diffusion model including a reduction of effective porosity due to salt crystallisation is necessary for reproducing the measured values in the case of the Na_2SO_4 saturated solution. The Boltzmann data transformed with the Boltzmann variable $\lambda = x \cdot t^{-1/2}$ obtained from the measured values of the saturated Na_2SO_4 solution are shown in *Figure 8* as a reference to demonstrate that the consideration of in-pore crystallisation

is required. The relationship of solution content and λ is not able to be formalised as one function because the data scattered, especially at high moisture content. *Figure 9* shows the identified solution diffusivity, D_{sol} as a function of solution content. The influence of salt on solution diffusivity is so large that the value for the saturated solution of NaCl is 0.17 times smaller than that of pure water. In addition, a higher solution concentration causes smaller solution diffusivity. This

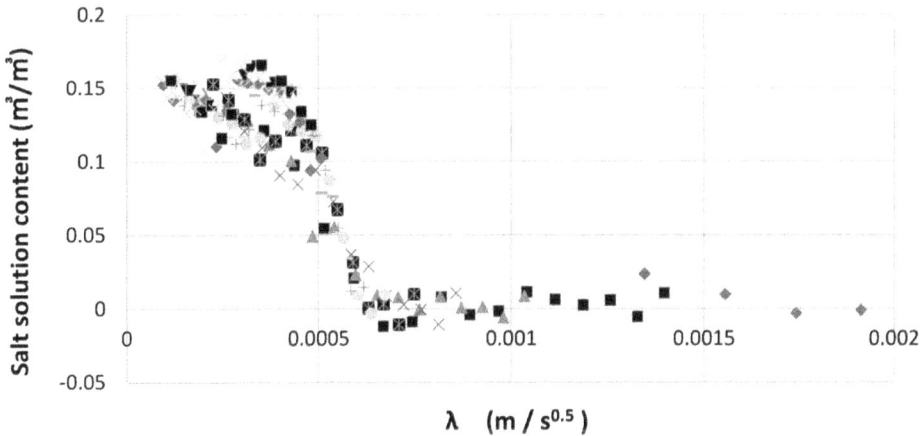

Figure 8: Boltzmann transformation of the measured solution content of the saturated Na_2SO_4 solution

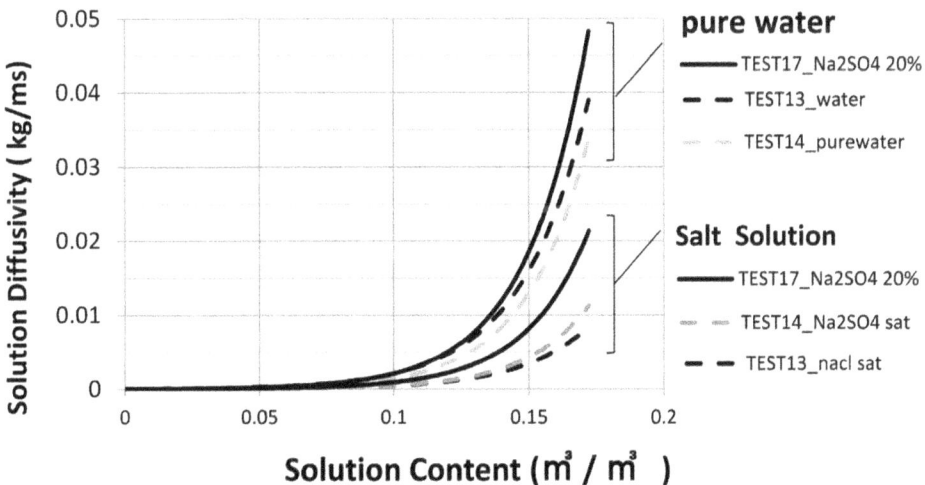

Figure 9: Identified solution diffusivity D_{sol} as a function of solution content

result can be explained by the rise in viscosity. From these results, we concluded that the moisture diffusivity model requires the consideration of both viscosity and in-pore crystallisation.

4. Conclusions

We conducted a solution uptake test of two kinds of sodium salts (NaCl, Na_2SO_4) which are believed to affect the deterioration of the inner walls of Hagia Sophia. Fired clay brick was used to investigate the influence of salt on solution diffusivity. The influence of salt on solution diffusivity is significant due to its viscosity and salt deposition characteristics. The uptake rate of a salt solution is, at most, approximately six times slower than that of pure water. It was observed from significant change of moisture behaviour that Na_2SO_4 easily crystallises in the material compared to NaCl. A numerical analysis of dissolved salt moisture transport was also conducted to identify solution diffusivity as a function of solution content. The solution diffusivities for the saturated solution of NaCl and the 20% saturated solution of Na_2SO_4 can be found by multiplying the moisture diffusivity of pure water by a constant including saturation degree of solution. The solution diffusivity model takes into account the change due to salt crystallisation, which is required to predict salt solution transfer with the influence of pore clogging, especially in the case of the high concentration Na_2SO_4 solution.

Acknowledgements

This work was supported by JSPS KAKENHI Grant Number 16H06363 (Grant-in-Aid for Scientific Research), Grant Number 26709043 (Grant-in-Aid for Young Scientists (A)) and Grant Number 26870897 (Grant-in-Aid for Young Scientists (B))

References

[1] E. Mizutani, D. Ogura, T. Ishizaki, M. Abuku, J. Sasaki, 'Degradation of the wall paintings of Hagia Sophia in Istanbul', Central European Symposium of Building Physics, Dresden, Germany, 2016.

[2] R. M. Espinosa-Marzal, G. W. Scherer, 'Impact of in-pore salt crystallization on transport properties', Environ. Earth Sci 69, 2013, 2657-2669.

[3] D. Ogura, M. Abuku, S. Hokoi, C. Iba, S. Wakiya, T. Uno, 'Measurement of salt solution uptake by ceramic brick using γ-ray projection', 3rd international Conference on Salt Weathering of Buldings and Stone Sculptures,2013, 529-532.

[4] J. Carmeliet, et al. 'Determination of the Liquid Water Diffusivity from Transient Moisture Transfer Experiments, Journal of Thermal ENV. & BLDG. SCI., Vol. 27, No. 4, 2004, 277-305.

[5] L. Pel, K. Kopinga, H. Blocken 'Moisture transport in porous building materials', HERON, Vol. 41, No. 2,1996,95-105.

[6] M. Matsumoto 'Energy conservation in heating cooling ventilating building: heat and mass transfer techniques and alternatives (ed. Hoeogendoorn C.J. and Afgan N.H.)', Washington: Hemisphere Pub. Corp.: 1978, 1-45.

[7] M. Abuku, S. Hokoi, S. Takada 'Heat and Moisture Transfer in Cloth Considering Salt Influences Part 2-Numerical model of simultaneous heat, moisture and salt transfer and analysis of moisture uptake and evaporation', Journal of heating, air-conditioning and Sanitary Engineers of Japan, 131 ,2008.

[8] O. Söhnel, P. Novotny, 'Density of Aqueous Solutions of Inorganic Substances', Elsevier, Amsterdam, 1985.

[9] C. Iba, 'A Study on Freezing-Thawing Processes in Porous Building Walls', Master thesis of Graduate School of Engineering, Kyoto University, 2002.

Local strain measurements during water imbibition in tuffeau polluted by gypsum

Mohamed Ahmed Hassine*, Kévin Beck, Xavier Brunetaud and Muzahim Al-Mukhtar
University of Orleans, INSA-CVL, PRISME - EA4229, Orleans, France
*Mohamed-ahmed.hassine@etu.univ-orleans.fr

Abstract

The research presented in this communication aimed to evaluate the mechanism of spalling generated by mechanical stresses and strains developed due to imbibition and the link with atmospheric pollution generating gypsum in the stone. Tests were carried out on a French limestone, called tuffeau. Local strains were measured using strain gage rosettes during water imbibition in polluted and unpolluted samples. Three rosettes were placed on a sample at different distances from the surface (1, 4, 7 cm) to measure locally the strain during water infiltration. These rosettes determine the strain in three directions 0°/45°/90°.
Results concerning the behaviour of polluted stone during imbibition are compared to unpolluted stone. Results concerning differential strains between

the surface and the core of the stone due to water infiltration demonstrate unusual behaviour not restricted to expansion alone: a local contraction zone and expansion zone in each direction separately for the stone. Pollution by gypsum has an effect on the strain measurements mainly on the first few centimetres of the stone.

Keywords: gypsum, spalling, local strain, imbibition, tuffeau

1. Research aim

Limestones constitute the main construction materials in historical monuments and are affected by several deterioration mechanisms, among which spalling.[1, 2] This decay is defined as a detachment of thick plates (1 to 3 cm thick) gradually formed on the surface of the stone *(Fig. 1)*. Once the plate falls off, the resulting stone surface turns to powder. Mineralogical analysis of the degraded stone, throughout its depth, shows the presence of gypsum ($CaSO_4.2H_2O$) located mainly within a crack network parallel to the surface at 1 or 2 cm depth.[3] Therefore, the study of the effects of pollution on these stones is one of main aims to understand the physico-chemical mechanism leading to spalling. There is also a consensus on the role of water, which is the source of physical changes and the dissolution/ crystallization of phases. In order to further our knowledge of the reaction of stones with respect to water migration and the effect of pollution,

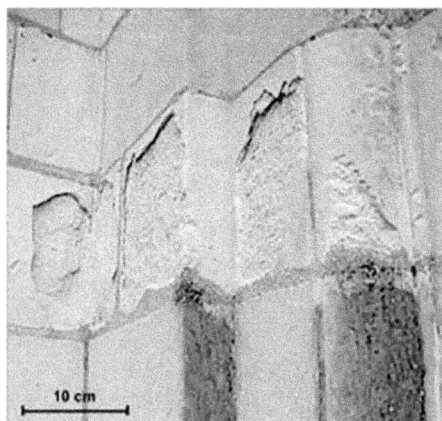

Figure 1: Example of spalling in tuffeau at the "Chateau de Chambord" (France)[3]

a

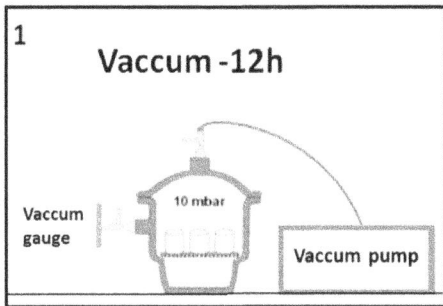

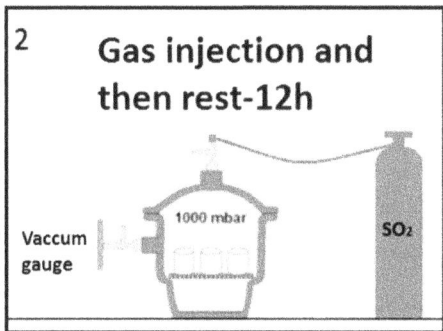

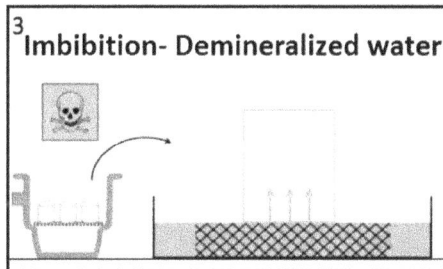

b

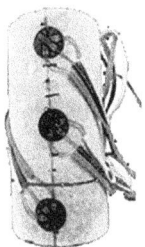

Figure 2: Sample pollution process (a), sample of tuffeau with strain gage rosettes (b)

this paper presents the results of an experimental campaign applied to a French limestone: tuffeau. The experiment consisted in subjecting stone samples (both polluted and unpolluted) to imbibition, and in monitoring strain at different heights in the samples. Pollution was ge-

nerated in the laboratory with the precipitation of gypsum in depth within the fresh stone samples. The work presented aimed to assess the mechanism of spalling generated by mechanical stresses and strains developed due to imbibition and the link with atmospheric pollution.

2. Studied stone and methodology

Tests were carried out on tuffeau, the stone used as building material in the construction of most cultural heritage buildings in the Loire Valley in France. It is of Turonian age (Upper Cretaceous period between 88-92 million years ago). It is mainly affected by spalling. Tuffeau is composed of a major calcite phase (50%), a high siliceous fraction (40%: opal Cristobalite-Tridymite and quartz) and a significant clay content (10%: glauconites, smectites, illite). It is very porous (45% of porosity), with a bi-modal (first peak at 8 µm; second peak at 0.01 µm) porous network.[4] The mechanical characteristics depend on the degree of water saturation and on the orientation with respect to the bedding plane.[5] The stone tested in this experimental campaign was extracted from the Usseau quarry, located in the Vienne department, in the Center-West of France. Two cylindrical stone samples, 40 mm in diameter and 80 mm in height, were cored in the direction parallel to the bedding plane (i.e. the cylinder axis is parallel to the bedding plane) *(Fig. 2)*. This direction was chosen to simulate a real in-situ imbibition process due to rain on the stonework.

For each sample, three strain gage rosettes were glued on the lateral surface of the cylindrical sample, at different heights: 10 mm (J1); 40 mm (J2); 70 mm (J3). Each strain gage rosette was composed of three strain gages, oriented at 0°, 45° and 90° depending on the eigenvectors of the loading, which correspond to the same directions as the axes of symmetry of the sample. The rosettes are „KFG 120" type from Kyowa and were glued using a cyanoacrylate glue (CC-35A). The diameter of each rosette is 1 cm. The error of measurement is about 1E-06 m/m and the quantification limit is about 1E-08 m/m. Before testing, the samples were dried in an oven at 60°C during 92 hours. Then, they were cooled down in a desiccator with a drying salt that maintains an almost zero relative humidity. One of the samples was subjected to pollution by gypsum inside the stone. Pollution was generated in the laboratory by artificial ageing through an injection of dry gaseous SO_2 in the porous network of the stone followed by a partial imbibition of samples with water *(Fig. 2)*. For the creation of gypsum in the porous network, the following chemical equations were used:

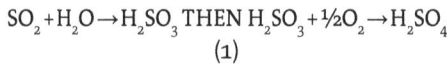

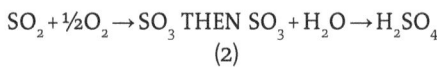

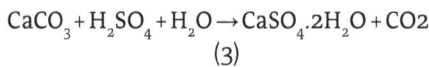

$$SO_2 + H_2O \rightarrow H_2SO_3 \text{ THEN } H_2SO_3 + \tfrac{1}{2}O_2 \rightarrow H_2SO_4$$
$$(1)$$

$$SO_2 + \tfrac{1}{2}O_2 \rightarrow SO_3 \text{ THEN } SO_3 + H_2O \rightarrow H_2SO_4$$
$$(2)$$

$$CaCO_3 + H_2SO_4 + H_2O \rightarrow CaSO_4.2H_2O + CO2$$
$$(3)$$

The first two equations (1) and (2) are two possible reactions to obtain sulfuric acid and then the reaction of sulfuric acid, water and calcite (dissolved from the stone) makes it possible to obtain gypsum in the porous network inside the stone and not only at the surface (3). The injection of sulfur dioxide (SO_2) was carried out in the dry state in samples previously dried under vacuum for about 12 hours. The sample was therefore saturated with SO_2 gas ($\approx$ 12 hours) and after opening of the desiccator, it was subjected to partial capillary imbibition ($\approx$ 2 cm) of distilled water. Afterwards, the samples were stored in an oven at 50°C for 15 days. The stone sample, in the dry state, was placed on a plastic grid that allows deminerali-

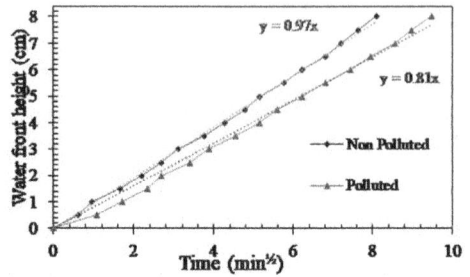

Figure 3: Imbibition curves of polluted and unpolluted tuffeau

zed water to be imbibed into the sample by a capillary process. During the test, the stone sample was subjected to imbibition in controlled conditions (20°C and 50% of relative humidity).

3. Results

First, we present the well-known imbibition curves based on visual measurements of water front height and the time of capillary rise. The curve of water front height (cm) over square root of time (min½) is linear *(Fig. 3)*. The slope represents an intrinsic property of the material.

The imbibition curves show that the hydric properties of tuffeau changed due to the presence of gypsum. The slope of the curve decreased from 0.97 to 0.81. The imbibition rate decreased by 16%. This is the first evidence for the effect of pollution.

Figures 4.a and 4.b show respectively the results of strain measurements of vertical gages for unpolluted and polluted tuffeau. V10, V40, V70 refer respectively to the vertical strains of gages at 10 mm, 40 mm and 70 mm of height.

While the water penetrates into the stone, local vertical strains change. During the capillary imbibition and depending on the relative location of the water with respect to the rosette, the strain may be a contraction, an exten-

a

b

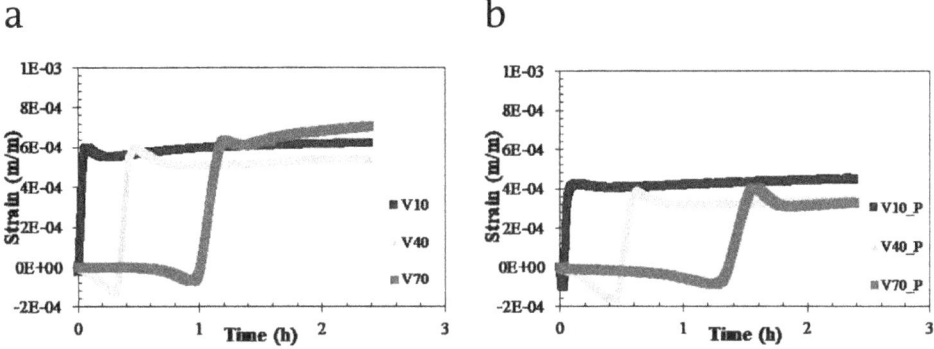

Figure 4: Vertical strains of unpolluted tuffeau (a) and polluted tuffeau (b)

sion, or a stabilization. When the water is below the rosette, the stone is locally contracted. The amplitude is estimated at 1.6E-04 m/m which represents 28% of total strain. Once the water reaches the rosette height, there is an abrupt extension up to a value of about 6E-04 m/m for unpolluted tuffeau and around 3.8E-04 m/m for polluted tuffeau. Then, a small contraction (relative to the amplitude of the extension) is measured, followed by a stabilization step. These stages are present for polluted and unpolluted tuffeau. Two differences were reported between the two states: The magnitude of total vertical strain was higher for unpolluted stone (around 6E-04 m/m against 4E-04 m/m), and the imbibition rate was higher for unpolluted stone. The pollution

by gypsum decreased the total strain by 33%, which is significant.

Figures 5.a and 5.b show respectively the results of strain measurements of horizontal gages for unpolluted and polluted tuffeau. H10, H40, H70 refer respectively to the vertical strains of gages at 10mm, 40mm and 70mm of height. The horizontal strains show monotonous evolution in extension for polluted and unpolluted tuffeau. The same reasoning was adopted for the water front. When the water front is below the rosette, stone extension begins. Then, a sudden extension occurs when the water front reaches the rosette. A second stabilization plateau is obtained when the water front exceeds the rosette. Pollution has no real effect on the total horizontal strain value.

a

b

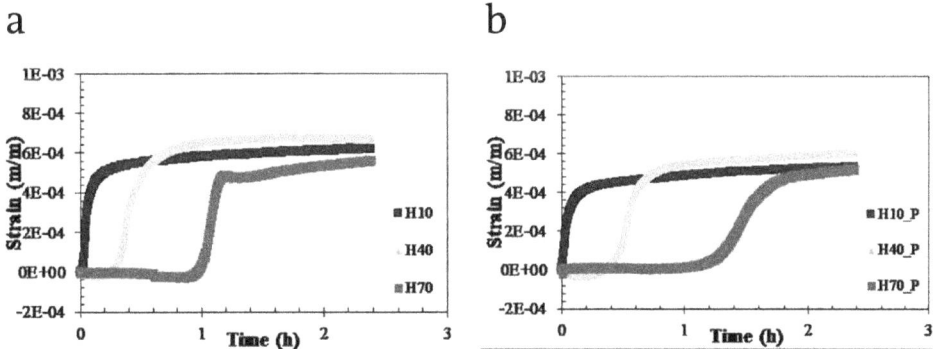

Figure 5: Horizontal strains of unpolluted tuffeau (a) and polluted tuffeau (b)

4. Discussion

As pointed out in the Results section, water has an effect on local strain changes. In vertical strains, four distinguishable stages are observed for tuffeau (polluted and unpolluted).

The first stage starts when the water front has not yet reached the rosette; there is vertical contraction and the onset of horizontal extension. This behaviour could be the consequence of the horizontal extension of the imbibed zone just under the rosette. Vertical contraction is due to the Poisson effect.[5] The horizontal tensile strain is accompanied by vertical compressive strain.

When the water front crosses the zone of the rosette, the extension is in all directions due to pore saturation. This phenomenon is the expected reaction of stone due to imbibition.[6–8] The total vertical strain of tuffeau is higher than the total vertical strain due to the high quantity of clays in tuffeau (10%).[7] The presence of gypsum salt has an effect on the extension of tuffeau and reduces the maximum strain value by 33%. In polluted sample, the gypsum which replaced the calcite acted as a local reinforcement of the stone structure. Gypsum is a soluble salt with a low solubility (2 g/l) but moves slightly during an imbibition. The main problem during partial imbibition is the accumulation zones of gypsum due to tidal effect with the movement of water. In these zones, gypsum content could be very high, and this effect of reduction of strain could be higher.

Then, the third stage begins when the water front is above the rosette. There is a slight contraction in vertical strain and a continuity of horizontal extension. This effect is due to two phenomena: the continuity of capillary imbibition in capillary pores and the slower saturation of small pores which decrease the saturation of the capillary pores. This stage is restricted because of the bi-modal porosity network of tuffeau as explained by Hassine.[5] The stone has the same behaviour even in the presence of gypsum. The final stage concerns the stabilization of strain in all directions, so all porosity is fully saturated by water. *Table 1* sums up the main phenomena affecting tuffeau stone during imbibition.

The main interesting feature of the results concerns the effect of pollution during imbibition on the local vertical strain. As shown by Hassine et al.[5], the effect of imbibition is not limited to monotonous extension, as was previously thought.[7, 8] Firstly, the existence of vertical contraction with significant amplitude may contribute to spalling. Secondly, the presence of salt (gypsum) in the stone stiffens the skeleton and creates a difference in stiffness between polluted and unpolluted areas. Subsequently, the polluted zone (the most rigid one) creates restraint stresses and a shear plane between the two zones. By comparison with in situ conditions of stones in a wall submitted to a rain event where pollution is located on the first few centimeters, the vertical contraction observed during the experiments corresponds to hori-

Vertical strain	- No effect on contraction phase
	- Decreases the maximum strain value by 33%
Horizontal strain	- No effect on strain changes
Imbibition kinetics	- Slows the imbibition rate

Table 1: Effect of gypsum pollution on tuffeau behaviour during imbibition

zontal contraction perpendicular to the surface of the stone. This contraction is restrained by the polluted area, resulting in tensile stress, situated just after the water front, i.e. around 1 to 2 cm depth. The amplitude of this tensile stress is not negligible compared to the low tensile strength of tuffeau. This could be the origin of a crack parallel to the surface at 1 to 2 cm depth, which corresponds to the description of spalling.

5. Conclusion

This paper evaluated the effect of pollution during water migration on the strain of a French limestone, tuffeau, and proposed an interpretation for the degradation process causing spalling. The local strains were measured by using strain gage rosettes. On tuffeau, which is a homogeneous stone, the strain measurements during an imbibition on several samples of unpolluted tuffeau showed reproducible results.[5] The evaluation of strain measurements in a sample polluted and unpolluted is reproducible. The tests have been repeated 3 times in each case and the error is about 1E-06 m/m.

Gypsum does not affect the contraction phase in the vertical direction, and decreases the maximum vertical strain value by 33%. However, no change is detected horizontally. The imbibition kinetics of the tuffeau are slowed down.

The presence of vertical contraction just above the water front can contribute to the discussion about the cause of spalling. This vertical contraction may correspond to horizontal contraction for in-situ stones in a wall exposed to rain. If the displacement of the stone in the direction of depth is restrained by harder stone (polluted area), it could result in significant tensile stress, able to generate a crack parallel to the surface. As the typical depth of water imbibition during a rain event in tuffeau is between

1 and 2 cm, the depth of the crack would be around those values. Hence, the presence of gypsum and the presence of vertical contraction observed on a sample submitted to imbibition could be a key element in the clarification of the origin of spalling. Indeed, this study is the first step for understanding the mechanisms of spalling of tuffeau and this hypothesis must be supported by further tests.

This work will be pursued on a second stone, Richemont stone, used as a replacement stone during restoration work at the end of the 20th century. The behavior of the two stones will be compared in order to understand the effect of gypsum pollution during imbibition and to explain why no degradation by spalling is observed on Richemont stone. Finally, research on both tuffeau and Richemont stone will be extended to investigate cycles of water imbibition – drying with and without gypsum pollution to better understand their long-term behavior.

References

[1] V. Vergès-Belmin (Ed.), Illustrated glossary on stone deterioration patterns, (2008) http://www.icomos.org/publications/monuments_and_sites/15/pdf/Monuments_and_Sites_15_ISCS_Glossary_Stone.pdf.

[2] C. Walbert, J. Eslami, A.-L. Beaucour, A. Bourges, A. Noumowe, Evolution of the mechanical behaviour of limestone subjected to freeze–thaw cycles, Environ. Earth Sci. 74 (2015) 6339–6351. doi:10.1007/s12665-015-4658-2.

[3] S. Janvier-Badosa, K. Beck, X. Brunetaud, M. Al-Mukhtar, The occurrence of gypsum in the scaling of stones at the Castle of Chambord (France), En-

viron. Earth Sci. 71 (2014) 4751–4759. doi:10.1007/s12665-013-2865-2.

[4] A. Al-Omari, X. Brunetaud, K. Beck, M. Al-Mukhtar, Effect of thermal stress, condensation and freezing-thawing action on the degradation of stones on the Castle of Chambord, France, Environ. Earth Sci. (2013) 1–13. doi:10.1007/s12665-013-2782-4.

[5] M.A. Hassine, K. Beck, X. Brunetaud, M. Al-Mukhtar, Strain changes during the progress of water infiltration in tuffeau stone, in: Int. RILEM Conf. Mater. Syst. Struct. Civ. Eng. Conf. Segments Hist. Mason., Lyngby, 2016.

[6] E. Colas, J.D. Mertz, C. Thomachot-Schneider, V. Barbin, F. Rassineux, Influence of the clay coating properties on the dilation behavior of sandstones, Appl. Clay Sci. 52 (2011) 245–252. doi:10.1016/j.clay.2011.02.026.

[7] J. Berthonneau, P. Bromblet, F. Cherblanc, E. Ferrage, J.-M. Vallet, O. Grauby, The spalling decay of building bioclastic limestones of Provence (South East of France): From clay minerals swelling to hydric dilation, J. Cult. Herit. 17 (2016) 53–60. doi:10.1016/j.culher.2015.05.004.

[8] Dessandier D., Role of the clays in the modifications of hydric properties: consequence on the mechanism of plates exfoliation of Tuffeau., in: Proc. 9th Int. Congr. Deterior. Conserv. Stone, Venice, 2000.

Assessment of the durability of lime renders with Phase Change Material (PCM) additives against salt crystallization

Loucas Kyriakou, Magdalini Theodoridou* and Ioannis Ioannou
Department of Civil and Environmental Engineering, University of Cyprus, Nicosia, Cyprus
* mtheodo@ucy.ac.cy

Abstract

Energy consumption in buildings is mostly associated with the use of heating and cooling systems. Renders with the addition of Phase Change Materials (PCMs) have the ability to absorb and release thermal energy, when the temperature changes accordingly, thus enhancing the thermal comfort and energy efficiency of buildings. Nevertheless, the performance of such renders with traditional binders lacks international experimental data, especially regarding their durability against salt weathering.

This paper focuses on the effect of different percentages of commercial microencapsulated powder PCMs on the properties of hydrated and hydraulic lime renders, investigating at the same time the durability of the end-products against salt crystallization. The aim is to produce energy efficient and durable lime-based renders for the upgrading of contemporary buildings, as well as for conservation purposes.

The modified composites have significantly lower thermal conductivity and increased specific heat capacity at 90 days after laboratory production, thus confirming the great potential of PCMs in enhancing the thermal performance of the aforementioned renders. Comparative tests show that the addition of PCMs has an adverse effect on the mechanical properties of the renders and a noteworthy reduction of their bulk density. Nevertheless, the salt crystallization resistance of the modified renders improves with the percentage of PCM addition, when assessed both quantitatively and qualitatively following 15 full immersion wetting and drying cycles in Na_2SO_4 solution.

Keywords: lime, composites, PCMs, salt crystallization

1. Introduction

Buildings nowadays account for about 40% of the global energy consumption.[1] The continuously augmenting energy consumption in buildings is directly linked to thermal comfort, as well as to the growth rate of population.[2] Therefore, it is imperative that efficient construction technologies and materials are used or developed for the reduction of energy consumption in the built environment and for the preservation of energy resources.

Conventional thermal insulating building materials are usually inappropriate for traditional building envelopes, since normally they are used in thick layers and most of the times they cannot offer the desired results without collateral hitches, such as load bearing problems.[3] The application of latent heat storage is probably one of the most efficient methods for enhancing the thermal behaviour of buildings, since it is based on the phase change enthalpy of a material, which can store heat within a temperature range.[4] Phase Change Materials (PCMs) are widely known due to their ability to absorb and store energy and release it back to the environment, under certain condi-

Mixture	Aggregates	Binder		Additive (% w/w of solids)	W/B	Workability (mm)
		A	H			
REFA	3	1	-	-	1.06	174
PCMA5	3	1	-	5	1.00	176
PCMA10	3	1	-	10	1.13	174
PCMA20	3	1	-	20	1.50	175
REFH	3	-	1	-	0.83	174
PCMH5	3	-	1	5	0.87	176
PCMH10	3	-	1	10	1.05	180
PCMH20	3	-	1	20	1.34	176

Table 1: Mix design of laboratory composites. All quantities are measured by mass. A: hydrated (aerial) lime; H: hydraulic lime.

tions, during the melting and solidifying process respectively.

While the introduction of Phase Change Materials in structural elements often involves the dispersion of PCMs into the matrix, encapsulation of the raw material can prevent leakage without altering the thermal energy transfer efficiency.[5] The incorporation of microencapsulated PCMs in composite construction materials, such as renders, plasters and concrete [6, 7, 8, 9], shows the inordinate interest of the research community in investigating the potential of these additives towards the enhancement of the energy performance of buildings.

Nevertheless, although PCMs have been used as additives in renders, the durability of PCM-enhanced renders has not yet been investigated. Furthermore, additional research is needed on the effects of PCM addition on the physico-mechanical properties of renders, since existing related data is rather inconclusive and contradicting (e.g. some studies report a negative impact on the mechanical properties, whereas others suggest a positive effect).[6, 10]

This study focuses on the effects of PCMs on the performance of lime-based renders, and includes research on the durability of PCM-enhanced lime renders against salt weathering. Due to the lack of a suitable standardized methodology to study the salt weathering phenomenon on renders with traditional binders under laboratory conditions, the concept of full immersion in sodium sulphate (Na_2SO_4) solution and subsequent drying is adopted. It should be noted that this test is considered to be highly aggressive and inconsistent with natural environmental conditions, let alone the fact that in this study it is performed on relatively weak construction materials, such as the lime renders under investigation.

2. Materials and methods

2.1. Materials and sample preparation

Eight lime-based mixtures were designed and produced in the laboratory; two reference mixtures and six optimized ones with the addition of PCMs at different percentages (5%, 10% and 20% w/w of solids). The binder consisted of either hydrated (aerial) lime (A: CL80 supplied by Hellenic Mining Public Co.) or natural hydraulic lime (H: NHL 3.5 supplied by Lafarge). The aggregate fraction consisted of Latouros sand, a local calcarenite fine aggregate with 0-2 mm particle size.[11] A commercial PCM (Micronal DS 5038 X supplied by BASF) was also used; this is microencapsulated in powder (dry) form and has its main melting and crystallization peaks in the range of 24-25 °C. The selected binder/aggregate ratio was 1:3 by weight, based on the prevalence of this ratio in ancient and traditional composites in Cyprus and other areas of the world.[12, 13] The water demand of each mixture was subsequently estimated after achieving a constant workability in the range of 175±5 mm according to EN 1015-3.[14] The water to binder ratio (W/B) along with all the other mix design details are given in *Table 1*.

2.2. Experimental tests

2.2.1. Physico-mechanical and thermal properties

A series of laboratory tests and analytical techniques were used for the determination of the physical and mechanical performance of the samples. All the tests were carried out at different curing times (28, 56 and 90 days); this paper, however, presents only the results recorded 90 days after casting, due to page limitation. Fresh hydrated mortars were stored in a room with constant temperature (23±5) °C and humidity (50±5%), while hydraulic ones were cured in closed plastic containers at constant temperature (23±5) °C and high humidity conditions.

Mercury Intrusion Porosimetry (MIP) was carried out on bulk samples to evaluate porosity (p_o), apparent density (p_a), average pore diameter and pore size distribution. Moreover, the capillary absorption coefficient (s) was measured on prismatic specimens (40x40x160 mm), using water as the wetting liquid. Prismatic specimens were also used for testing the materials under flexural load (three-point bending). Compression strength tests were executed on the fragments of each specimen resulting from the flexural tests. The analysis of the temperature response of the test specimens (175x50x30 mm) to heat flow impulses, following a standardized bulk measurement in accordance with ASTM E1530-11 [15], was used for the determination of the thermal properties of the renders investigated. The test was carried out in the curing room, at a stable temperature of 23±5 °C; this temperature is lower than the phase change temperature of the PCM particles (25 °C).

Level	Description
Perfect	Specimen intact
Good	Very minor damage or minor cracks
Moderate	Rounding of corners and several cracks or detachment of small fragments
Bad	Specimen with several major cracks or broken
Fail	Specimen in pieces or disintegrated

Table 2: Five-level rating classification for visual inspection of the materials resistance to salt crystallization.

2.2.2. Salt crystallization

For the assessment of the renders' resistance to salt crystallization, six specimens (50x50x50 mm) from each mixture were subjected to a maximum of 15 wetting and drying cycles using a 10% solution of Na_2SO_4. The 16th cycle corresponded to a full immersion of the specimens in deionised water for 24 hours and then drying until constant mass was achieved. The test was carried out when the samples had completed at least 90 days of curing. The durability of each specimen after each crystallization cycle was assessed based on two parameters: (i) visual inspection (qualitative evaluation) and (ii) percentage loss in mass (quantitative evaluation). With regard to the qualitative evaluation, a five-level rating classification was devised *(Table 2)*.

3. Results and discussion

The results of the thermal tests *(Table 3)* demonstrate the potential of PCMs in enhancing the thermal performance of lime-based renders.[16, 17] PCM addition causes a significant reduction in the thermal conductivity of the modified end-products, compared to the reference ones. Higher reductions correspond to the renders with the higher percentage of PCM addition.

Increase in the specific heat capacity is further recorded for the modified renders *(Table 3)*; this is also proportional to the PCM content in the mix design. Other researchers[16] also noted a positively correlation between the specific heat capacity and the PCM addition.

The thermal diffusivity of the PCM-enhanced composites is notably lower *(Table 3)*. As in the case of thermal conductivity, this decrease is related to the amount of PCM in the mixture.

Higher open porosity and lower apparent density values are observed *(Table 3)* for the mixtures with the PCM addition, compared to the reference composites (REFA and REFH); the fluctuation is generally proportional to the percentage of the PCM additives in the mix design. The

Mixture	P_o	P_a	Av. Pore Diam.	s	FS	UCS	λ	Cp	a
	(%)	(g/cm³)	(nm)	(mm/min$^{1/2}$)	(MPa)		(W/mK)	(10²J/kgK)	(10⁻⁷ m²/s)
REFA	30.8	1.65	300.3	2.01	1.32	2.21	0.575	8.147	4.040
PCMA5	32.1	1.53	231.9	1.19	1.03	3.31	0.452	9.200	3.110
PCMA10	34.8	1.51	133.7	0.88	0.91	3.08	0.387	9.663	2.690
PCMA20	33.1	1.44	76.9	0.64	0.72	3.22	0.316	10.361	2.250
REFH	28.8	1.66	69.8	0.82	2.85	10.76	0.687	8.336	4.740
PCMH5	35.9	1.57	90.5	0.06	2.28	8.70	0.570	9.351	3.970
PCMH10	38.1	1.34	108.3	0.23	1.36	5.75	0.426	10.395	2.970
PCMH20	39.0	1.32	96.3	0.39	1.32	3.92	0.309	10.760	2.230

Table 3: Physico-mechanical and thermal properties of laboratory composites measured at 90 days after casting. po: open porosity, pa: apparent density, s: water capillary absorption coefficient, FS: flexural strength, UCS: uniaxial compressive strength, λ: thermal conductivity, Cp: specific heat capacity, a: thermal diffusivity.

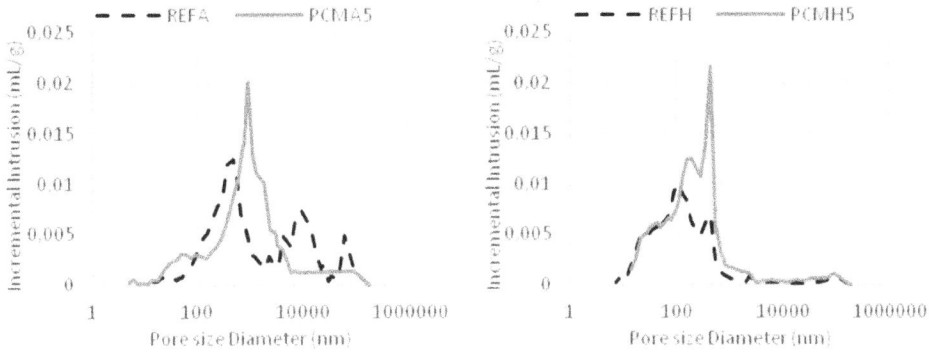

Figure 1: MIP pore size distributions for REFA & PCMA5 (left) and REFH & PCMH5 (right).

relatively higher porosity values can be attributed to the higher water demand of the PCM-enhanced mixtures in order to attain the desired workability[16, 18]; this may, in turn, be associated with the increased volume fraction of solids, the fineness of the microcapsules and the hydrophylic nature of their polymeric wall.

Regarding the capillary absorption results *(Table 3)*, the addition of PCMs seems to lead to significantly reduced values in both the hydrated and hydraulic renders (over 65% and 90% respectively). Similar results were also recorded by other researchers.[19]

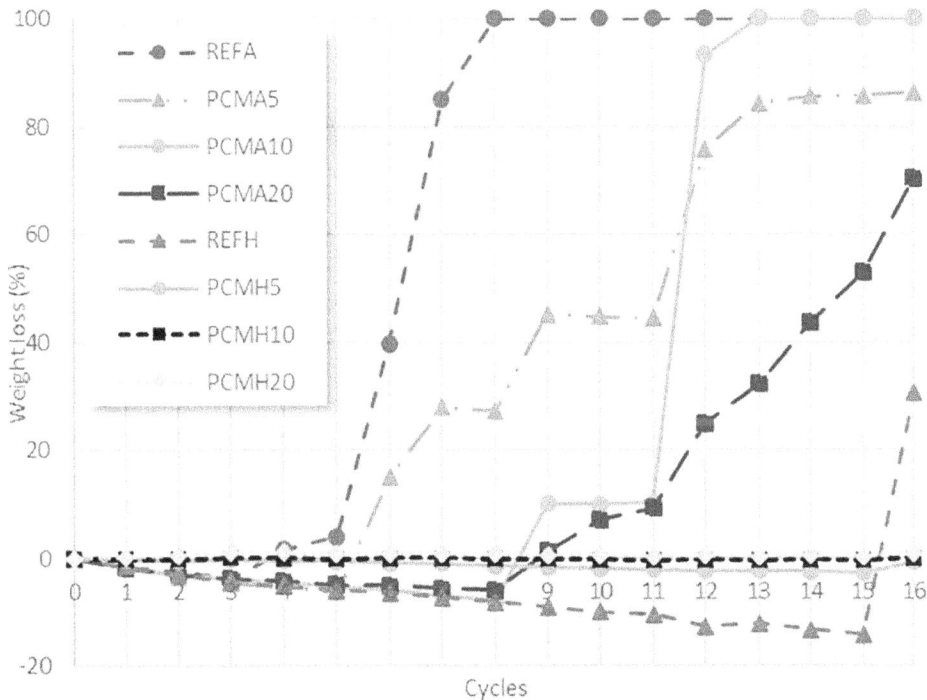

Figure 2: Weight loss of specimens after salt crystallization cycles.

The pore size distributions recorded by MIP *(Figure 1)* show changes in the pore structure of the mixtures with and without PCMs. More specifically, the PCM addition causes a shift of the main peak towards larger pore sizes in comparison with the reference mixtures; in the hydrated mixtures, the main peak shifts from ca. 500 to 900 nm (REFA and PCMA5 respectively) and in the hydraulic ones from ca. 100 to 430 nm (REFH and PCMH5 respectively). Furthermore, in the case of hydrated lime-based renders, the PCM addition causes a decrease of the larger pore volumes (i.e. peaks at ca. 9000 and 60000 nm). Other researchers[10] reported similar changes in the pore structure of PCM-enhanced hydrated lime composites.

Generally lower flexural and compressive strengths are reported for the hydraulic mixtures with the addition of PCMs, compared to the reference composites *(Table 3)*. This agrees well with the higher porosity values observed and may be attributed to the higher water demand of the PCM-enhanced mixtures. Lower flexural strength values are also recorded in the case of hydrated lime renders, compared to the reference materials *(Table 3)*. These observations are in line with other researches reporting on similar lime-based renders[17]. It is worth noting that, despite the fact that PCM addition leads to a decrease in flexural strength, it is found to enhance the compressive strength in the case of hydrated lime mortars. The latter may be attributed to the decrease of the bigger pore volumes recorded by MIP *(Figure 1)*.

The determination of the renders' resistance to Na_2SO_4 attack through quantitative evaluation is given in *Figure 2*, whereas the qualitative assessment of their condition following salt crystallization is presented in *Figure 3*.

Only six mixtures survived all 15 cycles of the artificial weathering test (PCMA5, PCMA20, REFH, PCMH5, PCMH10, PCMH20). Their noteworthy resistance to salt crystallization is primarily attributed to the binder nature (hydraulic lime is associated with enhanced durability

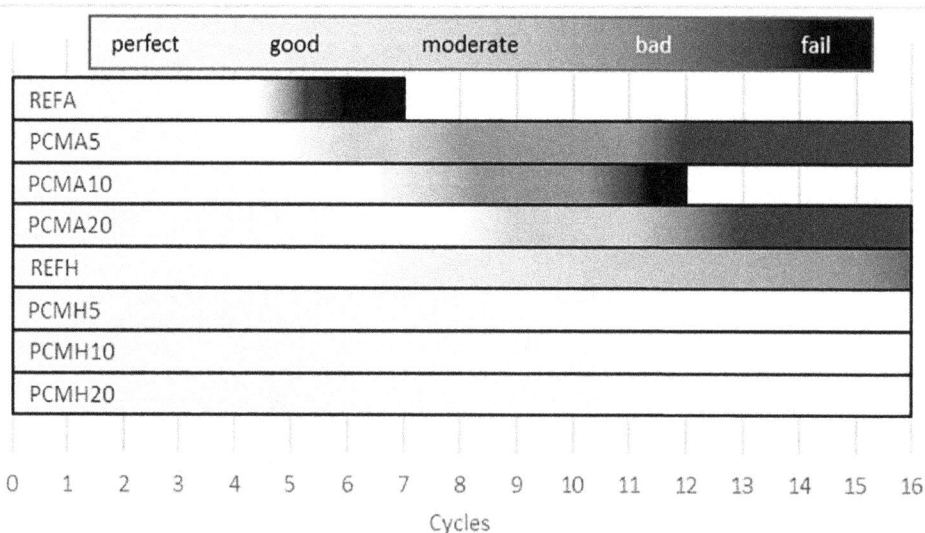

Figure 3: Qualitative evaluation of salt resistance by visual inspection of render mixtures. For explanation on the scale used see Table 2.

compared to hydrated lime). At the same time, PCM addition in the case of hydrated lime renders seems to lead to better durability results; it is worth noting that, in the case of hydraulic composites, the PCM-enhanced specimens completed the durability test without any weight loss. It is therefore evident that better durability results correspond to the PCM-enhanced renders, when compared to the reference mixtures. This is found to be in good agreement with the materials' capillary absorption results, as well as with their high values of porosity and the shift of the main peak of pore volume towards bigger pore sizes *(see Table 3 and Figure 1)*; this may have led to less damage linked to high crystallization pressures due to crystal growth within the pores.[20] It is worth noting that the increase observed in salt crystallization resistance is in constrast to the decrease reported in flexural strength; the latter is possibly attributed to localised flaws in the specimens containing PCMs, however further investigation using enhanced microscopy (e.g. SEM) is needed to confirm this.

The positive effects of PCM addition on the durability of the lime renders investigated in this study is confirmed by the outcomes of the visual inspection *(Figure 3)*. All PCM-enhanced hydrated renders show a better performance against salt weathering, compared to the reference. In the case of hydraulic renders, and despite the fact that the mixture without PCMs (REFH) survived until the end of the crystallization test, the addition of PCMs is associated with efficient damage prevention, since all relevant specimens (PCMH5, PCMH10, PCMH20) are found intact after the end of the test.

4. Conclusions

PCMs significantly reduced the thermal conductivity and thermal diffusivity, while they increased the specific heat capacity of the modified composites. The modification in thermal properties is proportionally linked to the amount of PCMs added to the mixture. This demonstrates the potential of PCMs in enhancing the thermal performance of lime-based renders.

The relatively higher open porosity and lower apparent density values of the PCM-enhanced mixtures are associated with the fact that mixtures with higher PCM content require higher amounts of water to reach the desired workability. Nevertheless, PCM addition significantly reduced the capillary absorption coefficient in both hydrated and hydraulic renders. Noticeable changes in the pore size distribution were also reported since PCM-modified renders were characterized by a shift of the main peak of the pore size distribution towards bigger size pores.

PCM enhancement had in general an inverse effect on the mechanical properties of the end products. However, PCM addition benefited the compressive strength in the case of hydrated lime mortars; this can be attributed to the decrease of pore volumes in the relatively larger pore sizes.

The alterations in several physical values (e.g. lower capillary absorption) due to the addition of PCMs led to noticeable changes in the salt crystallization resistance of the mixtures. Better durability results corresponded to the modified composites when compared with the reference mixtures. This is probably due to the reduced amount of salt absorbed owing to the low capillary absorption of the PCM-enhanced composites; a noteworthy increase of the salt crystallization resistance may also be attributed to the hydraulic binder. Visual inspection of the samples studied confirmed the enhanced performance of PCM modified renders.

The results of this research are overall positive in terms of producing innovative thermally upgraded, yet durable, lime renders. Therefore their incorporation in the construction industry of southern European countries for upgrading the energy efficiency of contemporary buildings and for conservation purposes is deemed promising.

References

[1] L. YANG, H. YAN, J. LAM, Thermal comfort and building energy consumption implications – A review, Applied Energy, (115), (2014), 164-173.

[2] M. LACHHEB, Z. YOUNSI, H. NAJI, M. KARKRI, S. NASRALLAH, Thermal behaviour of a hybrid PCM/plaster: A numerical and experimental investigation, Applied Thermal Engineering, (111), (2017), 49-59.

[3] R. BAETENS, B. JELLE, A. GUSTAVSEN, Phase change materials for building applications: A state-of-the-art review, Energy and Buildings, (42), (2010), 1361-1368.

[4] A. SHARMA, V.V. TYAGI, C.R. CHEN, D. BUDDHI, Review on the thermal energy storage with phase change materials and applications, Renewable and Sustainable Energy Reviews, (13), (2009), 318-345.

[5] L. COPPOLA, D. COFFETTI, S. LORENZI, Cement-Based Renders Manufactured with Phase-Change Materials: Applications and Feasibility, Advances in Materials Science and Engineering, (2016), 1-6.

[6] M. THEODORIDOU, L. KYRIAKOU, I. IO-ANNOU, PCM-enhanced lime plasters for vernacular and contemporary ar-chitecture, Energy Procedia, (97), (2016), 539-545.

[7] C. CARBONARO, Y. CASCONE, S. FAN-TUCCI, V. SERRA, M. PERINO, M. DUTTO, Energy assessment of a PCM-embedded plaster: embodied energy versus operational energy, Energy Procedia, (78), (2015), 3210-3215.

[8] M. KHERADMAND, M. AZENHA, J. AGUIAR, J. CASTRO-GOMEZ, Experimental and numerical studies of hybrid PCM embedded in plastering mortar for enhanced thermal behaviour of buildings, Energy, (94), (2016), 250-261.

[9] A. JAYALATH, R. SAN NICOLAS, M. SOFI, R. SHANKS, T. NGO, L. AYE, P. MENDIS, Properties of cementitious mortar and concrete containing micro-encapsulated phase change materials, Construction and Building Materials, (120), (2016), 408-417.

[10] S. S. LUCAS, V. M. FERREIRA, J. L. BAR-ROSO DE AGUIAR, Latent heat storage in PCM containing mortars – Study of microstructural modifications, Energy and Buildings, (66), (2013), 724-731.

[11] R. FOURNARI, I. IOANNOU, D. VATYLIO-TIS, A study of fine properties and their effect on the quality of cementitious composite materials, Proceedings of the IAEG XII Congress Engineering Geology for Society and Territory – Volume 5: Urban Geology, Sustainable Planning and Landscape Exploitation, G. LOLLINO, A. MANCONI. F. GUZZETTI, M. CULSHAW, P. BOBROWSKY, F. LUI-NO, Springer International Publishing, Switzerland 2015, 33-36.

[12] M. THEODORIDOU, I. IOANNOU, M. PHILOKYPROU, New evidence of early use of artificial pozzolanic material in mortars, Journal of Archaeological Science, (40), (2013), 3263-3269.

[13] A. MOROPOULOU, A. BAKOLAS, A. ANAGNOSTOPOULOU, Composite materials in ancient structures, Cement and Concrete Composites, (27), (2005), 295-300.

[14] European Committee for Standardization. Methods of test fot mortar for masonry. Determination of consistence of fresh mortar (by flow table), EN 1015-3, 2007.

[15] ASTM International. Standard Test Method for Evaluationg the Resistance to Thermal Transmission of Materials by the Guarded Heat Flow Meter Technique. ASTM E1530-11, 2016.

[16] L. VENTOLÁ, M. VENDRELL, P. GIRALDEZ, Newly-designed traditional lime mortar with a phase change material as an additive, Construction and Building Materials, (47), (2013), 1210-1216.

[17] J. VIEIRA, L. SENFF, H. GONÇALVES, L. SILVA, V. M. FERREIRA, J. A. LABRINCHA, Functionalization of mortars for controlling the indoor ambient of buildings, Energy and Buildings, (70), (2014), 224-236.

[18] S. CUNHA, J. AGUIAR, PACHECO-TORGAL F., Effect of temperature on mortars with incorporation of phase change materials, Construction and Building Materials, (98), (2015), 89-101.

[19] Z. PAVLÍK, M. PAVLÍKOVÁ, V. KAULICH, A. TRNÍK, J. ONDRUŠKA, O. ZMĚŠKAL, R. ČERNÝ, Properties of a New Type of Plaster Containing Phase-Change Material, IPCSIT, (28), (2012), 122-126.

[20] G. SCHERER, Stress form crystallization of salt, Cement and Concrete Research, (34), (2004), 1613-1624.

Salt crystallization tests

Salt crystallization tests: Focus on their objective

A. Elena Charola[1]*, Inge Rörig-Dalgaard[2], Jacek Chwast[3] and Jan Elsen[3]
[1] Museum Conservation Institute, Smithsonian Institution, Washington, DC, USA
[2] DTU, Department of Civil Engineering, Group of Construction Materials, Lyngby, Denmark
[3] KU Leuven, Department of Earth and Environmental Sciences, Leuven, Belgium
* charolaa@si.edu

Abstract

Many factors influence the durability of a building material, such as its mechanical resistance, exposure conditions and the presence of soluble salts in it. Since the latter interact with each other, it is difficult to relate any of them to the specific damage observed. Lubelli et al.[1] have recently summarized the shortcoming of some salt crystallization tests and of the mathematical models based on the accepted salt crystallization theories. The net result is that there is no single salt crystallization test that can provide all answers since crystallization kinetics, depending on specific circumstances, play a critical role in the induced deterioration. Nonetheless, specific tests have been developed which have proved to be practically viable in assessing particular material compatibility or potential damaging sources. Two such tests are described, one using sodium chloride to determine compatibility of restoration mortars, and another where the efflorescence of gypsum for brick masonry is evaluated. These methods have proven their reliability and lead to the conclusion that salt tests should be designed for specific objectives.

Keywords: sodium chloride test, sodium sulfate test, masonry materials, crystallization kinetics

1. Introduction

It has long been known that laboratory tests have limitations and rarely can reproduce the deterioration induced on building materials over time. Lubelli et al.[1] have recently summarized the shortcoming of these tests and of the mathematical models based on the accepted salt crystallization theories.

The deterioration induced by salt crystallization has been studied for over a century and the present paper aims to point out relevant issues with regards to the most common salts present in monument and structures, such as the high damage inducing sodium sulfate, the ubiquitous sodium chloride, and the less damaging gypsum. Because sodium sulfate induces such fast and intense deterioration in materials, it has been the salt most used for crystallization tests to evaluate mechanical resistance (durability) of materials. However, one of the problems with this test is that the conditions under which it is undertaken strongly influence the results obtained.[1] Therefore, results from different laboratories are practically impossible to compare.

Meanwhile, other crystallization tests have been developed for specific materials, such as the sodium chloride for replacement mortars, and the gypsum efflorescence for evaluation of both brick as well as mortar formulations. Tests with a more focused objective have proven more useful than the generalized sodium sulfate crystallization test.

2. Crystallization tests

2.1. Sodium sulfate (Na$_2$SO$_4$)

The oldest salt crystallization test uses a Na$_2$SO$_4$ solution to impregnate the samples and then subject them to wet-drying cycles[2,3] since it can result in intense damage to the substrate. It was used to test durability and mechanical resistance. The deterioration of the substrate can be attributed to the crystal growth that may occur both during the drying as well as the wetting step[4] and that both thenardite, the anhydrous form, and mirabilite, the decahydrate, can crystallize simultaneously, one at the expense of the other, thus inducing repeated crystallization cycles.[5]

From the study of the actual behavior of salts as they crystallize out, both Pühringer [6] and Zehnder and Arnold[7] show that crystal size and habit will depend on the relative humidity (or moisture availability) during the crystallization process and strongly related to crystallization kinetics.

Rodriguez Navarro et al. [8] showed that under normal conditions, i. e., non-equilibrium, 40% RH and 20ºC and relative fast evaporation, the anhydrous thenardite will crystallize directly from solution (although it is below the 32.4ºC thenardite-mirabilite transition point) inducing significant deterioration to the substrate and thereby also reflecting the importance of crystallization kinetics. Furthermore, no direct hydration of thenardite occurred, with mirabilite being produced only from the dissolution of thenardite and its reprecipitation, as previously shown [5]. For materials with a relatively homogeneous pore system, those with finer pores tend to suffer more deterioration from salt crystallization than those with larger pores. Further studies by Cultrone and Sebastián[9] confirmed the increase of smaller pores (<1 μm) after salt crystallization tests is possibly caused by

microfissures formation. Benavente et al.[10] characterized material parameters of several different rocks and applied a non-standard Na$_2$SO$_4$ crystallization test to determine their resistance to it. Using a principal component analysis they correlated the resulting salt weathering to the pore structure of the stone, its water transport properties, as well as its mechanical strength finding that the latter was the most important parameter in resisting this particular salt weathering in agreement with previous tests carried out on bricks.[11] It would appear that this non-standard crystallization test proved to be more useful than the accepted standard test.

2. Sodium chloride (NaCl)

Sodium chloride, halite, is probably one of the most ubiquitous salts originating both from marine environments and the use of deicing salts in northern countries.[12] This salt induces deterioration such as powdering, which is more likely to develop when isometric crystals grow, as observed on Venetian brick shown in *Figure 1* left; or flaking, which in our experience can be attributed to columnar crystals growth as shown on another Venetian brick sample, *Figure 1* right. This particular sample was collected in Venice but only analyzed some years later. As shown in the photomicrograph, initially NaCl formed the cubic crystals, but with the change in environmental conditions (closed sample holder) over years, the slow drying out resulted in the change of habit of the crystal growth following the pattern described by Zehnder and Arnold[7] and clearly showing that crystals grow from the bottom up.

Sodium chloride is far less damaging than sodium sulfate. When tested in immersion/drying cycles far more cycles are needed to produce damage[13] than with Na$_2$SO$_4$, even though studies have

Figure 1: Left: A brick sample collected from the S.Stefano church in Venice and analyzed some years later show the change in crystal growth habit resulting from the changed environmental conditions. Right: Cubic halite crystals growing on brick from the Abbazia da Misericordia in Venice (ca. 1980), where brick powder can be seen being pushed away by the growing crystals.

shown that NaCl also tends to crystallize in smaller pores (>1μm).[14] Lubelli[15] presents an overview of salt weathering tests and concludes that sodium chloride crystallization damage is enhanced by repeated dissolution/crystallization cycles induced by fast drying of the moistened sample or by changing RH conditions. Based on these premises, a test[16] that uses this latter approach was developed by Prof. Henriques and his team[17, 18] to evaluate replacement mortars for historic buildings, so as to ensure that the replacement mortar will be compatible with both the remaining original mortar and the weathered masonry material. The test is briefly described below.

The test[16] uses regular shaped specimens for easy visualization of any changes. The specimens are dried at 105 ± 5°C for 24 hours and weighed. Once cooled to room temperature they are immersed in a supersaturated solution of sodium chloride at 20-25ºC for 24 hours. They are left to dry for 1 hour and then dried at 105 ± 5°C until constant weight, i.e., when the weight between two subsequent 24-hour is equal or less than 0.1% of the sample weight. Once this has been achieved, they are cycled in a climate chamber between dry (<40% RH) and humid (>90%RH), every 12 hours. After each cycle, or number of cycles, the samples are weighed and visually evaluated, using the following scale: U unaltered; SD surface disaggregation; HD half disaggregate; and, D disaggregated.

The test has been used extensively and has proved useful, particularly for evaluating restoration mortars for situations where sodium chloride is the main deterioration factor.[18]

2.3. Gypsum (CaSO$_4$·2H$_2$O)

Most of the papers dealing with gypsum address the black crusts formed on calcareous stones from air pollution and the deterioration they caused. Fewer papers deal with the efflorescence of gypsum. These are currently considered as being mainly an aesthetic problem, rather than the potential deteriorating agent it will prove to be with time. When in the 1960s the main monuments and buildings in Paris were cleaned of their black crusts using an aggressive sanding procedure, the reaction of the conservation community was that some black crusts could be considered "protective". It took some years until the phenomenon of black crust formation and its deteriorating effect was clearly understood.

An interesting example of the deterioration that gypsum can induce was reported when a Romanesque mural painting was uncovered in Austria.[19] It had been covered by a thick plaster and had suffered quite some damage in the lower part by rising damp problems. Once uncovered, the mural paintings were desalinated by poulticing. The desalination extracted the more soluble NaCl whose presence significantly increases the solubility of the gypsum.[20, 21] However, its removal reduces the solubility of the gypsum that started "bloom" lifting up the paint layers. Therefore, further conservation measures were required to stabilize the mural painting.[19]

A more prosaic of gypsum efflorescence was found on a brick retaining wall in the garden at the Cowper Hewitt Museum as shown in *Figure 2*. The gypsum, probably found in the soil of the garden on the other side of the wall, migrates solubilized in water (from watering plants and rain) through it and crystallizes out as a white efflorescence at about the top soil level on the garden side.

It is interesting to compare the gypsum crystals formed from direct crystallization with a fairly constant supply of moisture, to those formed by reaction of air pollution with a calcareous material, such as limestone. *Figure 3* shows the well-formed crystal shapes that resulted

Figure 2: Detail of the brick wall showing the gypsum efflorescence formed on the parallel wall below grade to that seen in the garden at right.

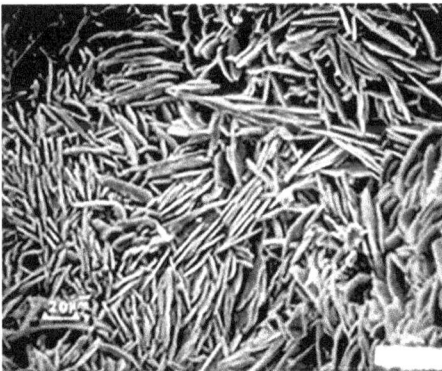

Figure 3: SEM photomicrographs of the gypsum efflorescence on brick, left (see Fig.2 left), and the framboidal growths resulting from air pollution on limestone of the Cowper Hewitt Museum, right.

from growth in the presence of sufficient moisture[7] and those formed as a result from air pollution reacting with the limestone of the Cowper Hewitt Museum *(see Figure 2)* with limited moisture, highlighting the importance of crystallization kinetics in producing significantly different crystallization patterns and degrees of deterioration, as discussed elsewhere.[8] The poorly crystallized gypsum in the black crusts will migrate into the substrate when enough moisture is available for its dissolution. Once in the subsurface it will re-crystallize into larger and better formed crystals, thus slowly inducing more damage to the material with time as described by Domaslowski.[22]

Over the past 35 years, gypsum efflorescence was found to form on new construction in the UK, The Netherlands and Belgium.[23-25] The origin of this efflorescence was attributed mainly to the used masonry mortar formulations, mostly ready-to-use based on Ordinary Portland Cement, and containing additives such as air-entrainers, surfactants among others. The development of tests to study this issue was first undertaken by Bowler and collaborators.[23, 26] However, these tests had several drawbacks as discussed by Chwast[27] who then tried to improve the testing procedure, as discussed below.

A wick-test setup *(see Figure 4)* was adapted to evaluate the risk of gypsum efflorescence formation on brick masonry.[27] The test uses a thoroughly desalinated brick core sample (ø = 30mm, height = 62mm) which serves as the transport medium by being kept in constant contact with the test solution (that represents the source of the salt). Evaporation takes place only over the brick core sample promoting continuous efflorescence formation. Gypsum can be found both in brick[23] but also in carbonating cement paste[27], so to include this latter source carbonated cement paste, and if desired, mortar admixtures, can be added. For this purpose, both powdered brick and hardened cement paste are added to the solution. The sulfate and Ca^{++} ions migrate then into the brick core and will precipitate efflorescing or subflorescing as gypsum on the drying surface. The ef-

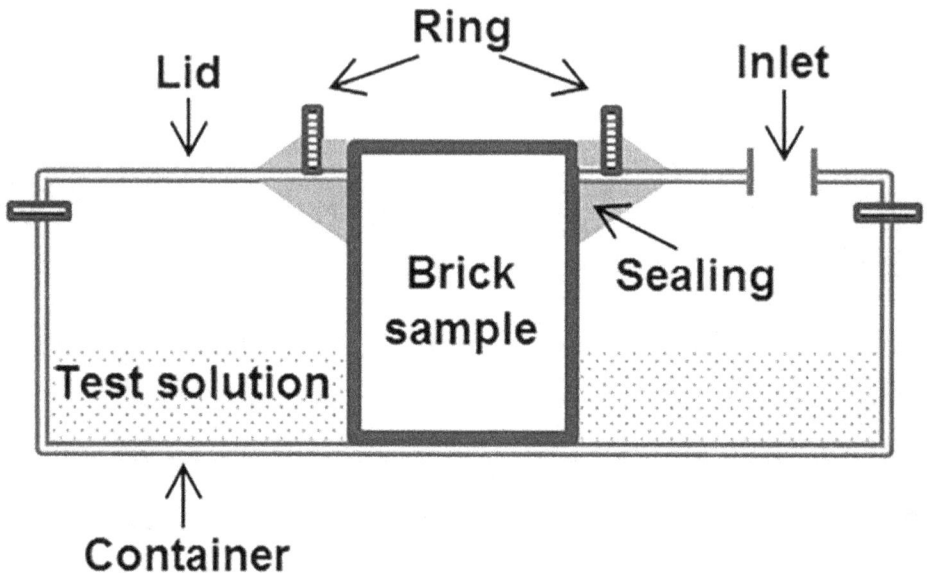

Figure 4: A sketch of the optimized wick-test set up.

fect of admixtures can be evaluated alone (mixed with a gypsum solution), as a component of carbonated cement paste, or added to the brick powder. The amount of the tested sample is calculated to quantitatively reflect the mortar to brick ratio in real masonry. The efflorescence appearance and thus human appreciation of its extent substantially depend on the light conditions and surface dryness. For this reason, high quality efflorescence photographs are collected under well controlled conditions from dry samples allowing for a reliable evaluation of the discoloration extent by means of visual and digital image analysis.

Furthermore, efflorescence composition is determined by XRD and HCl tests, while its effect on the wick performance is evaluated by a drying rate analysis.[27]

specific material and cannot be used to evaluate different ones. Finally, the fact that salt distribution in porous materials is not homogeneous[29] makes this an even more complex situation.

The above considerations lead to the conclusion that salt tests should be designed for specific objectives, such as the two test examples mentioned for sodium chloride and gypsum. By focusing on the dominating deteriorating effect, as done by Henriques[16], relatively simple tests can be elaborated that will provide the information required. However, the complexity of the salt induced deterioration as outlined above, suggests that it cannot be expected that one single test provide all the information that would be desirable.

3. Discussion and conclusion

Salts induce deterioration in materials when they migrate into it and crystallize out. Some salts induce far more damage than others, sodium sulfate being the prime example. Therefore it has been used to evaluate material durability. But the durability of a material depends on the material itself as well as on the conditions under which the salt crystallizes. Since the former may be heterogeneous, as is the case for bricks[28]; and the latter cannot be predicted in practice, the results of standard crystallization tests are not completely reliable.

Sodium chloride will also induce damage, albeit at a slower rate, while gypsum is probably the salt that is slowest in inducing deterioration. Crystallization kinetics will vary depending on environmental conditions; therefore, the same amount of salt will produce different deterioration patterns, which will also depend on the nature of the material. This suggests that the use of a single salt test needs to be specifically developed for a

References

[1] Lubelli B., van Hees R.P.J., Nijland T.G., Salt crystallization damage: how realistic are existing ageing tests? 3rd International Conference on Salt Weathering of Buildings and Stone Sculptures, ed. De Clercq H., KIK-IRPA, Royal Institute for Cultural Heritage, Brussels, (2014), 259-273.

[2] Schmölzer A., Zur Entstehung der Verwitterungsskulpturen an Bausteinen. Chem. Erde (10), (1936), 479-520.

[3] de Quervain F., Jenny V., Verhalten der Bausteine gegen Witterungseinflüsse in der Schweiz. In Beiträge zur Geologie der Schweiz, Geotechnische Serie, 30 Lieferung, ed. Schweizerische Geotechnische Kommission, Kümmerly & Frey, Geographischer Verlag, Bern, (1951), 1-66.

[4] Steiger M., Asmussen S., Crystallization of sodium sulfate phases in porous

materials: The phase diagram Na2SO4–H2O and the generation of stress. Geochimica and Cosmochimica Acta (72), (2008), 4291–4306.

[5] Charola A.E., Weber J., The hydration-dehydration mechanism of sodium sulphate. In: 7th International Congress on Stone Deterioration and Conservation, eds. Delgado Rodrigues J., Henriques F., F.Telmo Jeremias F., Laborátorio Nacional de Engenharia Civil, Lisbon, (1992), 581-590.

[6] Pühringer J., Salt disintegration. Salt migration and degradation by salt – a hypothesis. D15:1983. Swedish Council for Building Research: Stockholm, (1983).

[7] Zehnder K., Arnold A., Crystal growth in salt efflorescence. Journal of Crystal Growth (97), (1989), 513-521.

[8] Rodriguez-Navarro C., Doehne E., Sebastian E., How does sodium sulfate crystallize? Implications for the decay and testing of building materials. Cement and Concrete Research, (30), (2000), 1527-1534.

[9] Cultrone G., Sebastián E., Laboratory simulation showing the influence of salt efflorescence on the weathering of composite building materials. Environmental Geology, (56), (2008), 729-740.

[10] Benavente D, Cueto N., Martínez-Martínez J., García del Cura M.A, J.C. Cañaveras J.C., The influence of petrophysical properties on the salt weathering of porous building rocks. Environmental Geology, (52), (2007), 215–224.

[11] Somsiri S., Zsembery S., Ferguson J.A., A study of pore size distributions in fired clay bricks in relation to salt attack resistance. In Proc.7th International Brick Masonry Conference, eds. McNeilly T., Scrivener J.C., eds. University of Melbourne, Melbourne, (1985), 253-260.

[12] Lewin S.Z. and Charola A.E., The physical chemistry of deteriorated brick and its impregnation technique. In Il Mattone di Venezia. Laboratorio per lo Studio della Dinamica delle Grande Masse del CNR and Universitá di Venezia, Venice, (1979), 189-214.

[13] Binda L., Baronio G., Charola A.E., Deterioration of porous materials due to salt crystallization under different thermohygrometric conditions. I. Brick. 5th International Congress on Stone Deterioration and Conservation, Félix G., ed. Presses Polytechniques Romandes, Lausanne, (1985), 279-288.

[14] Amadori M.L., Lazzarini L., Massa S., Il deterioratmento da sodio cloruro di rocce compatte e porose a Venezia. La conservazione dei monumenti nel bacino Mediterraneo. 1st International Symposium 1989, ed. Zezza F., Grafo, Brescia, (1990), 83-89.

[15] Lubelli B.A., Sodium chloride damage to porous building materials. Ph.D. Thesis. Technical University of Delft, (2006).

[16] Henriques F.M.A., Pedras naturais e artificiais. Análise de argamassas. Determinação da resistência a acção dos cloretos. Universidade Nova de Lisboa, Caparica. (1996), Fe 12.

[17] Henriques F.M.A., The effects of soluble salts on a vaulted ceiling. Proc. 7th Int. Congress on Deterioration and Conservation of Stone, eds. Delgado Rodrigues J., Henriques F, Telmo Jeremias F. Laboratório Nacional de Engenharia Civil, Lisbon, (1992), 1355-1362.

[18] Faria P., Henriques F.M.A., Rato V., Com-

parative evaluation of lime mortars for architectural conservation. Journal of Cultural Heritage, (9), (2008), 338-346.

[19] Leitner H., The treatment of wall paintings affected by salts: An interdisciplinary task as seen from a conservator's perspective. Restoration of Buildings and Monuments, 11[6] (2005) 365-380, and, European Research on Cultural Heritage 5, ARCCHIP Workshops, eds. Simon S., Drdácký M. Institute of Theoretical and Applied Mechanics: Prague (2006), 201-314.

[20] Charola A.E., J.Pühringer J., Steiger M., Gypsum: a review of its role in the deterioration of building materials. Environmental Geology, (52), (2007), 339-352.

[21] Larsen P.K., The salt decay of medieval bricks at a vault in Brarup Church, Denmark. Environmental Geology, (52), (2007), 375-383.

[22] Domaslowski W. La conservation préventive de la pierre. Musées et Monuments, XVIII, UNESCO, Paris, (1982), 51-57.

[23] Bowler G.K., Winter N.B., Investigation into causes of persistent efflorescence on masonry. Masonry International, (11), (1997), 15–18.

[24] Brocken H., Nijland T.G., White efflorescence on brick masonry and concrete masonry blocks, with special emphasis on sulfate efflorescence on concrete blocks. Construction and Building Materials, (18), (2004), 315–323.

[25] Chwast J., Todorovic J., Janssen H., Elsen J., Gypsum efflorescence on clay brick masonry: Field survey and literature study. Construction and Building Materials, (85), (2015), 57-64.

[26] Bowler G.K., Sharp R.H., Testing of various brick/mortar combinations for mortar durability, efflorescence potential and resistance to rain penetration. Proc. British Masonry Society, (1998), 31–36.

[27] Chwast J., Gypsum efflorescence on clay brick masonry. Ph.D. Thesis. KU Leuven, (2017).

[28] Rörig-Dalgaard I., Ottosen L.M., Hansen K.K., Diffusion and electromigration in clay bricks influenced by differences in the pore system resulting from firing, Construction and Building Materials, (27), (2012), 390-397.

[29] Arnold A., Zehnder K., Monitoring Wall Paintings Affected by Soluble Salts. The Conservation of Wall Paintings, Carter S. ed. The Getty Conservation Institute, Los Angeles, (1991), 103-135.

Mitigating salt damage in lime-based mortars with mixed-in crystallization modifiers

Sanne J. C. Granneman[1], Barbara Lubelli[1]* and Rob P. J. van Hees[1, 2]
[1] *Delft University of Technology, Delft, The Netherlands*
[2] *TNO Technical Sciences, Delft, The Netherlands*
*B.Lubelli@tudelft.nl

Abstract

This paper presents the most important results of a research project which focused on the use of crystallization modifiers mixed in lime mortar to mitigate salt crystallization damage. The research focused on two of the most damaging salts, sodium chloride and sodium sulfate, and suitable crystallization modifiers (sodium ferrocyanide and borax). We report the major findings related to the effectiveness of the modifiers when mixed in the mortar, and the results of characterization of the additivated mortars in comparison with reference mortars. Moreover, the durability of the developed mortars to salt decay is discussed, based on the results of an accelerated salt weathering test carried out in laboratory. No major effects of the modifiers on the fresh and hardened mortar properties were observed which might restrain the application of crystallization modifiers in restoration mortars. Additionally, the mortars with mixed-in modifiers showed a considerable improvement of the salt resistance when compared to reference mortars. Considering these results an outlook for future research pathways is given.

Keywords: crystallization modifiers, self-healing, lime mortar

1. Introduction

Salt crystallization damage in porous building materials is a ubiquitous threat to our built cultural heritage. Low mechanical strength makes lime-based mortars especially susceptible to salt damage. In restoration or renovation works, replacement of renders and plasters often constitutes a large part of the total costs of the project. Current solutions, such as using a stronger binder or changing the moisture transport properties of the mortar, usually have a limited resistance to salt decay and low compatibility with the existing materials.[1, 2]

Alternatively, the use of crystallization modifiers has been proposed.[3] Crystallization modifiers do not aim to alter the material properties, but the damaging mechanism itself. Using crystallization modifiers in porous building materials has gained wide research interest in the last years (see e.g.[3-5]). However, the use of modifiers in a fresh mortar, thereby giving the mortar "self-healing properties", is relatively new. By mixing modifiers in during mortar production, they can become active as soon as the damaging salts enter the porous material. Promising results have already been obtained in a pilot study.[6]

A research project was started to further study the feasibility of the use of crystallization modifiers to mitigate salt crystallization damage. This project had the following aims:

i) Identify suitable modifiers for two of the most abundant and damaging salts: sodium chloride and sodium sulfate,

ii) Study the modifier-salt interaction in bulk solution to elucidate the working mechanism and find a suitable concentration,

iii) Test the effect of modifier addition on mortar properties (which might limit their application), and iv) Test the durability of additivated mortars in an accelerated salt weathering test.

From literature research, sodium ferrocyanide and borax (sodium tetraborate decahydrate) were identified as potential modifiers for sodium chloride and sodium sulfate, respectively.[7] Ferrocyanide is a well-known modifier for sodium chloride. Therefore, this research focused on the study of its effect on fresh and hardened mortar properties and on its capability of mitigating salt decay in mortar. Ferrocyanide could reduce crystallization damage in two ways:

i) it keeps the salt longer in solution favouring transport to the drying surface, and

ii) it alters the crystallization habit of sodium chloride from cubic to dendritic, fact which also favours transport to the surface and enhanced drying.

Both factors lead to enhanced transport to the surface, where subsequently the salt can crystallize as harmless efflorescence.[8] Borax on the other hand is less well-known and was therefore first studied in bulk solution experiments. In this research, the effect of borax addition on solution properties and its effect on crystallization of sodium sulfate were studied. As has been reported in Ref.[9], depending on the starting concentration of the solution, two different phases of borax can precipitate, each having a different effect on sodium sulfate crystallization. One phase favours the crystallization of hydrated sodium sulfate (mirabilite) at or near saturation, meaning that no or lower crystallization pressures can develop. The other phase modifies the habit of anhydrous sodium sulfate (thenardite) to elongated needles. This habit modification can be supposed to lead, thanks to a larger evaporation surface, to enhanced salt transport to the drying surface. Similar to ferrocyanide, this would favour the formation of harmless efflorescence.[8]

In this paper the major findings relating to the effect of modifier addition on mortar properties and to the durability of additivated mortars are reported. First, an experiment to test whether borax' effectiveness is affected by the carbonation process is described. Then, the effect of modifiers on mortar properties such as workability, water absorption and drying and flexural and compressive strength are reported. Finally, the durability of the additivated mortars with respect to salt crystallization damage is discussed based on the results of an accelerated salt weathering test.

2. Materials and Methods

2.1. Mortar characterization

Two types of specimens were prepared: lime only specimens and mortar (lime + sand) specimens. The first, used to study the effect of borax on sodium sulfate crystallization, were prepared by mixing calcium hydroxide powder (Sigma-Aldrich, ≥ 96% purity) with distilled water only or with water additivated with the modifier. After carbonation, a blank specimen was treated with borax solution. Then, the blank and the two borax specimens contaminated with sodium sulfate solution. After drying, the specimens were broken and the cross section was studied using

SEM. By comparing specimens additivated with borax prior to carbonation and those to which borax was added later on, the effect of carbonation on the effectiveness of borax as modifier of sodium sulfate crystallization was assessed. Full details of this experiment can be found in Ref.[10]

The second type of specimens, used to test fresh and hardened mortar properties and assess the durability towards salt decay, was prepared according to EN1015-2. Commercial air lime (Supercalco90) and standard sand (EN 196-1, sieved to select the grain size between 0.25 and 1.0 mm, were used. The modifier was added to the water used to prepare the specimens: 0.94 wt% sodium ferrocyanide and 3.2 wt% borax were added with respect to the lime. Several fresh and hardened mortar properties were tested according to standard procedures or techniques: workability (EN1015-3), water absorption and drying (EN1015-18), porosity and pore size distribution (Mercury Intrusion Porosimetry) and flexural and compressive strength (EN1015-11). Full details on the preparation of the mortars and the testing methods can be found in Ref.[10]

2.2. Accelerated salt weathering test

The salt crystallization resistance of the reference and additivated mortar specimens was tested with a custom designed salt weathering test, shown in *Figure 1*, simulating circumstances found in practice. 80 RH% is above the equilibrium relative humidity of sodium chloride (RHeq = 75%), but below that of the sodium sulfate phases. Consequently, the sodium chloride crystals will deliquesce when the humidity goes up and recrystallize when the humidity goes down again. This ensures multiple crystallization cycles for sodium chloride, a requirement for this specific salt for damage develop-

ment. Recrystallization of sodium sulfate was obtained only by rewetting with liquid water at the end of a cycle. To have a comparable test protocol, the specimens with sodium chloride were also rewetted with liquid water after each cycle.

Before the test, the specimens were contaminated with salt solution via capillary absorption from the bottom. A precise amount of solution was used to ensure contamination with 1 wt% sodium sulfate or 2 wt% sodium chloride with respect to the mortar. In total 3.46 (reference NaCl), 3.91 (ferrocyanide), 1.77 (reference Na_2SO_4) or 1.91 (borax) gram salt was added to the specimens during the complete test. These values include the brushed off efflorescences during the test. For each mortar type, 3 replicas were tested.

After each complete cycle, all specimens were rewetted with demineralized water equal to the initial amount used to contaminate the specimens. After 3 cycles (63 days), salt solution (exact amount to obtain again 1 wt% sodium sulfate or 2 wt% sodium chloride) was

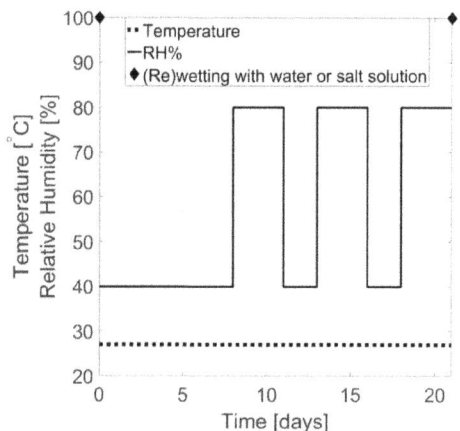

Figure 1: Temperature and RH cycles used in the accelerated salt weathering test. This entire cycle was repeated 5 times (in total 105 days). The diamonds correspond to wetting by capillarity at the start (day 0) or after each 3 week cycle (day 21) at 22.9°C ± 0.3/29.1RH% ± 2.4. At day 0 and day 63 salt solution was used, the other times demineralized water.

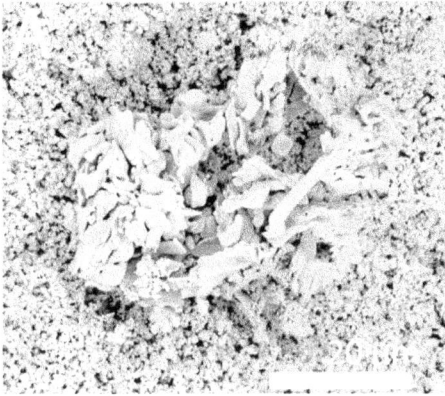

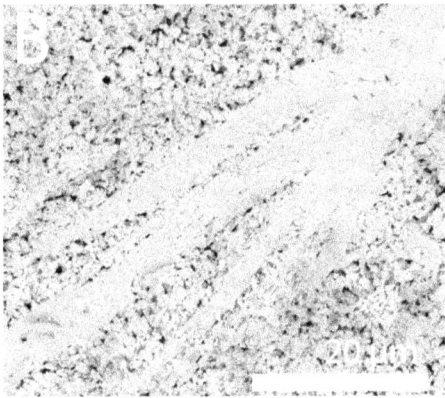

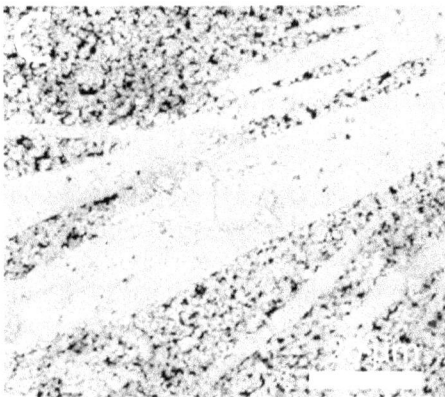

Figure 2: SEM images of lime-only specimens contaminated with sodium sulfate. A: reference specimen; B: specimen with 3.2wt% borax mixed in during preparation (thus before carbonation); C: Specimen additivated with borax after full carbonation of the specimen. A clear difference in crystal habit can be observed between the reference specimen and both specimens with borax. Contrarily, no distinction can be observed between B and C, meaning that the carbonation process has no effect on the borax effectiveness as modifier of sodium sulfate crystallization.

used for rewetting, in order to replenish the brushed off salt. After rewetting, any loose material was brushed off and the specimens started a new cycle. The brushed off material was washed and dried in order to separate the salt efflorescences from the debris. The debris was weighed and the amount of salt calculated by the difference. In total the specimens were tested for 5 cycles. Full details of the experiment can be found in Ref.[11]

3. Results and discussion

3.1. Mortar properties

The effect of borax on sodium sulfate crystallization can be observed in the SEM pictures in *Figure 2*. When *Fig. 2a* is compared to *2b/c*, it is clear that the crystal habit of sodium sulfate without borax is different from the crystal habit in the presence of borax. If subsequently *Figures 2b and 2c* are compared, a similar crystal habit can be seen in both figures. This means that the carbonation process of the mortar has no effect on the effectiveness of borax as modifier for sodium sulfate. With this experimental technique it is not possible to investigate the nature of the precipitated phases, but this would be interesting for future work.

A selection of the measured fresh and hardened mortar properties is summarized in *Table 1* (additional characterization results can be found in Ref.[10]). When the values for additivated and reference specimens are compared, no notable differences can be observed. It can therefore be concluded that there are no negative consequences to mixing these quantities of modifiers in the mortar during production, as the addition of these modifiers does not negatively affect the mortar properties.

Property	Method	Reference	Ferrocyanide	Borax
Workability	Flow table test	170 mm	161 mm	161 mm
Water content	-	15.95 wt%	15.14 wt%	14.59 wt%
WAC [kg/m²h^{1/2}]	Capillary rise	8.05	7.62	7.84
Density [kg/m³] Open porosity (%V/V)	Saturation at atm pressure	1943 26.7 ± 0.19	1964 25.9 ± 0.24	1933 27.1 ± 0.06
Bulk density [g/ml] Open porosity (%V/V)	Mercury Intrusion Porosimetry	1.977 25.1 ± 0.11	1.971 25.4 ± 0.36	1.961 25.1 ± 0.55
Tensile strength (N/mm²)	-	0.79 ± 0.11	0.85 ± 0.03	0.92 ± 0.11
Compressive strength (N/mm²)	-	2.01 ± 0.33	2.08 ± 0.18	2.61 ± 0.22

Table 1: Fresh and hardened mortar properties of the different 1:3 lime:sand mortar mixtures. Results previously reported in [10]

3.2. Salt durability[11]

During the accelerated salt weathering test, the specimens were monitored both visually and gravimetrically. The weight loss of material (with respect to the mortar) is plotted in *Figure 3*, and the weight loss of salt is visualized in *Figure 4*. It is clear that for both salts, the reference specimens suffer considerable material loss after 5 cycles. Contrarily, the additivated mortars show no or only minor material loss. The ferrocyanide stimulates efflorescence of the salt, i. e. crystallization outside the material, as does borax but to a lesser extent. This is consistent with the hypotheses on damage reduction proposed in the introduction. *Figure 5* compares specimens contaminated with sodium chloride at the start and end of the test. The reference specimen shows considerable surface loss at the end of the test. Contrarily, the specimen with ferrocyanide shows no material

loss, but extensive efflorescence, which developed very fast already just after rewetting via capillarity and brushing of the specimens. In *Figure 6* the specimens contaminated with sodium sulfate are compared. At the end of the test, both specimens show damage at the surface, but this is in the case of the specimen additivated with borax considerably less than in the reference specimen. Both the material loss and the visual observations show that both sodium chloride and sodium sulfate have the potential to cause considerable damage in the reference specimen. However, when the mortars are additivated with modifiers, damage does not occur or is considerably less.

Two additivated specimens were desalinated at the end of the test and the boron and iron content of the desalination water was analysed via ICP (Inductively Coupled Plasma) spectrometry. Approximately 10 wt% (borax) and 1 wt% (ferrocyanide) of the original modifier

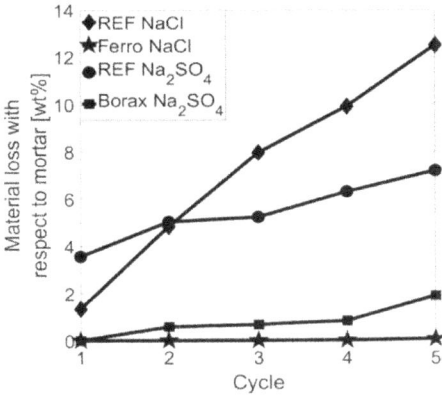

Figure 3: Cumulative material loss, comparison between reference and additivated specimens.

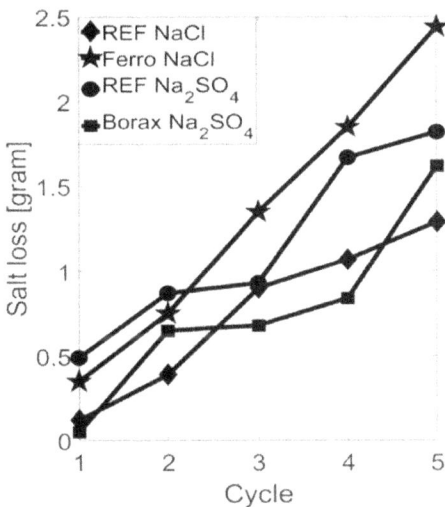

Figure 4: Cumulative salt loss, comparison between reference and additivated specimens.

4. Conclusions

The additivation of mortars with crystallization modifiers during production has been proposed here to mitigate salt crystallization damage in porous building materials. Suitable crystallization modifiers for sodium chloride (sodium ferrocyanide) and sodium sulfate (borax) were identified to be mixed in a mortar during production. In this paper, at first the effectiveness of borax as a modifier for sodium sulfate crystallization when mixed in lime was assessed and confirmed. In a next step, additivated mortars were characterized and compared to reference mortars in order to identify potential (negative) effects on fresh and hardened mortar properties. None of the tested properties was affected by the addition of the modifiers, meaning that there are no contra-indications to mixing them in the mortar in the used concentrations during production.

Finally, the salt crystallization resistance of the additivated mortars was assessed using a custom designed accelerated salt weathering test. The mortars with mixed-in modifiers showed a considerable improvement of the salt resistance when compared to reference mortars. Combining all these results it can be concluded that additivating mortars with crystallization modifiers during their production is a feasible method in order to mitigate salt crystallization damage in porous building materials.

amount was still present. These values indicate that either the modifiers have leached out (together with the salt efflorescence) or that they are (partially) tightly bound to the mortar structure. Modifier leaching is an important parameter to take into account in future applications, whereas the binding of the modifier could have implications for its effectiveness and working mechanism.

5. Outlook

The research presented in this paper shows the viability of using crystallization modifiers to mitigate salt weathering damage in porous building materials. Although the proof-of-principle has been shown on the laboratory scale, more research is needed to develop the material into a commercial product, suitable for

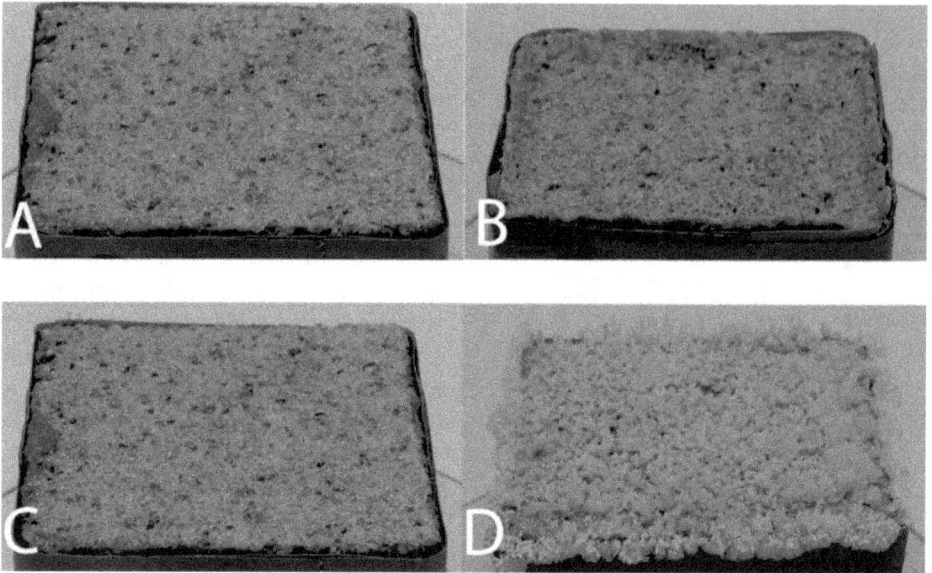

Figure 5: Comparison between reference (A/B) and specimens with mixed-in ferrocyanide (C/D), both contaminated with sodium chloride. A/C show the specimens at the start of the test, B/D show the specimens ~ 15 minutes after brushing after the 5th cycle. The reference specimen (B) shows sanding of the surface. The specimen with ferrocyanide shows no surface damage, only a large amount of efflorescence.

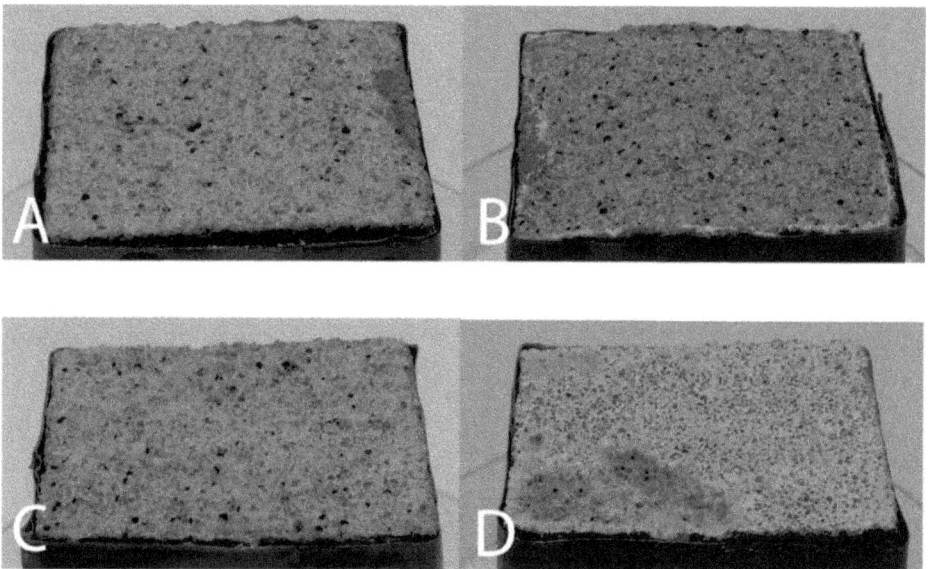

Figure 6: Comparison between reference (A/B) and specimens with mixed-in borax (C/D), both contaminated with sodium sulfate. A/C show the specimens at the start of the test, B/D show the specimens ~ 15 minutes after brushing after the 5th cycle. The reference specimen (B) shows clear damage at the surface. The specimen with borax shows only minor surface damage at the lower left corner (D).

renovation or restoration works. Interesting research paths to further develop the mortar designed in this project are:

- Studying the effect of modifier in mortars with different composition (e.g. cement-based).

- Studying the speed of modifier leaching and if necessary developing possible solutions, e.g. encapsulation.

- Assessing the effect of the identified modifiers on other salts and on salt mixtures and the possibility of combining different modifiers.

- Assessing the effectiveness of the developed mortar (on test panels) in situ.

Acknowledgements

This research has been financed by the Dutch IOP program on Self-Healing Materials, under Grant number SHM012018.

References

[1] B. Lubelli, R. P. J. van Hees, and C. J. W. P. Groot. Sodium chloride crystallization in a „salt transporting" restoration plaster. Cement and Concrete Research, 36: 1467–1474, 2006.

[2] C. Groot, R. van Hees, and T. Wijffels. Selection of plasters and renders for salt laden masonry substrates. Construction and Building Materials, 23: 1743–1750, 2009.

[3] C. Selwitz and E. Doehne. The evaluation of crystallization modifiers for controlling salt damage to limestone. Journal of Cultural Heritage, 3: 205–216, 2002.

[4] B. Lubelli and R. P. J. van Hees. Effectiveness of crystallization inhibitors in preventing salt damage in building materials. Journal of Cultural Heritage, 8: 223–234, 2007.

[5] C. Rodriguez-Navarro and L. G. Benning. Control of crystal nucleation and growth by additives. Elements, 9: 203–209, 2013.

[6] B. Lubelli, T. G. Nijland, R. P. J. van Hees, and A. Hacquebord. Effect of mixed in crystallization inhibitor on resistance of lime-cement mortar against NaCl crystallization. Construction and Building Materials, 24: 2466–2472, 2010.

[7] S. J. C. Granneman, E. Ruiz-Agudo, B. Lubelli, R. P. J. van Hees, and C. Rodriguez-Navarro. Study on effective modifiers for damaging salts in mortar. In Proceedings of the 1st International Conference on Ageing of Materials and Structures, 2014.

[8] S. J.C. Granneman, B. Lubelli, and R. P. J. van Hees. Mitigating salt crystallization damage with mixed-in modifiers – a review. Manuscript in preparation.

[9] S. J. C. Granneman, N. Shahidzadeh, B. Lubelli, and R. P. J. van Hees. Effect of borax on the wetting properties and crystallization behavior of sodium sulfate. CrystEngComm, 19: 1106–1114, 2017.

[10] S. J. C. Granneman, B. Lubelli, and R. P. J. van Hees. Characterization of lime mortar additivated with crystallization modifiers. Manuscript submitted 07-2017.

[11] S. J.C. Granneman, B. Lubelli, and R. P. J. van Hees. Salt resistance of lime mortars additivated with crystallization modifiers. Manuscript in preparation.

[12] S. J. C. Granneman, B. Lubelli, and R. P. J. van Hees. Mitigating salt damage in lime-based mortars with mixed-in crystallization modifiers. In Proceedings of the 4th WTA International PhD Symposium, 2017.

Studies for conservation issues

Efficiency of laboratory produced water repellent treatments on limestone

Cleopatra Charalambous and Ioannis Ioannou*
Department of Civil and Environmental Engineering, University of Cyprus, Cyprus
* ioannis@ucy.ac.cy

Abstract

A number of cultural and architectural heritage structures all over the world are built with natural stone. Although this material is considered to be one of the most durable geomaterials, many existing stone buildings and monuments show clear evidence of decay and weathering. The deterioration of stone is strongly related to the presence and movement of water within its pore network. Therefore, hydrophobic surface treatments are usually adopted to protect existing or new stonework. Such treatments, however, should not affect the breathability of stone; else, there is a risk of enhancing possible decay mechanisms, such as salt crystallization.

Natural limestones appear to have a degree of inherent water repellency. This has been confirmed through multiple measurements of capillary absorption at different temperatures, using water and organic liquids. The measurements were carried out on several building and decorative limestones, showing in each case an anomalously low water sorptivity. This natural water repellency of limestones was attributed to the presence of organic contaminants, such as fatty acids, in the pore network of the materials under investigation.

In this paper, the natural water repellency of Cypriot limestones is exploited to develop several water repellent surface treatments, based on oleic acid. The aforementioned laboratory produced treatments were applied on a Cypriot calcarenite with proven poor durability characteristics. The results suggest that all treatments can permanently reduce the wettability of the stone under investigation, without modifying its composition or appearance.

In order to investigate the durability of the treated stone, wetting/drying cycles were performed. The results provide strong evidence that treatment with oleic acid positively affects the durability of the stone under study. Consequently, the aforementioned surface treatment may be potentially used in practice to protect stone facades in buildings and cultural heritage sites.

Keywords: limestone, oleic acid, water repellency, wetting, drying

1. Introduction

The decay and weathering of historical stone masonries is one of the most common and severe problems the construction industry is facing all over the world. Although natural stone is considered to be one of the most durable geomaterials, stone structures are susceptible to water-mediated decay processes, such as salt crystallization, induced by alternate cycles of wetting and drying.

Salt weathering is, in fact, considered to be one of the most important degradation mechanisms that a porous stone may undergo at or near the Earth's surface.[1] The slow process of absorption of water into a porous structure and its subsequent evaporation may lead to the gradual deposit of salts in a wall. The masonry acts as a filter system for impure water into a structure, as the various soluble salts are drawn into the wall and then left behind.[2-3] Upon changes in the environmental conditions, salts can crystallize either at the surface (efflorescence) or inside the pores (subflorescence) of a material.[4-5] Crystallization of salts into a porous stone causes loss of coherence between the grain and the matrix. Weight loss, changes in the size of grains and pores, splitting and visible surface deterioration can therefore be produced by salt dissolution-crystallization cycles. Thus, salt crystallization may modify the properties of the affected porous stone, leading to the reduction of the lifetime of a stone building or monument.[6]

The need to protect existing stonework in buildings and cultural heritage sites from decay and weathering mechanisms, such as salt crystallization, has led several scientists to use hydrophobic surface treatments.[7] Those treatments usually include alkyl silicone products (i.e. alkyl siliconates, alkyl silanes, silicone resins), water-based fluoralkylsiloxane, polydimethyl siloxanes, phosphoric and polymaleic acid products.[7-10] However, when applied, they may restrict the breathability of the treated material by suppressing stage I drying, thereby affecting its microstructure by inducing pore clogging.[4] Thus, suitable alternative surface treatments need to be developed; these must confer hydrophobicity without affecting the breathability of the material.[8]

Limestones have an unpredictable behavior when water repellents are applied to them.[11] Furthermore, despite the fact that calcite surfaces are naturally hydrophilic[12], many researchers[13-14] state that when carbonate minerals are exposed to natural environments, they acquire organic contaminants, which induce to them a degree of natural resistance to capillary water absorption. In fact, these organic contaminants reduce the affinity

of calcite for water, and therefore modify its wettability. Ioannou et al.[14] confirmed the partial wettability of limestones to water through a series of measurements of capillary absorption at different temperatures, using water and organic liquids.

The most severe modification of calcite surfaces is due to the absorption of carboxylic and especially fatty acids.[15] As quoted in the literature, the strongest affinity for carbonate surfaces is shown by medium-to-long chain fatty acids and carboxylate polymers.[15] Consequently, surface treatments using fatty acids can modify the wettability of calcium carbonate surfaces, without affecting their inherent composition. In this paper, the efficiency of three laboratory produced water repellent surface treatments based on oleic acid is thoroughly investigated.

2. Experimental Work

In the experimental study, three light grey (according to Munsel Soil-Color charts) freshly quarried Cypriot limestones were used. They originate from the area of Agios Theodoros in Cyprus, which belongs to the Pachna Geological Formation. Those stones are packstones or poorly washed biosparites, with microsparry calcite and a small portion of micrite as their intergrain cementing material.[16] They also have a complex mineralogy; they show a rather high percentage of calcite, with significant amounts of quartz, aragonite, zeolites (analcime), pyroxenes (augite), plagioclases (anorthite, andesine, orthoclase, albite) and clay minerals (chlorite, montmorillonite). *Table 1* summarizes the mineralogical composition of the stone variety under study.

Initially, the sorptivity (S) of all three specimens was measured at different temperatures using both water and organic liquids (i.e. ethanol, 2-propanol, n-heptane) in accordance with EN 1925.[17]

Stone variety	Petrological family	Colour (Munsell Chart)	Mineralogy (XRD)
Agios Theodoros	Calcarenite	5Y 7/2 Light Grey	calcite (65-69%), quartz (6-7%), clinopyroxene (4-5%), chlorite (4-5%), plagioclase (5-6%), K-feldspar (1-2%), aragonite (9-11%), montmorillonite and analcime (traces)

Table 1: General details and mineralogical composition of Agios Theodoros stone.

The results were plotted against (σ/η) ½, where σ [Nm⁻¹] is the surface tension and η [Nsm⁻²] the viscosity of the wetting liquid at each temperature. From the graphs and using equation (1), the so-called intrinsic sorptivity (S_i) and the water wetting index (β) of each sample was estimated.[18]

$$S = S_i \left(\beta * \sigma/\eta\right)^{1/2} \qquad (1)$$

Following the initial sorptivity measurements, the surface of each sample was treated with a different in-house developed water repellent treatment, aiming to control its wettability. All treatments were based on oleic acid (i.e. a fatty acid with 18 carbon atoms in a chain). As quoted in the literature[19], when oleic acid is used as an organic additive, it can control the nucleation and growth of calcium carbonate ($CaCO_3$), and it can modify the wettability of its surface. In fact, back in 2012, in a research carried out by Walker et al.[8], the use of oleic acid induced additional hydrophobicity to the calcite surface of the York Minister Cathedral.

The first treatment included an ethanol solution of oleic acid (concentration 0.5% w/w). The second treatment was performed using an aqueous solution of sodium oleate (concentration 0.5% w/w). Sodium oleate is the unsaturated metal soap of oleic acid; it has an equally high affinity for carbonated mineral surfaces and has been shown to act as a kind of surfactant/plasticizer in renders applied on limestone substrates.[20] During the setting of the render, a reaction with calcium ions in limestones takes place and the oleate is transformed to a hydrophobic metal soap. Because of this reaction, sodium oleate is considered a 'Reactive Hydrophobic Agent'. It is worth noting that sodium oleate shows no gelling effect, due to its high content of unsaturated fatty acids.[20] The third treatment was performed using calcium oleate (i.e. the calcium soap of oleic acid), produced by mixing sodium oleate and calcium carbonate in an aqueous solution. Once again, the concentration of calcium oleate solution was 0.5% w/w. The three solutions were applied to the top surface of the specimens by brushing. Following this, the samples were allowed to dry at room temperature. In order to observe the effect of the treatment, the sorptivity of the treated samples was again determined through capillary absorption experiments at different temperatures using water and organic liquids.

The durability of the aforementioned surface treatments was investigated by subjecting the test specimens to wetting/drying cycles. Wetting was performed using vacuum saturation, while drying took place in an oven at 105°C, until constant weight was reached. Ten cycles of wetting/drying were performed. After each cycle, the sorptivity of the sample was determined through water capillary absorption tests.

(I) 1,5

S (mm min$^{-1/2}$)

10

20

$(\sigma/\eta)^{1/2}$ (m$^{1/2}$ s$^{-1/2}$)

········ water line

– – – organic liquids line

(II) 1,5

S (mm min$^{-1/2}$)

10

20

$(\sigma/\eta)^{1/2}$ (m$^{1/2}$ s$^{-1/2}$)

········ water line

– – – organic liquids line

(III) 1,5

S (mm min$^{-1/2}$)

10

20

$(\sigma/\eta)^{1/2}$ (m$^{1/2}$ s$^{-1/2}$)

········ water line

– – – organic liquids line

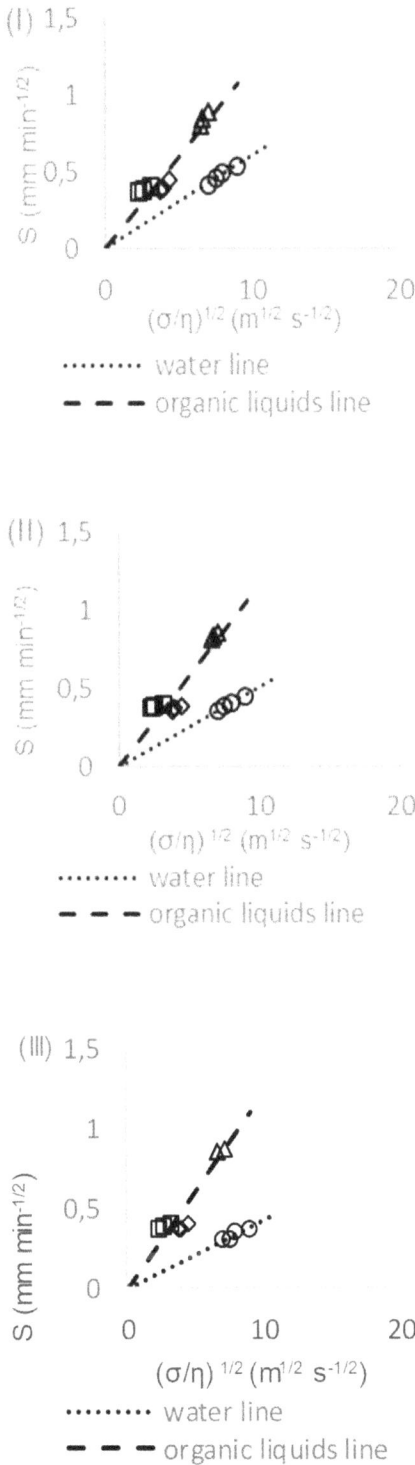

Figure 1: Sorptivity values S versus $(\sigma/\eta)^{1/2}$ for the three limestones under study. (I): Agios Theodoros I. (II): Agios Theodoros II. (III): Agios Theodoros III.

3. Results and Discussion

3.1. Capillary Absorption Measurements before the Treatments

The results of capillary absorption measurements before the treatments showed a linear correlation between the cumulative absorption per unit surface area i and the square root of time $t^{1/2}$, as expected. Consequently, the sorptivity S ($=i/t^{1/2}$) of each sample under study was derived from the slopes of the respective graphs. When the sorptivity values were plotted against $(\sigma/\eta)^{1/2}$, the data points fell into two groups, which lied on separate straight lines *(Fig. 1 I, II, III)*. Despite the fact that both the organic liquids and the water sorptivity values increased linearly with $(\sigma/\eta)^{1/2}$, the water data lied on a line with a lesser slope. This is in line with previous relevant work[14] and confirms the partial wettability of limestone specimens, which is attributed to natural organic contamination, due to the presence of a low-energy adlayer on the specimen's surface.[14, 18] From the results of *Fig. 1*, the water wetting indices of each specimen were calculated using equation (1) *(Table 2)*.

3.2. Capillary Absorption Measurements after the Treatments

The results of the capillary absorption experiments after the application of the surface treatments are shown in *Figure 2* for each specimen. A linear behavior between the sorptivity S and $(\sigma/\eta)^{1/2}$ is once again noted. The organic liquids line remains generally unchanged after all treatments (an indication that no significant chemical or structural changes have occurred in the test specimens[14]), whereas the water line has an even lower slope. This, strongly indicates that oleic acid and its by-products adsorb well on the calcite surface of the samples under

Treatment	Intrinsic Sorptivity ($\times 10^{-4}$ mm$^{\frac{1}{2}}$)	Specimen Water Wetting Indices		
		Original	After surface treatment	After 10 cycles of wetting/drying
Oleic acid	4.83	0.26	0.004	0.03
Sodium oleate	4.74	0.19	0.10	0.04
Calcium oleate	5.84	0.13	0.02	0.01

Table 2: Water wetting indices of test specimens before (original) and after each treatment.

study and induce to them additional hydrophobicity. The latter is attributed to the deposition of the resulting Ca(C$_{17}$H$_{33}$COO)$_2$ onto the calcite (CaCO$_3$) surface of the test specimens.

3.3. Capillary Absorption Measurements after the Wetting/Drying Cycles

The results of the soprtivity tests carried out on each specimen after the completion of the ten wetting/drying cycles are also shown in *Figure 2*. Tests with pure organic liquids were further performed after the completion of the wetting/drying procedure.

From the results, it is evident that the organic liquids line, and consequently the intrinsic sorptivity of the samples, continues to remain unchanged. The results also indicate that the samples were not affected by the wetting and drying cycles. In fact, in the cases of sodium and calcium oleate, there is a further significant reduction (ca. 50-60%) in the water wetting index of the samples after the wetting/drying procedure *(Table 2)*. Whilst for sodium oleate this may be attributed to enhanced diffusion of the treatment into the sample, for calcium oleate the reason is not clear and needs to be further investigated.

It is worth noting that sodium oleate chemically adsorbs on the CaCO$_3$ surfa-

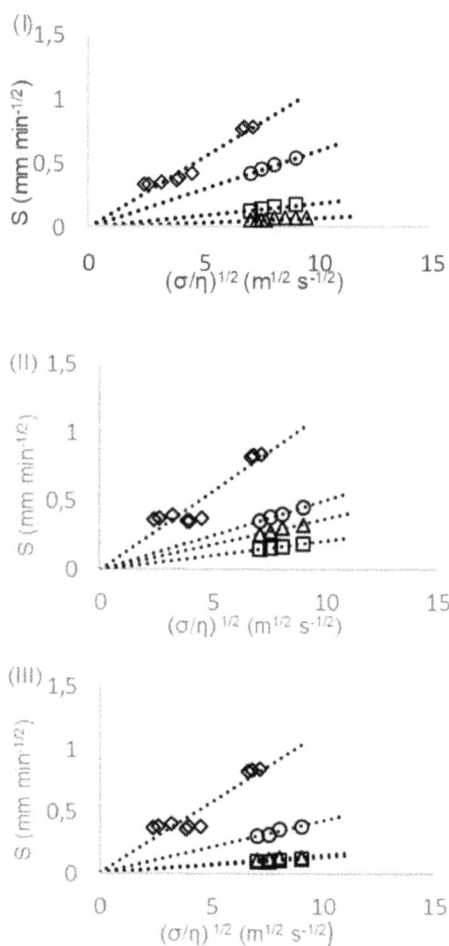

Figure 2: Sorptivity values S versus $(\sigma/\eta)^{1/2}$ for water and organic liquids before and after all treatments. (○) water before treatments, (◊) organic liquids before and after all treatments, (△) water after water repellent surface treatments, (□) water after 10 cycles of wetting and drying. (I): Agios Theodoros I, (II): Agios Theodoros II, (III): Agios Theodoros III.

ce of the stone, which interacts in itself with the oleate anions, thus producing a calcium oleate product. Since the latter is water insoluble, it remains on the CaCO$_3$ surface, thus inducing further water repellency.

Even in the case of oleic acid, where there appears to be an increase in the water wetting index of the sample after the ten wetting cycles *(Table 2)*, the surface of the stone is still nearly water repellent at the end of the experimental procedure (β=0.03). These results, therefore, provide strong evidence that oleic acid and its by-products (sodium and calcium oleate) adsorb well on calcite surfaces, without modifying their chemical/structural composition or appearance *(Figure 3)*. Hence, the aforementioned treatments may be adopted in conservation projects for the protection of stone masonries.

4. Conclusions

The inherent water repellency of limestones originating from the Agios Theodoros area in Cyprus has been confirmed by the experimental work presented in this paper. When the samples were subjected to capillary absorption experiments with water and organic liquids, at different temperatures, a significant differentiation in the respective S vs $(\sigma/\eta)^{\frac{1}{2}}$ graphs was observed; the water data consistent-ly fell on a line of lesser slope, compared to the organic liquid data. This was attributed to the presence of a hydrophobic natural organic contaminants adlayer below the specimen's surface, inducing a degree of inherent water repellency to the stone.

The natural water repellency of the Agios Theodoros stone has been exploited through chemical modification of the surface of three test specimens using different treatments based on the use of oleic acid. All three surface treatments (i.e. oleic acid, sodium oleate, calcium oleate) proved successful in further reducing the water wetting index of the limestones under study.

The efficiency of the aforementioned treatments was further investigated through wetting/drying cycles. Data from capillary absorption measurements after these cycles revealed that oleic acid and its salts and soaps adsorb very well on the calcite bearing surface of the limestone under study. Despite the fact that, there was an increase in the water wetting index of the sample treated with oleic acid after wetting/drying, the stone retained its hydrophobicity. The other two samples showed a further reduction in their water wetting indices after ten cycles of wetting and drying.

This work has practical significance, since the durability of stone masonry is largely controlled by processes mediated

Figure 3: Physical appearance of the stone surfaces before/after each treatment. A0: Untreated stone, A1: Oleic acid treated stone, A2: Sodium oleate treated stone, A3: Calcium oleate treated stone

by water, such as salt crystallization. The results strongly suggest that the in-house developed water repellent treatments with oleic acid, sodium and calcium oleate not only manage to reduce the water absorption of the treated stone, but they are also durable against wetting/drying cycles. Therefore, the aforementioned treatments may be potentially used in conservation projects for the protection of stone facades in buildings and cultural heritage sites.

References

[1] Benavente D., Garcia del Cura M.A., Fort R., Ordonez S. Durability estimation of porous building stones from pore structure and strength. Engineering Geology (74), 2004, 113-127.

[2] Benavente D., Martinez J., Cueto N., Garcia den Cura M.A. Salt weathering in dual-porosity building dolostones. Engineering Geology (94), 2007, 215-226.

[3] Rirsch E., Zhang Z. Rising damp in masonry walls and the importance of mortar properties. Construction and Building Materials (24), 2010, 1815-1820.

[4] Ioannou I., Hoff W.D. Water repellent influence on salt crystallization in masonry. Construction Materials (161), 2008, 17-23.

[5] Franzoni E., Bandini S., Graziani G. Rising moisture, salt and electrokinetic effects in ancient masonries: From laboratory testing to on-site monitoring. Journal of Cultural Heritage (15), 2014, 112-120.

[6] Benavente D., Garcia del Cura M.A., Bernabeu A. Ordonez S. Quantification of salt weathering in porous stones using an experimental continuous partial immersion on method. Engineering Geology (59), 2001, 313-325.

[7] Charola A.E., Water Repellents and other "Protective" Treatments: A critical review. Proceedings of Hydrophobe III, Hannover, 2001, 3-2001.

[8] Walker R., Wilson K., Lee A., Woodford J., Grassian V., Baltusaitis J., Rubasinghege G., Cibin G., Dent A. Preservation of York Minister historic limestone by hydrophobic surface coatings. Scientific Reports. (2:880), 2012, 1-5.

[9] Thompson M., Wilkins S.J., Compton R.G., Viles H.A., Polymer coatings to passivate calcite from acid attack: polyacrylic acid and polyacrylonitrile. Journal of Colloid and Interface Science (259), 2003, 338-345.

[10] Carmona P.M., Panas I., Svensson J.E., Johansson L.G., Blanco M.T., Martinez S. Protective performance of two anti-graffiti treatments towards sulfite and sulfate formation in SO2 polluted model environment. Applied Surface Science (257), 2010, 852-856.

[11] Charola A.E., Water repellent treatments for building stones: A practical overview. APT Bulletin (24), 1995, 10-17.

[12] Okayama T., Keller D.S., Luner P. The wetting of calcite surfaces. The Journal of Adhesion (63), 1997, 231-252.

[13] Thomas M., Clouse J.A., Longo J.M. Adsorption of organic compounds on carbonate minerals: 1. Model compounds and their influence on mineral wettability. Chemical Geology (109), 1993, 201-213.

[14] Ioannou I., Hoff W.D., Hall C. On the role of organic adlayers in the anomalous water sorptivity of Lepine limestone. Journal of Colloid and Interface Science (279), 2004, 228-234.

[15] Zullig J.J., Morse J.W. Interaction of organic acids with carbonate mineral surfaces in seawater and related solutions: I. Fatty acid adsorption, Geochimica et Cosmochimica Acta (52), 1988, 1667-167.

[16] Modestou S., Theodoridou M., Fournari R., Ioannou I. Physico-mechanical properties and durability performance of natural building and decorative carbonate stones from Cyprus. Geological Society London, Special Publications (416), 2015, 145-162.

[17] EN 1925:1999 Natural stone test methods – Determination of water absorption coefficient by capillarity, European Committee for Standardization 1999.

[18] Taylor S., Hall C., Hoff W.D., Wilson M. Partial wetting in capillary absorption by limestones. Journal of Colloid and Interface Science (224), 2000, 351-357.

[19] Zhang L., He R., Gu H. Oleic acid coating on the monodisperse magnetite nanoparticles. Applied Surface Science (253), 2006, 2611-2617.

[20] Stolz H.J. Oleochemicals – Important additives for building protection. ZKG International, 2008.

Environmental control for mitigating salt deterioration by sodium sulfate on Motomachi Stone Buddha in Oita prefecture, Japan

Kyoko Kiriyama[1]*, Soichiro Wakiya[2], Nobumitsu Takatori[1], Daisuke Ogura[1], Masaru Abuku[3] and Yohsei Kohdzuma[2]
[1] *Kyoto University, Japan*
[2] *Nara National Research Institute for Cultural Properties, Japan*
[3] *Kindai University, Japan*
* *kiriyama.kyoko.24n@st.kyoto-u.ac.jp*

Abstract

This study aimed to develop an appropriate interior environment to mitigate the deterioration by sodium sulfate crystallization on Motomachi Stone Buddha in Oita prefecture of Japan. This Stone Buddha is carved on the fragile welded tuff and has been covered by a shelter for the preservation. On this site, salt crystallization, especially in winter, is a major deterioration factor of this Buddha. In previous studies, the crystallization of sodium sulfate and calcium sulfate were identified by XRD. The temperature and relative humidity were measured both indoor and outdoor. Water quantity analysis of the groundwater near the Stone Buddha has been conducted revealing a higher solute content near the Buddha than in the general groundwater in Japan. According to the temperature and humidity measured inside the shelter in winter, it was suggested that the solubility of sodium sulfate decreased greatly with dropping in interior temperature and phase change from mirabilite to thenardite (sodium sulfate anhydrate) caused the deterioration of the statue. Hence, the shelter was improved to reduce ventilation frequency and block the direct solar radiation in order to reduce evaporation and decreasing interior temperature in winter. In this study, we have conducted environmental research and salt crystallization experiments under controlled temperature and relative humidity to assess the improvements of the shelter. After improvements of the shelter, the indoor temperature and relative humidity increased in winter. According to our experimental results, the deterioration of the Stone Buddha is reduced highlighting the efficiency of the improvements.

Keywords: salt deterioration, sodium sulfate, shelter

1. Introduction

Salt crystallisation is one of the most common cause of stone deterioration.[1] In the field of conservation , it is well known that repeated cycles of dissolution and recrystallization of salts inside a stone can provoke serious damage. Especially for sodium sulfate, with both hydrated and anhydrous phases, is one of the most dangerous salts for stone.[2-5] The solubility depends on temperature and decreases as the temperature falls. The effects of temperature and relative humidity on in-situ salt weathering have been investigated in literature.[6,7] However, the mechanism of in-situ salt weathering remains controversial.

At the Motomachi stone-cliff Buddha to prevent collapse of the stone, polymers have been used. But they have strengthen the stone surface causing more degradation. A tunnel has been dug to lower the groundwater levels and limits the salt crystallization. However, the issue of an ideal environment to restrain salt deterioration awaits further investigation. Many recent studies focus on mixed salt behaviour. However, it is too complicated and controversial to discuss the matter in this paper. Therefore, this paper attempts to investigate the effects of the improvements of the shelter at the Motomachi stone-cliff Buddha and aims to consider methods to mitigate salt deterioration, especially in sodium sulfate.

2. Background of the Motomachi stone-cliff Buddha

The Motomachi stone-cliff Buddha is located in Oita prefecture, Japan. It was engraved onto a cliff of soft welded tuff around the 11th to 12th centuries and was covered by a shelter which was built during in the 20th century *(Figure 1 and Figure 2)*. The entrance of the shelter faces east. Salt crystallisation especially in winter is the most important cause of deterioration. Powdering and scaling occur at the surface of the stone. A previous investigation identified thenardite, mirabilite and gypsum by X-ray diffractometry (XRD).[8,9] Furthermore, a past study showed that sulphate, calcium and sodium ions, that are the origins of thenardite, mirabilite and gypsum, are detected in the groundwater near the site. And the content is higher than in the general groundwater in Japan.[9] Although a drainage tunnel and well were built in order to lower the groundwater levels *(Figure 3)*, an increase in the amount of salt was reported after this construction. A past study also suggested that drop in inside temperature in winter due to door-opening of the shelter and water leaking into the high-water-content stone could be causes of salt crystallisation at the Motomachi stone-cliff Buddha.[9] Therefore, the shelter was improved by setting door closer, double glass windows and boards to reduce ventilation frequency and to block the direct solar radiation to the Buddha in November 2015 by the Oita-city board of education.

Figure 1: Motomachi stone-cliff Buddha

Figure 2: Exterior of the shelter

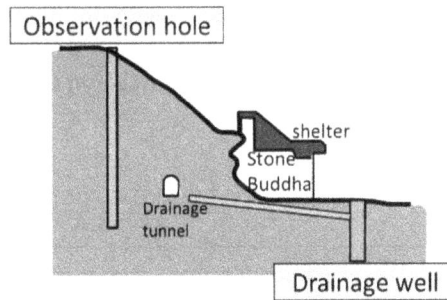

Figure 3: Cross sectional view

3. Methods

3.1. Environmental research

In order to assess the effects on the shelter improvements, the indoor temperature, relative humidity, ventilation frequency and local meteorological data have been measured before and

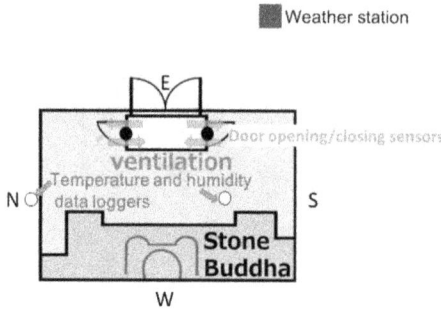

Weather station

Figure 4: Schematic representation of the inside with location of data loggers

after the shelter improvements. Temperature and relative humidity have been measured inside and outside the shelter with data loggers (HOBO Prov2) from November 2014 *(Figure 4)*. The counts of door-opening/closing have been measured with door-opening/closing sensors (HOBO UX90 State Data Logger) from February 2015 *(Figure 4)*. A weather station was installed in February 2015 which measured local meteorological data (temperature, relative humidity, wind direction, wind speed, solar radiation and pressure). All measurements have been done at 10 minutes intervals and the data were collected roughly once in three months.

3.2. Experiments

First, Aso soft welded tuff, which is thesame stone of the Motomachi stone-cliff Buddha, was selected for the experiment. The stone with the porosity of 30.4% and the density of 1.64(g/cm³) was cut into cylinders (50mm x 20mm). Second, to saturate the samples in order to have less effect on the depth of salt distribution at different relative humidity, all samples were immersed in distilled water for 3 days. Finally, the bottom face of the water-saturated samples were put into a 20ml/10w%Na_2SO_4 solution under controlled temperature and relative humidity (Table 1).

The conditions of 20°C, 10°C, 75% RH and low RH were chosen to reproduce field conditions in summer and winter in order to assess the effects of temperature and humidity conditions on crystallization of mirabilite and thenardite *(Table 1)*.

To prepare for samples A and B, an incubater was programmed to 20°C and 75% humidity. Sample A was enclosed in a desiccator containing silica gel, used to create a low humidity environment. Sample B was placed in an open desiccator. Both desiccators were placed into

	A	B	C	D
Temperature (°C)	21.7±1.0	21.0±1.1	10.4±0.2	10.8±0.2
Relative Humidity	59.8±14.5	77.2±5.0	27.8±29.8	73.1±1.4

Table 1: Experimental conditions

	Average temperature (°C)
Nov. 2014 - Feb. 2015	9.9±2.8
Nov. 2015 - Feb. 2016	13.3±3.2

Table 2: Average indoor temperature before and after improvements in winter ns

the incubator. To prepare for samples C and D, the other incubator was programmed to 10°C and 75% humidity. Sample C was placed in the desiccator containing silica gel and sample D was placed in the open desiccator. Both of these desiccators were placed in the second incubator. The experiments consisted of 5 dissolution/recrystallization cycles at intervals of at least 4 days. At the beginning of the cycles, 20ml distilled water was added to saturate the samples. After 5 cycles, stone powder was gently removed from the samples with a brush and collected. The residue was filtrated with 1μm pore size quantitative filter papers to remove the salt. The filter papers and the residue were immersed in 50ml of distilled water, which was replaced twice at an 1 hour interval. The filter papers were dried at 60°C until they reached constant weight. The percentage of damage was calculated by using equation (1).

Percentage of damage =
$(M_{amount\ of\ residue}/M_{drystone\ initial})$ (1)

4. Results and Discussion

4.1. Effects on the improvements of the shelter

The doors closing periods are given in *Figures 4a, b. Figure 4a* shows the door closing periods before improvements, on the other hand, *Figure 4b* indicates the state after the improvements. Before setting the door closers, the doors could stay open for a few days because visitors could keep opening and closing the doors freely. After the improvements, the doors became usually closed except during an operation periods of inside the shelter. Therefore, the results indicate that the ventilation frequency was reduced by the improvements.

Figure 5 illustrates a phase diagram for sodium sulfate and temperature and relative humidity of daily average inside the shelter. The continuous lines indicate the boundaries of the stable phases.[4] This diagram suggests that the interior environment before the improvements (November 2014-October 2015) mainly varied in both thenardite and mirabilite stable conditions. Previous studies showed the damage by sodium sulfate occurs because thenardite dissolution can produce solutions highly supersaturated with respect to mirabilite, so that precipitation of this phase can lead to large crystallization pressures.[5,10] Thus, it is important for the conservation of the Buddha that crystallization of sodium sulfate should be reduced, cycles of crys-

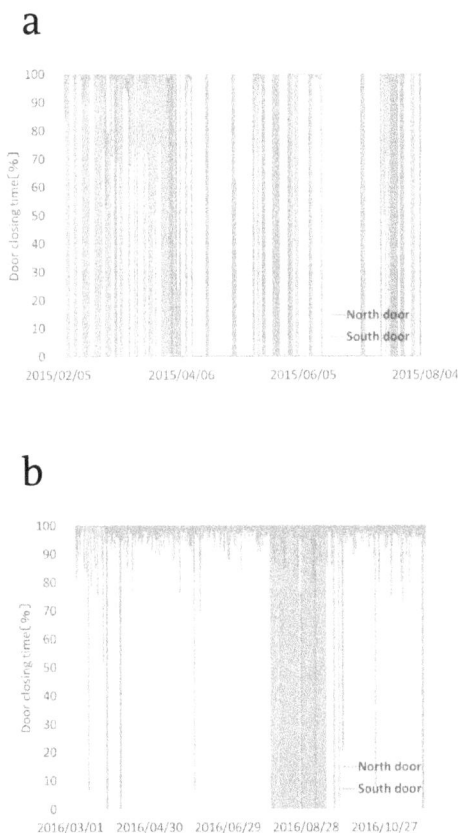

a

b

Figure 5: Door closing time a) before the improvements b)after the improvements

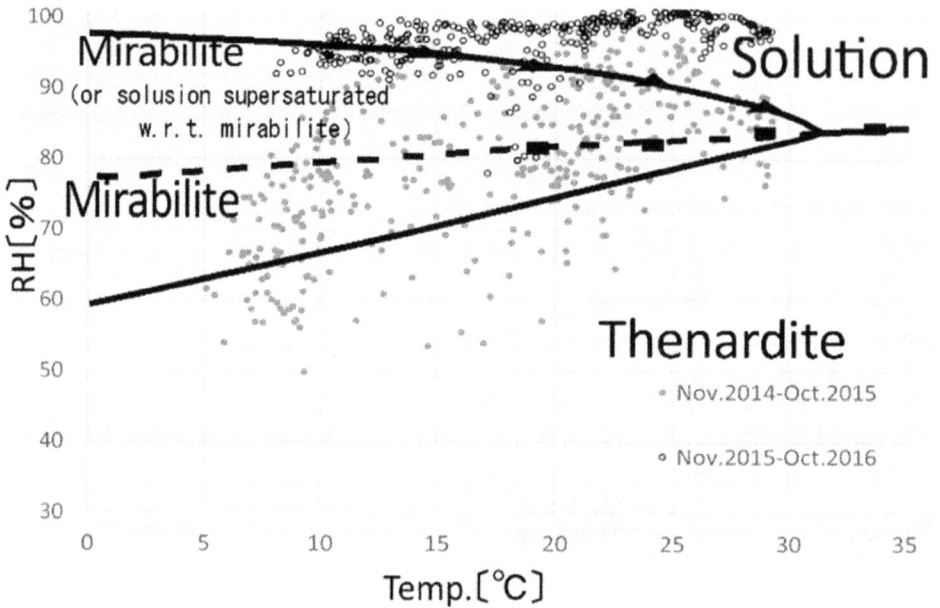

Figure 6: Phase diagram for sodium sulfate and temperature and relative humidity of daily average inside the shelter (based upon data from[4] Blue dots are before improvements Temp. and RH, on the other hand white dots are after improvements.

Rate of deterioration (%)	A	B	C	D
	0.0034±0.0009	0.0009±6E-05	0.0035±0.0026	0.0003±8E-05

Table 3: Percentage of damage from equation (1)

tallization and dissolution of sodium sulfate should be avoided and phase change to thenardite should be suppressed even if mirabilite crystallizes. *Figure 5* indicates that the interior environment after the development of the shelter (November 2015-October 2016) were mostly under solution of sodium sulfate stable phase, even though sometimes in mirabilite stable phase. This results shows that the improvements reduced the risk of precipitation of thenardite. Furthermore, the indoor temperature before and after improvements in winter is shown in *Table 2*. The temperature in winter increased 3.4°C after the renovation, which could lead to inhibit decreasing solubility of

sodium sulfate. Consequently, the interior environment has changed to reduce crystallization of sodium sulfate.

The amount of salt crystallization could not be measured quantitatively because usual salt crystallized area in winter was covered with Japanese papers in order to remove salts. However, according to observations, the amount of salt crystallization in winter decreased obviously. Therefore, it is considered that deterioration by powdering on the stone surface was reduced. On the contrary, on the other part of the stone surface, new salt crystallization was confirmed after the improvement. The salt was identified as gypsum by XRD. Furthermore, molds were confirmed on wood parts of the shelter by observation after the improvement. The high water content caused by the rise in inside relative humidity may be considered as the causes of increasing gypsum and molds. These changes after the improvement were caused by suppressing the ventilation of the shelter throughout the year. Generally speaking, winter in Japan is cold and dry, on the other hand summer is hot and humid. Therefore, it is necessary to consider operation method of the shelter in each season in the future. This could lead to reducing molds or other microbiology attack caused by humid condition.

4.2. Deterioration comparison by sodium sulfate

The results of the experiments are provided in *Table 3*. The rate of deterioration of samples are calculated by using equation (1). The rate of deterioration of sample A and sample C were greater than sample B and sample D. It is inferred that phase change to Thenardite may have an effect on the rate of deterioration. It is also possible that Increasing in evaporation rate may promote further damage. As for the difference of temperature (sample B and sample D), significant difference did not be detected. These result indicate that to reduce lowering relative humidity is more important than to decrease drop in temperature. Furthermore, the depth of salt crystallization affected by evaporation rate did not be examined in this study. Hence, this is a subject for future analysis.

Conclusion

Salt crystallization, especially in winter, is a major deterioration factor of this Buddha. Although salt deterioration may have been caused by decreasing interior temperature and relative humidity in winter, environmental control could not have been carried out because of the poor airtight of the shelter. Then, the shelter was improved in order to reduce a drop in temperature and humidity in winter. To evaluate the shelter improvements' effect on the salt deterioration, environmental research and observation were carried out. After these shelter improvements which led to reduction of ventilation frequency, it is considered that deterioration by powdering on the stone surface was reduced. However, predictable problems by humid condition such as new crystallization of gypsum and mold were detected after the improvement. Furthermore, in order to set the target value of the environmental control, effects of temperature and humidity on salt deterioration were investigated by salt crystallization experiment. The results of the experiments indicate that to reduce lowering relative humidity is more important than to decrease a drop in temperature.

For future research, a target value of the interior environment in each season needs to be considered. Furthermore, by an appropriate operation of the shelter which increased the airtight and could control environment more easily, mitigating methods for not only salt crystallization in winter but also other deterioration problems of the Buddha should be considered continuously.

Acknowledgements

We are grateful to the Oita-city board of education for the image of stone-cliff Buddha (Figure5) and the groundwater table of the observation hole.

This work was supported by JSPS KAKENHI Grant Number 26709043[Grant-in-Aid for Young Scientists (A)]

References

[1] Goudie, A. and Viles, H., Salt Weathering Hazards, John Wiley, Chichester, (1997).

[2] Scherer, G. W., Stress from crystallization of salt George, Cement and Concrete Research, (34), (2004), 1613-1624.

[3] Steiger, M. and Asmussen, S., Crystallization of sodium sulfate phases in porous materials: The phase diagram Na-2SO4–H2O and the generation of stress, Geochimica et Cosmochimica Acta, (72), (2008), 4291-4306.

[4] Flatt, R. J., Salt damage in porous materials: how high supersaturations are generated, Journal of Crystal Growth, (242), (2002), 435-454.

[5] Desarnaud, J. and Shahidzadeh, N., Impact of the Kinetics of Salt Crystallization on Stone Damage During Rewetting/Drying and Humidity Cycling, Journal of Applied Mechanics, (80), (2013), 6-8.

[6] Bionda, D., Methodology for the preventive conservation of sensitive monuments : microclimate and salt activity in a church, Proceedings of the 10th International congress on deterioration and conservation of stone, Stockholm, (2004), 627-634.

[7] Zehnder, K. and Schoch, O., Efflorescence of mirabilite, epsomite and gypsum traced by automated monitoring on-site, Journal of Cultural Heritage, (10), (2009), 319-330.

[8] Oita city board of education, Kunishiteishiseki Oitamotomachisekibutsu Hozonsyurijigyo Hokokusho (in Japanese), oita, sohrinsha, (1996).

[9] Kiriyama, K., Wakiya, S., Takatori, N., Ogura, D., Abuku, M. and Kohdzuma, Y., The current state and the factors of salt deterioration at buddha statue carved onto a cliff at motomachi in oita prefecture of japan, Proceedings of the 13th international congress on the deterioration and conservation of stone : VOLUME II, Hughes, J. and Howind, T. (eds.), University of the West of Scotland, (2016), 1163-1170.

[10] Tsui, N., Flatt, R. J. and Scherer, G. W., Crystallization damage by sodium sulfate, Journal of Cultural Heritage, (4), (2003), 109-115.

Numerical analysis on salt damage suppression of the Buddha statue carved into the cliff by controlling the room temperature and humidity in the shelter

Nobumitsu Takatori[1]*, Daisuke Ogura[1], Soichiro Wakiya[2], Masaru Abuku[3], Kyoko Kiriyama[4] and Yoshei Kohdzuma[2]
[1] Department of Architecture and Architectural Engineering, Kyoto University, Japan
[2] National Institutes for Cultural Heritage Nara National Research Institute for Cultural Properties, Nara, Japan
[3] Faculty of Architecture, Kindai University, Osaka, Japan
[4] Graduate School of Advanced Integrated Studies in Human Survivability, Kyoto University, Japan
*takatori.nobumitsu.62a@st.kyoto-u.ac.jp

Abstract

Motomachi Sekibutsu is a Buddha statue that was carved into a cliff in Oita City, Japan, during the 11th or 12th century. It was designated as a national historic site in 1934. The stone statue is constantly affected by the penetration of heat and moisture into the cliff, and concerns have been raised about its deterioration. Various preservation measures have been taken to prevent this; however, the main cause of deterioration, salt damage, has not been eliminated. Here we develop a numerical analysis model to calculate the heat and moisture behaviour in the statue and its shelter. Using this model, we reproduce the shelter's hygrothermal environment before and after renovation and evaluate it with respect to damage caused by sodium sulphate. Our results show that the improvement in airtightness drastically contributes to decreasing the evaporation from the statue and suppressing the salt phase change; thus, the renovation of the shelter suppressed sodium sulphate salt damage to the statue.

Keywords: conservation environment, architectural environment, coupled heat and moisture transfer, sodium sulphate, phase change

1. Research aim

Motomachi Sekibutsu *(Fig. 1)* is a stone statue of Buddha that consists of tuff and was carved into a cliff in Oita City, Japan, during the 11th or 12th century. The statue is constantly affected by the hygrothermal environment in the room and heat and moisture transfer through the cliff (from which it cannot be separated). Therefore, the statue has suffered salt damage, exfoliation and growth of mould and bryophytes.[1] Salt damage, caused by so-

Figure 1: Motomachi Sekibutsu

Figure 2: Shelter exterior

dium sulphate, has long been a concern. To prevent these deteriorations, various preservation measures have been undertaken from 1986 to 1996 and again from 2011 onward. These include constructing a shelter *(Fig. 2)*, boring a tunnel behind the statue to reduce groundwater and renovating the shelter. Salt damage by sodium sulphate is considered to be halted as a result of the shelter renovation during November 2015.[2]

Our aim is to suppress the further degradation of the statue from salt damage by controlling the hygrothermal environment of the shelter that houses the Buddha statue. Previously, using calculations of the heat and moisture behaviour of the statue and considering the shelter's hygrothermal environment, we clarified that salt damage tends to occur at the knee of the statue under the room environment before the renovation in 2013.[3] In this study, we focus on how the hygrothermal environment in the shelter is produced, and we reproduce it using numerical analysis. In addition, we examine a method for coordination of the room environment, which suppresses the salt damage of the statue.

2. Motomachi Sekibutsu Room Environment

2.1. Shelter

At Motomachi Sekibutsu, the shelter was renovated twice, in 1995 and in 2015. In 1995, various construction works were done to improve thermal insulation, such as insulating the wall by a thermal insulation material and installing a windbreak room at the entrance.[1] In 2015, the renovations focused on salt damage due to sodium sulphate and aimed to protect from solar radiation, improve the performance of thermal insulation and improve the airtightness of the windows.[2] During that renovation, thermal insulation was improved, a door closure was mounted, the windbreak room was mounted, and the windows were covered with solar shading sheets. In this study, we refer to that renovation as 'the renovation'.

2.2. Room temperature, humidity and ventilation rate

We measured the hygrothermal environment inside and outside the shelter from November 2014. *Figures 3 and*

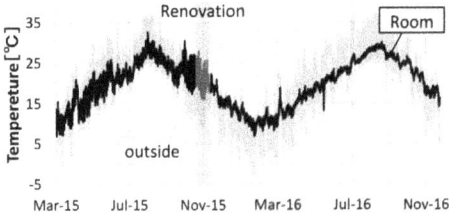

Figures 3: *Variation of the measured temperature inside and outside the shelter*

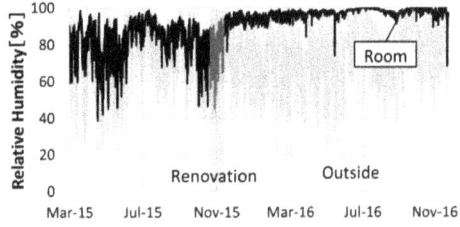

Figure 4: *Variation of the measured relative humidity inside and outside the shelter*

4 show the inner and outer temperature and relative humidity from March 2015 to November 2016. After the renovation, the daily fluctuation in room temperature decreased but the relative humidity maintained its high value. We measured the inside and outside concentrations of CO_2 from October 3 to October 4, 2016, and calculated the ventilation rate via the decrease in CO_2 concentration after opening or closing the window. The ventilation rate was 0.21 times per hour when the windows were closed, and it was 9.9 times per hour when they were opened.

3. Method for Examining Salt Damage

Salt damage to the porous statue material is caused by precipitated salt. The porous body is destroyed by volume expansion of salt, which is caused by crystallisation, thermal expansion and hydration.[4] To destroy the statue material, the following processes must occur: i) salt must precipitate in the material, and ii) the crystal pressure of salt must be high enough to break the material. Motomachi Sekibutsu consists of a tuff., the most prevalent damaging salt is sodium sulphate, and it is assumed that the pressure occurs when the precipitated salt ch-

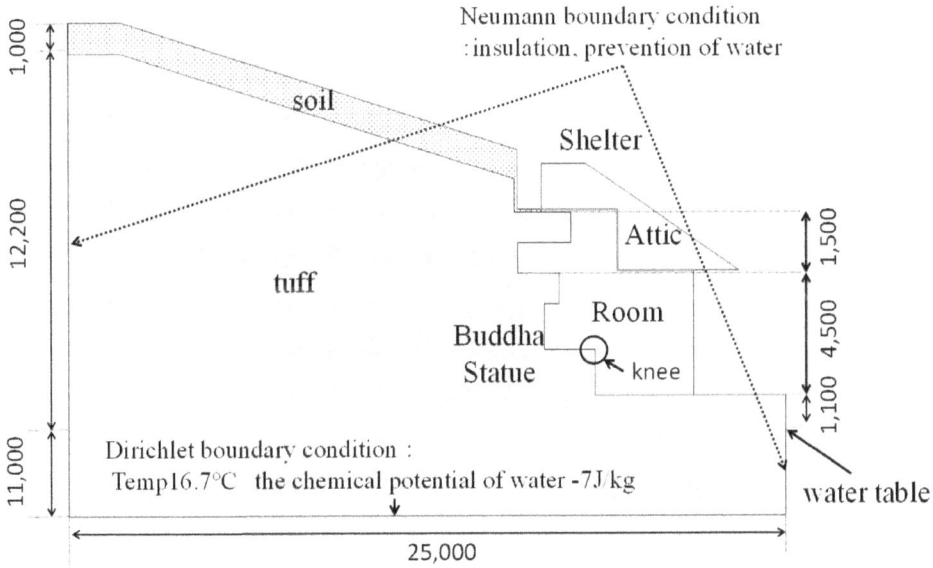

Figure 5: *Schematic of the numerical model of the cliff, including the statue*

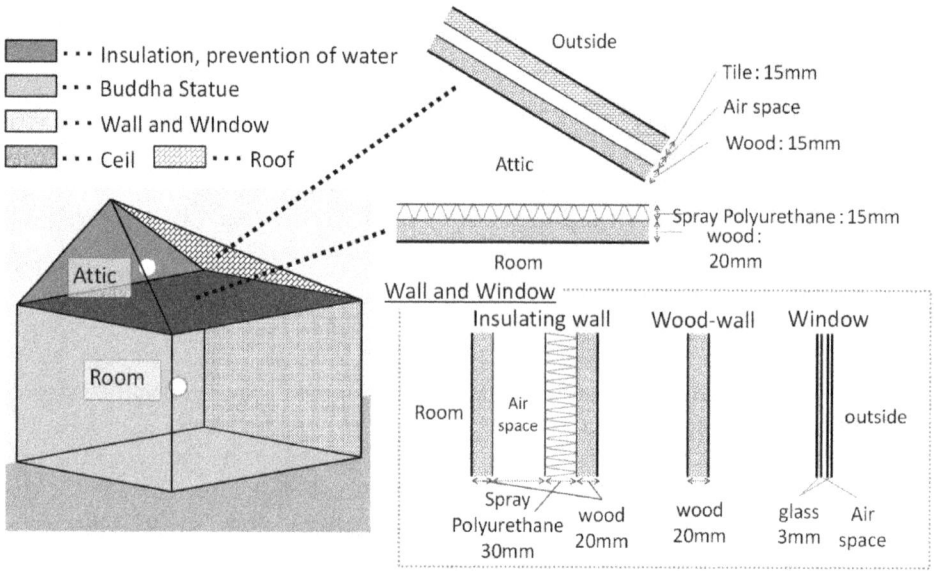

Figure 6: Model for analysing the room environment

anges its phase from thenardite (Na_2SO_4) to mirabilite ($Na_2SO_4.10H_2O$).[5] The salt precipitates when the solubility decreases with decreasing temperature and/or water evaporates.

4. Numerical Analysis

4.1. Methodology

We developed a numerical model to calculate the heat and moisture behaviour in the statue and shelter. The cliff, including the statue, is represented as a two-dimensional model, as used in our previous study[3] and shown in *Fig. 5*. The analytical model of the temperature and humidity in the shelter is shown in *Fig. 6*. We divided the shelter into room space and attic space. The shelter includes three kinds of walls: the insulated wall, the wooden wall and the glass window. The walls are oriented along the northeast, southeast and southwest directions. The roof is inclined at 30°. We calculated the

amount of solar radiation by considering the angle of inclination. The surface area of the statue was obtained by multiplying the length of each of the two-dimensional models by the depth, 10.8 m. The volumes of the room and the attic are 164.2 m² and 100 m³, respectively.

We used the heat and moisture balance equations [6] to analyse the heat and moisture behaviour in the statue and wall. The equations are written as follows:

Heat balance

$$c\rho \frac{\partial P}{\partial \rho} = \nabla \cdot \left\{ (\lambda + r\lambda'_{Tg}) \nabla g \right\} + \nabla \cdot (r\lambda'_{\mu g} \nabla \mu)$$

(1)

Moisture balance

$$\rho_w \left(\frac{\partial \psi}{\partial \mu} \right) \frac{\partial \mu}{\partial \mu} = \nabla \cdot [\lambda'_\mu (\nabla \mu - n_x g)] + \nabla \cdot (\lambda'_T \nabla \mu)$$

(2)

where the relation between the chemical potential of water μ and relative humidity h is given by

$$\mu = R_v T \ln h = R_v T \ln \frac{p}{p_{sat}} \qquad (3)$$

The boundary conditions for heat and moisture, respectively, are:

$$\alpha(T_0 - T_i) + r\alpha'_m(p_0 - p_i) + q_s =$$

$$-(\lambda + r\lambda'_{Tg})\frac{\partial T}{\partial n} - r\lambda'_{ug}\frac{\partial u}{\partial n} \qquad (4)$$

$$\alpha'_m(p_0 - p_i) + J_p = -\lambda'_u\left(\frac{\partial u}{\partial n} - n_x g\right) - \lambda'_T \frac{\partial T}{\partial n}$$

$$(5)$$

The boundary equations of the space are written as

Heat balance

$$c_A \rho_A V \frac{\partial T}{\partial t} = \sum S\alpha_i(T_w - T_i) + \sum c_A \rho_A V N (T_0 - T_i) + Q$$

$$(6)$$

Moisture balance

$$\rho_A V \frac{\partial A}{\partial t} = \sum S\alpha_i(P_w - P_i) + \sum \rho_A V N (A_0 - A_i)$$

$$(7)$$

where Q is the heat, which includes absorbed solar radiation in the space, transmitted through the windows.

4.2. Analysis conditions

In this analysis, the outside tempera-

ture and humidity, which are boundary conditions, were obtained from measured values (obtained from April 1, 2013 to March 31, 2014 and January 1, 2016 to December 31, 2016) [7]. The heat and moisture properties of each material were estimated from the literature.[8, 9] The coefficients of heat and moisture were 9.3 W/m2K and 2.85 × 10-8 kg/m² Pa for the room/attic space, respectively, and 23.3 W/m2K and 1.14 × 10-7 kg/m² Pa for the outside, respectively. Table 1 shows the calculation conditions. We aimed to clarify how the room's hygrothermal environment was changed by each renovation; therefore, we divided the renovation into the following factors to create our model.

i) Ventilation rate. Before the renovation, the ventilation rate between the room and the outside was 10.0 times per hour; however, after the renovation, it was 0.2 times per hour. Furthermore, the ventilation rate was 2.0 times per hour between the room and the outside and 0.1 times per hour between the room and the attic.

ii) Thermal insulation performance of the wall and window. Before the renovation, the U-value (the average transmission rate) of the wall was 4.55 W/m2K. After the renovation, the U-value was 1.23 W/m2K.

iii) Solar radiation shielding. Before the renovation, the rate of solar trans-

	Case 0	Case 1	Case 2	Case 3	Case 4	Case 5
Ventilation rate	A	B	A	B	B	A
Thermal insulation	A	B	B	A	B	A
Solar transmission	A	B	B	B	A	A

'A' means after the renovation, and 'B' means before the renovation

Table1: Analysis conditions

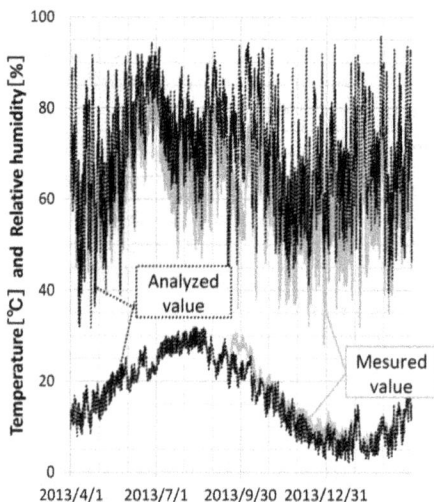

Figure 7: Comparison in Case 0

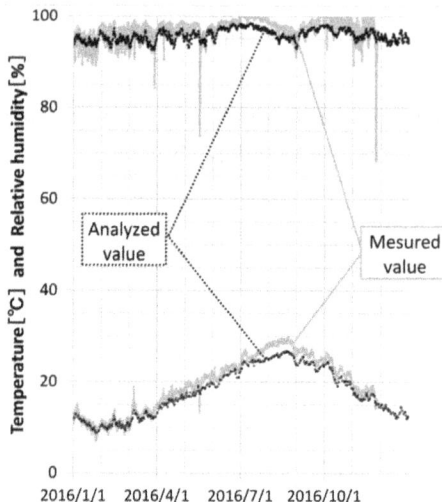

Figure 8: Comparison in Case 0

mission of the window was 0.75.[3] The amount of solar radiation incident on the statue was calculated by considering the shape of the roof and the position of the sun.[3] After the renovation, the rate of solar transmission was 0.6 and the amount of solar radiation that reached the statue was zero.

Using these conditions, we reproduced the room temperature and humidity in 2016 with Case 0 and reproduced from April 2013 to March 2014 with Case 1, respectively. In Cases 2–5, we considered in detail how each renovation affected the room environment.

5. Results and Discussion

5.1. Comparison of the measured and calculated values

Fig. 7 and Fig. 8 show the measured and calculated values of the temperature and relative humidity in the room for Case 0 and Case 1. In both cases, the annual fluctuation of the calculated value agrees very well with that of the measu-

red value, although there is a mismatch over a short period. Those errors can be mainly attributed to the constant ventilation rate throughout the year or the heat and moisture flux from the statue.

5.2. Water evaporation from the statue

We examined the trends in salt precipitation for Cases 1–5. Here we assumed that the moisture flux from the surface of the statue is equal to the evaporation in the statue. *Figure 9* shows the annual amount of water evaporation in the statue, and *Fig. 10* shows the monthly amount of evaporation. Comparing Case 1, before the renovation, to Case 2, it can be seen that water evaporation is suppressed by the decreasing ventilation rate. Similar results are obtained because of the improvement in insulation performance and prevention of solar transmission. These results show that after the renovation, evaporation and salt precipitation was drastically suppressed throughout the year. Here we focus on the effects of each renovation factor. The decreasing ventilation rate suppresses the evaporati-

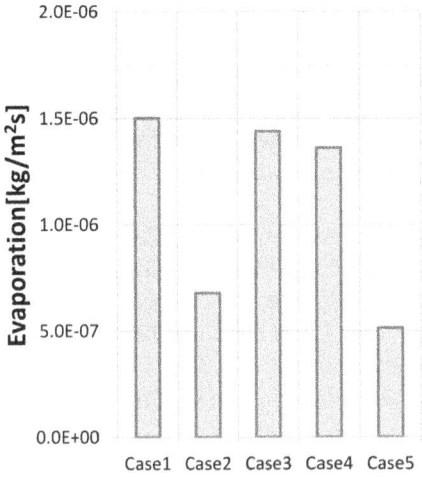

Figure 9: Annual evaporation from the surface of the statue

on and, in especially, contributes well in summer. The improvement in insulation performance suppresses the evaporation from winter to spring, but there is a possibility of promoting the evaporation from summer to autumn. The prevention of solar transmission suppresses the

evaporation, and the amount of evaporation does not change significantly across all seasons. The decreased ventilation rate has the greatest influence on the evaporation from the statue, and the improved insulation performance has the least influence.

5.3. Phase change of sodium sulphate at the statue surface

At Motomachi Sekibutsu, salt damage can be seen at the knee of the statue *(Fig. 11)*. In this section, we plot the calculated temperature and relative humidity at the surface of the statue on the phase diagram of Na_2SO_4 and examine the possibility of the resulting salt damage. *Figure 12* shows the results of Case 1, Case 3 and Case 4, and *Fig. 13* shows that of Case 1, Case 2 and Case 5. In Case 2 and Case 5, which are decreased-ventilation-rate cases, the phase change from Na_2SO_4 to Na_2SO_4-$10H_2O$ rarely occurs because the

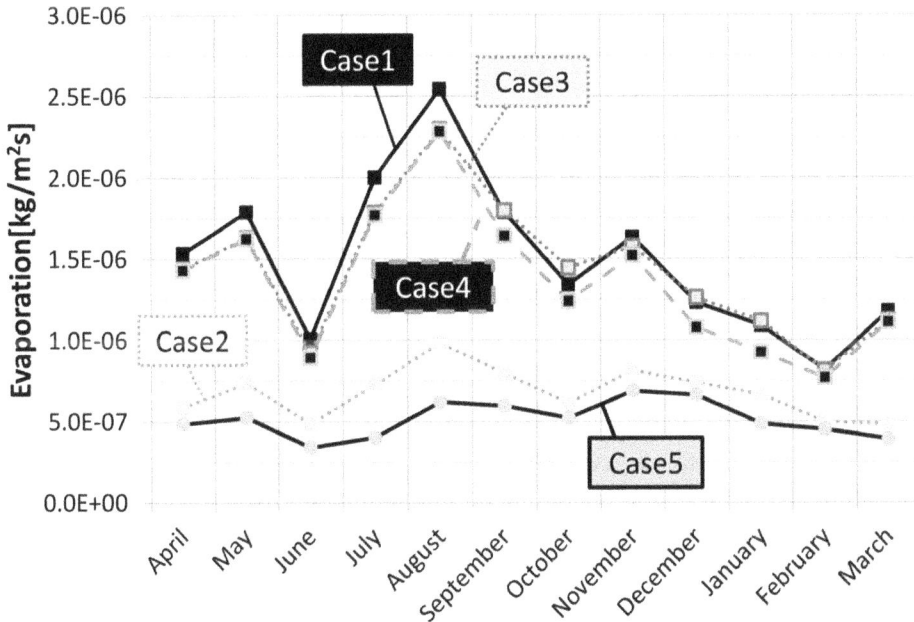

Figure 10: Monthly average evaporation from the surface of the statue

Figure 11: The knee of the statue

6. Conclusion

In this study, we developed a numerical analysis model to calculate the heat and moisture behaviour in the statue and shelter. Using this model, we reproduced the shelter's hygrothermal environment before and after the renovation and evaluated it with respect to sodium sulphate salt damage. The model was valid for recreating the hygrothermal environment in the shelter before and after the renovation. The renovation caused the evaporation from the statue to be suppressed by improving the airtightness and insulation performance and preventing solar radiation. The improvement in airtightness drastically decreased evaporation from the statue and suppressed the phase change of salt. Our results show that the renovation of the shelter suppressed sodium sulphate salt damage to the statue.

relative humidity is very high. However, in Case 1, Case 3 and Case 4, the temperature fluctuates around the low relative humidity, and phase change occurs easily. By focusing on the effect of each renovation factor, in Case 2, we see that the temperature fluctuates around the high humidity and the fluctuation of relative humidity is smaller in Case 2 than in Case 1 before the renovation.

The temperature and humidity in Case 3 are approximately equivalent to those in Case 1, but from winter to spring, the relative humidity is higher in Case 3 than in Case 1. Therefore, the phase change from $Na_2SO_4\text{-}10H_2O$ to Na_2SO_4 is restrained during these seasons. Considering the solar transmission, the temperature is lower and the relative humidity is higher in Case 4 than in Case 1 throughout the year. These results suggest that the improvement of airtightness strongly contributes to the suppression of salt damage to the statue.

References

[1] Oita City Board of Education, 1996, "Kunishiteishiseki Oita Motomachi Sekibutsu Hozonsyurijigyo Hokokusho (in Japanese)," Oita, sohrinsha.

[2] Oita City Board of Education, 2016, "Kunishiteishiseki Oita Motomachi Sekibutsu Hozonsyurijigyo Hokokusho" (in Japanese).

[3] N. Takatori, D. Ogura, S. Wakiya, M. Abuku, K. Kiriyama: Numerical analysis of the influence of hygrothermal variation on salt weathering of a Buddha statue carved into a cliff–Study on the conservation of a stone Buddha carved into a cliff at Motomachi PART1-, Journal of Environmental Engineering (Transactions of Architectural Institute of Japan), Vol. 733, pp. 215-225, 2017.

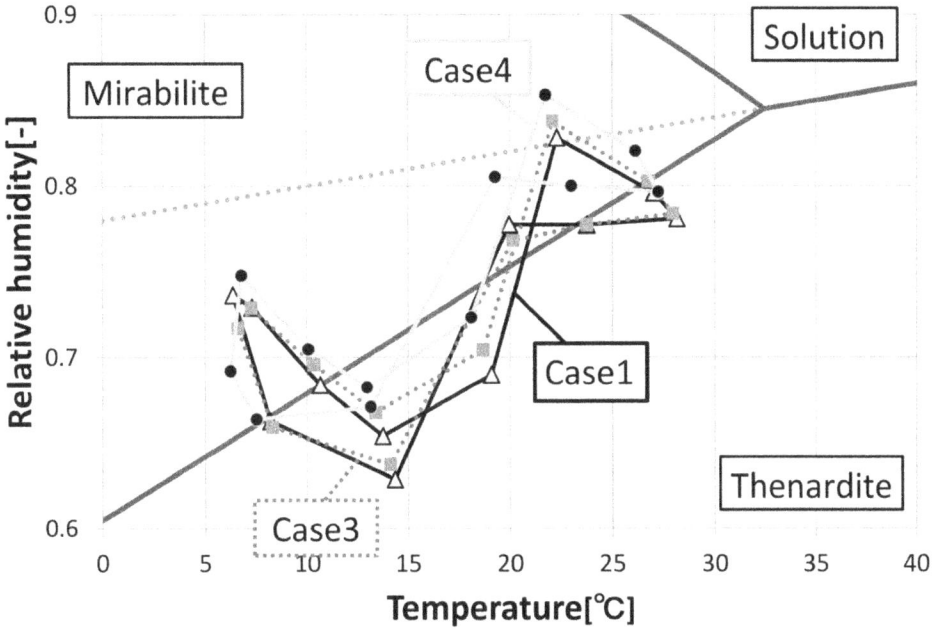

Figure 12: Phase diagrams for Case 1, Case 3 and Case 4

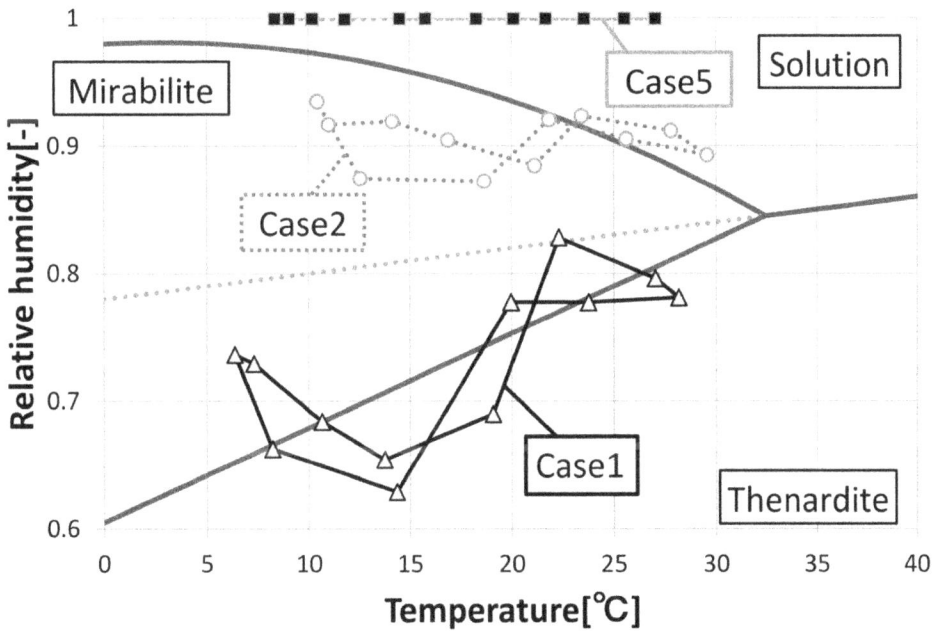

Figure 13: Phase diagrams for Case 1, Case 2 and Case 5

[4] A. S. Goudie, H. A. Viles: Salt Weathering Hazards, Wiley, 1997.

[5] R. J. Flatt: Salt damage in porous materials: How high supersaturations are generated, Journal of Crystal Growth, Vol. 242, pp. 435-454, 2001.

[6] M. Matsumoto: Energy conservation in heating cooling ventilating building: Heat and mass transfer techniques and alternatives (ed. Hoeogendoorn C.J. and Afgan N.H.), Washington: Hemisphere Pub. Corp., pp. 1-45, 1978.

[7] Japan Meteorological Agency, 2017, refer to the past weather data, <http://www.jma.go.jp/jma/index.html>.

[8] AIJ: Air and Moisture Transfer Through New and Retrofitted Insulated Envelope Parts (Hamtie), Final Report, Vol. 3, TASK 3: Material Properties, AIJ, 2001.

[9] M. K. Kumaran: Thermal and Moisture Transport Property Database for Common Building and Insulating Materials, Final Report from ASHRAE Research Project 1018-RP, 2002.

Management of sodium sulfate damage to polychrome stone and buildings

David Thickett* and Bethan Stanley
English Heritage
*david.thickett@english-heritage.org.uk

Abstract

English Heritage holds large collections of polychrome stone. Often the polychromy only remains as small fragments, loosely adhered to the stone surface with any original binders having converted to oxalate. The sixteenth century limestone Thetford Tomb fragments are amongst the most significant part of the collection. Salt analyses indicated very high concentrations of sodium sulfate present, (up to 1.5% by mass of the stone). Considering the very fragile nature of the remaining polychromy and the aggressive nature of this salt, strict preventive conservation is needed to ensure survival of the information. Theoretically, keeping the RH below the thenardite transition line (the RH is temperature dependant) should avoid any damaging salt transitions. Monitoring with acoustic emission in the storage environment, provided a direct tracing of salt transitions. When plotted against temperature, it became clear that transitions were occurring at between 3 and 7% lower RH than expected from theory. Tests with pure sodium sulfate powder coincided with the theoretical values. The most likely reason is the effect of the pore structure, particularly fine pores. The values are consistent for a single piece of stone and vary between pieces. A good correlation was observed between acoustic emission events and small pieces appearing on the imaging plate of a prototype particle deposition analyser. Loss of material from two interior building surfaces was monitored using a similar automated camera

system. Inspection of the environmental data when material fell, shows similar depression of the RH below the theoretical values for sodium sulfate.

Keywords: Sodium sulfate, polychromy, acoustic emission, image analysis, preventive conservation

1. Introduction

English Heritage holds large collections of polychrome stone. Often the polychromy only remains as small fragments loosely adhered. Such material is extremely susceptible to salt damage and evidence of polychromy can easily be lost. Sodium sulfate is an extremely damaging salt and used in several salt crystallisation tests.[1, 2] It undergoes an over 300% expansion when converting from the anhydrite, thenardite to the decahydrate, mirabilite. The RH that this conversion occurs at is strongly temperature dependant.[3] Controlling temperature and RH can prevent the damage. However, the exact parameters are needed to design effective control.

The sixteenth century limestone Thetford tomb fragments are amongst the most significant part of the collection. They formed the focus of a major recent historical research project.[4] The tomb was destroyed during the English Reformation (1530s) and the small (less than 24 cm) pieces were exposed to the elements for almost half a century before acquisition into the National collection. In these circumstances it is highly likely that the original organic binder present with the pigment has converted to oxalate. This makes the pigment stone bond extremely weak.

Two techniques have been used to directly monitor salt transitions or material loss caused by them. Acoustic emission allows a direct measurement of certain phenomena. It has been used to monitor sodium sulfate transitions[5] and stone decay.[6] The data can be of sufficient time resolution to allow direct correlation with environmental data. Automated image analysis of a glass plate was used to monitor material falling from the carved surfaces of the pieces. Various other analytical techniques were used to determine the nature of the lost material.

2. Methods and Materials

Salt analyses were undertaken with 0.4mm drilled samples from break surfaces. The first 2mm of the sample from the surface was discarded and sampling continued for another 10mm after that. The sample volume was over thirty times the maximum grain volume of the fine grained limestone determined by eye. The samples were dried at 110°C, and extracted with 18.2MOhm/cm water. The filtered solutions were analysed with a Dionex DX600 ion chromatograph with; AS14A column and an eluent of 8mM sodium carbonate and 1mM sodium bicarbonate for anions and CS12 column with a 1mM methane sulfonic acid eluent of for cations. The solution ion concentrations were converted to percentages by mass based on the dry mass of stone.

Five pieces were analysed with acoustic emission. Initially, the performance of two instruments was compared. A Physical Acoustics Pocket AE system was used with WD sensors (1-1000kHz, non resonant) with and without Slyglide couplant. A Hanwell Woodwatch was also assessed with the supplied sensor (1-

1000kHz, 150kHz resonant), clamped to the stone and with couplant present as well. Test blocks (6cm by 6cm by 6cm) of Caenstone impregnated to give approximately 1 and 0.2% sodium sulfate by weight, were exposed to increasing RH. The RH was generated in a polycarbonate chamber with a commercial Rh generatore, (Preservatech MiniOne unit). Slyglide couplant gel was found to be required to achieve sufficient sensitivity for the Woodwatch sensors with the 0.2% sulfate blocks. The Woodwatch system generated signals for the 1% sulfate blocks, but this was not deemed sensitive enough. Two cycles with the 0.2% sulfate blocks generated loss from their surfaces. More of the Woodwatch units were available to the project, although the use of couplant is not ideal. Removal of the couplant was investigated. Acetone was found to remove all visible residues of the Slyglide from the Caenstone. Staining with iodine vapour indicated no remaining organic material on the stone surface or in the interior (the blocks were cut).

A set of preliminary experiments were also undertaken to determine the effect of the other small concentrations of ions present on the acoustic emission detection of the hydration transition. Salts (sodium, calcium and potassium sulfate, sodium nitrate and chloride) were mixed by grinding to obtain a composition close to the soluble extracts from the stone object with highest percentage of non sulfate and sodium ions (object number 78101879), hence referred to as impure sodium sulfate. Water was added to form a solution and this was dried at 60°C in a Bakelite top. When dry the top and impure sodium sulfate was placed on top of a WD acoustic emission sensor. These were placed in an environmental chamber and the temperature set to 15, 20, 25 and 30°C. At each temperature the RH was increased from 5% below the pure sodium sulfate transition RH at 0.2% intervals using glycerol solutions.[7]

Object number	Ion concentration (% mass/mass of dry stone)					
	chloride	nitrate	sulfate	sodium	potassium	calcium
78101879	0.022	0.020	1.354	0.485	0.004	0.054
78101845	0.015	0.014	1.335	0.478	0.001	0.043
78101881	0.008	0.009	1.251	0.449	0.002	0.046
78101843	0.008	0.009	1.247	0.448	0.000	0.032
78101880	0.012	0.011	1.224	0.433	0.003	0.041

Table 1: Ion concentrations extracted from drillings of Thetford Tomb limestone

The temperature and RH was measured with a Rotronic Hygrolog logger. The RH at which significant acoustic emission occurred was taken to indicate the transition. Acoustic emission sensors were attached to the five pieces for twelve months in their storage environment. The temperature and RH was monitored beside each block with a Rotronic Hygrolog D logger with hygrclip probe, calibrated with UK National Measurement Accreditation Service traceable standards. Triaxial shock loggers (MSR145) were attached to two pieces to assess shock events that could be detected with the sensors and, potentially mis-interpreted, as salt activity. Acoustic emission monitoring of wooden objects has been limited by large amounts of noise.[8]

Further tests were undertaken with two pieces of stone (4.2 by 2.3 by 1.8 and 2.2 by 4.5 by 2.0 cm). The acoustic emission, T, RH (using a calibrated Meaco system with Rotronic hydroclip probes) and vibration were monitored. Each stone was placed above a prototype image analysis system, such that any material lost, fell onto a glass plate. A digital camera was focused through magnification onto the plate and recorded images with a circumferential LED lighting system every 30 minutes. The side LED lighting and camera/magnification combination were

found to reliably detect particles down to 3μm diameter.[9] The system works well in dark or very low light surroundings. This was facilitated by running the experiments in closed cupboards. Any fall of material onto the plate was determined by subtracting sequential JPEG images using Image J software. A particle appearing in the difference image was assumed to have been lost, in the 30 minute time interval from the stone surface. The position of the particles lost from the polychrome stone and falling onto the glass plate was noted. They were collected with a chemically sharpened needle under magnification every 3 months and analysed with a Renishaw System 1000 Ramascope confocal Raman system, 632.8 nm laser, Nicolet Inspect FTIR microscope and Eagle III XRF system.

A second set of monitoring was carried out with a sodium sulfate laden group of brick casemates at Fort Brockhurst, on the South coast of England. The prototype image analysis system was placed under a mortar join exhibiting powdering and loss. Salts had been observed mainly to act at the mortar and not the brick surfaces. The T and RH were measured with a Meaco system. The casemates are dark when not in use. The increase in percentage coverage of the glass plate was measured with image analysis (Image J) every

30 minutes. Readings were discarded for 72 hours after the room had been accessed due to dust raised and deposited. Previous experiments had shown over 99.9% of the small amount of dust settled in this period. No activities took place during the monitoring period that would raise large amounts of dust.

A basement store room in Rangers House, London was losing plaster from several areas of the walls. Salt analysis, extracted both from the losses and from drillings indicated sodium sulfate was the predominant salt present in the wall and in one area this made up over 99% of the soluble ions detected. The automated image analysis system was placed below an area on an internal wall' to capture loss and the room temperature and RH measured with a calibrated Meaco system. Access to the store was restricted for the monitoring period, so that re-depositing dust did not cause significant interference.

3. Results

The salt analyses for the Thetford polychrome stone are shown in *Table 1*.

Errors are in the order of 3% of the calculated value. The errors were calculated from the reproduceability of multiple standard injections on the ion chromatograph, the calibration graph [10] and the stated error in the measuring balance and measuring pipettes used to prepare the solutions. The major anion is sulfate, which makes up over 97% of the anions present. The levels are very high, over 1.2% by mass of the stone. Sodium is the predominant cation, making up over 88% of the cations present.

The temperature and RH recorded when acoustic emission events were observed, are shown in *Figure 1*. The mirabalite to thenardite transition for pure sodium sulfate (pure) is marked on the figure. *Figure 1* also shows three of the four points determined for the impure

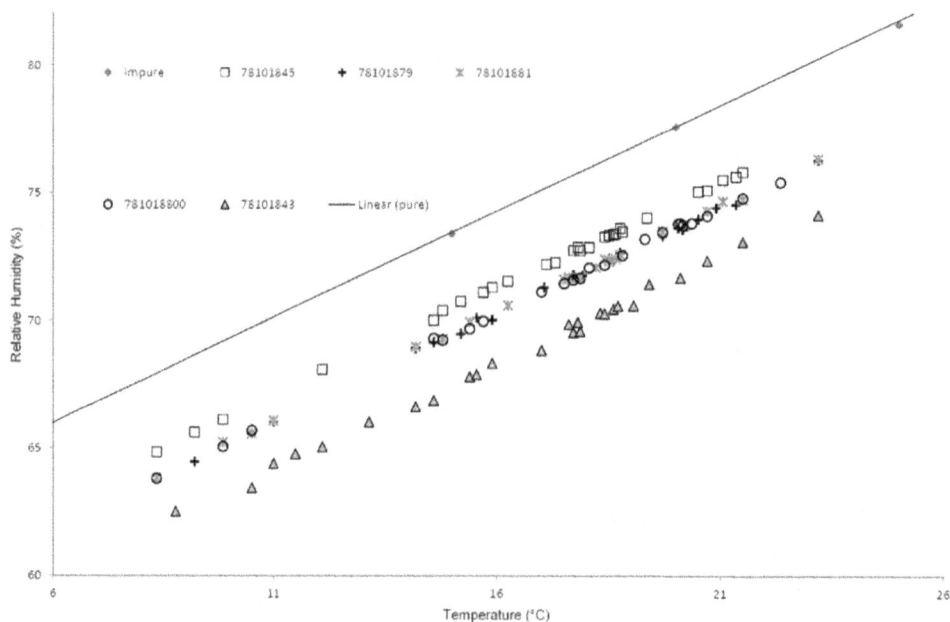

Figure 1: Temperature and RH values at which acoustic emission was observed.

sodium sulfate (labelled impure). They are coincident within the experimental parameters (1% RH). The fourth, higher temperature point was not plotted, to allow better reading of the graph, it also was coincident with the pure transition value.

Acoustic emission is occurring a few percent RH below the theoretical transition line in all instances. The stated accuracy for the probes is 0.03ºC and 0.8%. All RH values were within 1% during the three point calibrations. The RH is plotted to one decimal place. The digital output from the probe give a much higher precision but was not used with no way to check the calibration to that level of accuracy. The traceable standards are only valid to one decimal place. The depression is consistent for a particular piece of stone. Fragment 78101843 showed a value of approximately 6%. Fragments; 78101879, 78101881 and 781018800 showed approximately 4% and 78101845 showed 3%. The store in which the monitoring took place had temperatures mainly between 15 and 24ºC during the 7 month measurement period. Lower temperatures were experienced in the last 60 days.

The number of particles lost from stone 78101843 are shown in *Figure 2.*

Powder was observed falling from the objects, generally coincidentally with acoustic emission events detected (within the same 30 minute measurement slot). Three instances were observed with acoustic emission, but no powder detected, generally with lower numbers of acoustic emission events detected. Two instances of powder were observed with no acoustic emission. Examination of the vibration logger data indicated high shock levels coincident with two events. This was probably caused by unusual activity related to transferring files in the library above the room used for the experiments. Similar data was produce with 78101879. The Raman and FTIR microscope analysis showed approximately 30%

of the powder samples were pigments (hematite, carbon black, calcite), over 62% of the powder samples were sodium sulfate anhydrite (the phase information being lost due to the time delay before analysis) and the remainder, limestone powder. The Thetford limestone included approximately 2% iron, which XRF distinguished from calcite used as a pigment, that had no detectable iron present.

Two reasons for this observed RH depression were postulated. Slow ingress of RH and T into the stone, could retard the transition behind the prevailing ambient conditions. The presence of fine porosity could reduce the critical RH of the transition through the Kelvin effect.[11] This has been postulated for sodium sulfate, but this is the first experimental evidence.[12] Examination of the T/RH traces before the observed transitions indicated a large proportion, over 25% where if this was due to a lag in heating or RH transmission, then the transition would have been expected to be observed higher than the theoretical values, due to the environmental data around those instances.

The monitoring experiments at Fort Brockhurst and Rangers House showed powder falling from the mortar and plaster, coinciding with 4 and 6% RH below the thenardite to mirabilite transition line.

4. Discussion and Conclusions

It has been shown that the hydration transitions of sodium sulfate in some porous matrices can be lowered by several percent RH from the expected equilibrium values. This has also been shown to relate to the amount of loss observed. Care needs to be exercised when interpreting both acoustic emission and material loss data. Several other phenomena can cause acoustic emission, and its use for monitoring wooden objects has been impacted by significant noise.[8] Ca-

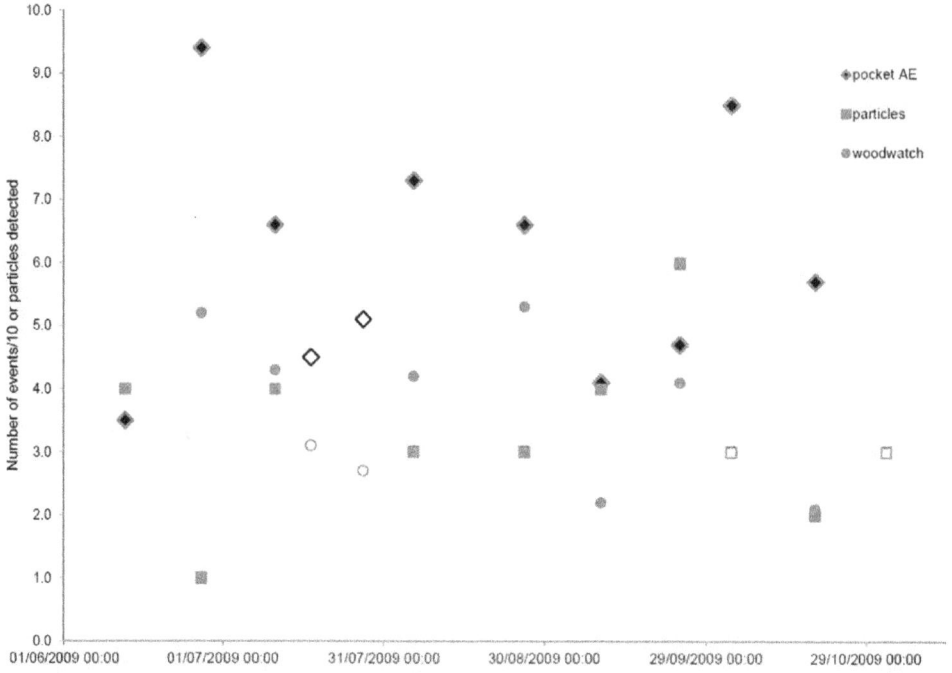

Figure 2: Acoustic emission and particle loss observed from polychrome limestone fragment 78101843

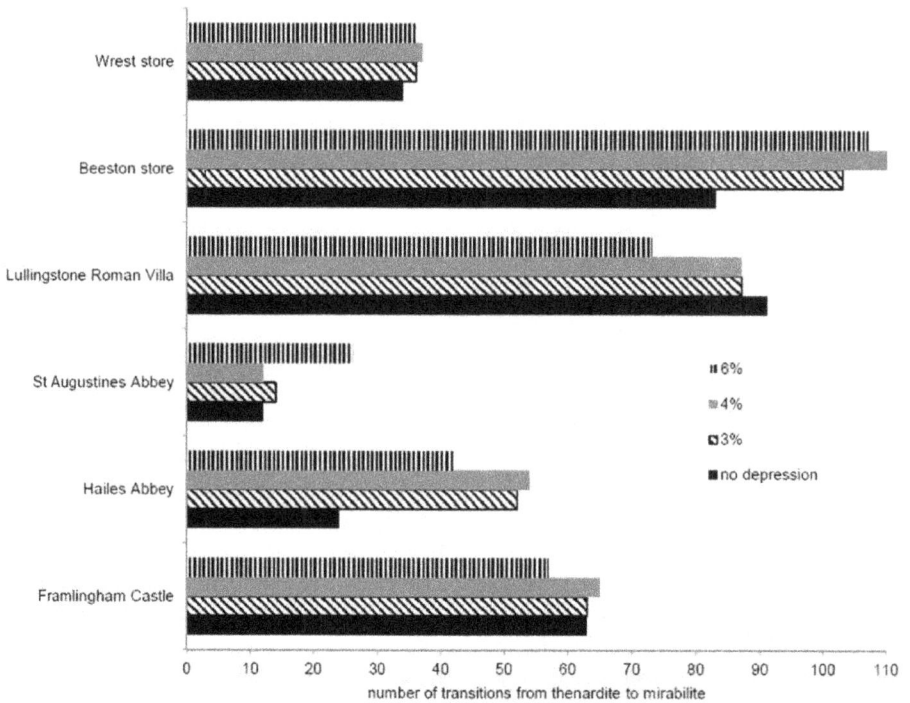

Figure 3: Number of thenardite to mirabilite transitions calculated from environmental data of English Heritage locations with polychrome stone.

reful control of the environment reduced this in these three instances. When salt expansion occurs, loss of material may not be instantaneous and loose, hardly bonded material is often observed, on surfaces. This could fall at a later date. Small numbers of events were detected with acoustic emission and no powder loss and two instances of powder loss with no acoustic emission due to large shock events. However, the overall excellent correlation between the two sets of results gives a reasonable degree of confidence, that the events detected fit the interpretation presented.

This effect has important ramifications for designing environments to control such phenomena. Display of some of the Thetford pieces was achieved in a dehumidified showcase, controlling the RH below 60% RH using a Miniclima unit. This value was determined by the research presented and knowledge of the minimum temperature likely to be encountered. Without the research, a value of 70% would probably have been used, reducing the conditioning load, but potentially allowing further damage to occur. The showcase was designed with a flat dark metal panel below the polychrome objects on display so any loss would be readily visible. This was examined every two years with magnification and any suspect particles analysed, as described previously. No pigment or limestone particles were detected during eight years on display. Depending on the situation, it is often not possible to achieve sufficient environmental control to totally stop an effect and the result is often expressed in a calculation of number of transitions. With the often complex environments in heritage buildings, even small changes in parameters can affect the number of transitions observed. *Figure 3* shows the number of events where the transition line was crossed from the thenardite, into the mirabalite region at English Heritage sites with polychrome

stone. For each location, four values are shown, those calculated from the equilibrium line and those, 3, 4 and 6% below.

In some instances, (Beeston store 4% and St Augustines Abbey 6%), the number of transitions increases dramatically, due to the nature of the environment. It was proposed to move the Thetford material from its existing store to a newly created store. The existing store environment was assessed using damage functions developed from this work and compared to the anticipated new store environment.[13]

References

[1] RILEM, Essais recommandees pour l'alkteration des pierres et evaluer l'efficacite des methodes de traitement, Materiaux et Constructions (17-75), 1980, 216-220.

[2] Lazzarini, L. and Laurenzi-Tabasso, M., Il Restauro della Pietra, Cedam, Padua 1986.

[3] Bionda, D., Modelling indoor climate and salt behaviour in historical buildings, PhD Thesis, Swiss Federal Institute of Technology, Zurich, 2006.

[4] Lindley, P.G., Representing Re-Formation: Reconstructing Renaissance Monuments, http://representingreformation.net/ accessed 25 May 2017

[5] Grossi, C.M., Esbert, Rosa Maria, Suarez del Rio, L.M., Montoto, Modesto, Laurenzo-Tabasso, M., Acoustic emission monitoring to study sodium sulphate crystallization in monumental porous carbonate stones, Studies in Conservation, (42), 1997, 115-125.

[6] Esbert, R.M, Grossi, C., Suarex del Rio, L.M., Calleja, L., Ordaz, J. and Montoto,

M., Acoustic emission generated in treated stones during loading. In Sixth International Congress on Deterioration and Conservation of Stone Proceedings, ed J.Ciabach, Turon, 1988, 403-410.

[7] Miner C.S. and Dalton N.N., Glycerol, Reinhold, New York, 1953.

[8] Jakieła, Sławomir, J. Bratasz, Ł. and Kozłowski, R., Acoustic Emission for Tracing the Evolution of Damage in Wooden Objects, Studies in Conservation (52-1), 2007, 101-109.

[9] Bowden, D. and Brimblecombe, P., Monitoring dust at Ickworth House with the dust-bug, Views, (42), (2005), 25–27.

[10] ISO11095 Linear calibration using reference materials, 1996.

[11] Camuffo, D., Microclimate for cultural heritage, Elsevier, Amsterdam, 1998.

[12] Charola, A. E., and Weber. J. The hydration-dehydration mechanism of sodium sulphate. In Seventh International Congress on the Deterioration and Conservation of Stone, Laboratorio Nacional de Engenharia Civil, Lisbon, 1992,. 581–590.

[13] Xavier-Rowe, A., Newman C., Stanley B., Thickett D., and Pereira Pardo L. A new beginning for English Heritage's archaeological and architectural stored collections. In ICOM-CC 17th Triennial Conference Preprints, Melbourne, ed. J. Bridgland, International Council of Museums, Paris, 2014, art. 1517, 10 pp. (ISBN 978-92-9012-410-8).

Conservation of marble artifacts by phosphate treatments: influence of gypsum contamination

Enrico Sassoni[1]*, Gabriela Graziani[1], Elisa Franzoni[1] and George W. Scherer[2]
[1] *Dept. of Civil, Chemical, Environmental and Materials Engineering (DICAM), University of Bologna, Italy*
[2] *Dept. of Civil and Environmental Engineering (CEE), Princeton University, USA*
*enrico.sassoni2@unibo.it

Abstract

The use of ammonium phosphate solutions has proven to be very promising for protection and conservation of marble. However, all the studies carried out so far have been performed on uncontaminated marble. Unfortunately, this is rarely the case in the field, because marble artifacts exposed outdoors are often affected by sulfation, i.e. formation of a gypsum crust on the surface. Because gypsum is much more soluble than calcite, the outcome of the ammonium phosphate treatment is expected to be sensibly altered by the presence of gypsum. Therefore, in this study the nature and morphology of the new calcium phosphate phases formed by reacting gypsum with aqueous solutions of diammonium hydrogen phosphate (DAP) were investigated. In particular, the effect of DAP concentration, ethanol addition (aimed at reducing gypsum solubility), and pH were explored. The result is that phase formation can be controlled by suitably tuning the above mentioned parameters. Phases with low solubility (such as tricalcium phosphate and hydroxyapatite) can be obtained by increasing the ethanol concentration, the DAP concentration or the pH. However, their formation is associated with diffused cracking, likely because of excessive growth of the new phases. Among the investigated formulations, treatment with a 0.1 M DAP solution with 30 vol% ethanol at pH=8 seems to be the most suitable one, as it leads to formation of brushite (about 30 times less soluble than gypsum), without cracking, so that a reduction in gypsum solubility in rain is expected.

Keywords: marble, gypsum, black crusts, hydroxyapatite, protection

1. Introduction

Aqueous solutions of diammonium hydrogen phosphate (DAP, $(NH_4)_2HPO_4$)) have proven highly promising for protection and consolidation of marble.[1-4] Thanks to the reaction between the phosphate solution (also containing a calcium source) and the substrate, new calcium phosphate (CaP) phases are formed[5]. These new phases are able to improve marble resistance to dissolution in rain (thanks to their lower solubility than calcite)[2,5] and marble cohesion (thanks to their bonding action at grain boundaries).[1,4] Ideally, the new calcium phosphate should be hydroxyapatite (HAP, $Ca_{10}(PO_4)_6(OH)_2$), which is the least soluble CaP phase in aqueous solutions at pH > 4.[5,6] However, depending on the reaction conditions (e.g., pH[3] or addition of external calcium sources[5]), different CaP phases may form alongside HAP, such as octacalcium phosphate (OCP, $Ca_8(HPO_4)_2(PO_4)_4 \cdot 5H_2O$), β-tricalcium phosphate (β-TCP, $β-Ca_3(PO_4)_2$), brushite ($CaHPO_4 \cdot 2H_2O$), monocalcium phosphate monohydrate (MCPM, $Ca(HPO_4)_2 \cdot H_2O$) and monocalcium phosphate anhydrous (MCPA, $Ca(HPO_4)_2$). These phases have sensibly different solubility in water[1], hence the exact nature of the new CaP phases is fundamental for success of the treatment.

As pointed out by some recent applications of the DAP-based treatment to some real artifacts[1,7], the possible presence of contaminants on the surface to be treated can have a significant impact on the nature of the new calcium phosphate phases formed after treatment. In fact, in the field marble is frequently contaminated with gypsum, resulting from marble sulfation induced by the high concentrations of sulphur dioxide in the atmosphere in the past decades. An example of marble affected by surface sulfation is illustrated in *Figure 1*. Gypsum formed on the surface of marble elements cannot always be completely removed before marble treatment. Because gypsum has a much higher solubility in water than calcite (~2.4 and ~0.014 g/l, respectively), the treatment outcome is expected to be sensibly affected by the presence of gypsum.[1]

A few studies have investigated the effects of treating gypsum with aqueous solutions of DAP.[7-10] Depending on the reaction conditions (e.g. DAP concentration, pH, duration), phases with low solubility (such as β-TCP, OCP, HAP) and more soluble phases (such as brushite) have been found.[7-10] However, when high DAP concentrations were used (3.0 M[9] and 3.8 M[8]), the formation of cracks and fissures in the new calcium phosphate layer was observed.[8,9]

The present study is aimed at investigating and optimizing the nature and the morphology of the new CaP phases formed starting from gypsum, by promoting the formation of phases with low solubility and preventing cracking. To this aim, the influence of several parameters (DAP concentration, pH, and addition of ethanol to reduce gypsum solubility) was investigated.

2. Materials and methods

2.1. Samples

To investigate the effect of gypsum on the nature of the new CaP phases, a simplified system was considered. Tests were carried out on specimens of gypsum pastes, produced by mixing bassanite with water (water/bassanite weight ratio 0.5). Prisms with $4 \times 4 \times 16 \, cm^3$ size were cast and then, after hardening, cubes with 1 cm edge length were sawn and used for the tests.

Figure 1: Example of marble decoration affected by sulfation (Monumental Cemetery in Bologna, Italy, XIX century). Beneath the gypsum surface layer (also incorporating some particulate matter), marble exhibits grain disaggregation

2.2. Treatments

Samples were treated by full immersion for 24 hours in 200 ml of aqueous solutions of DAP with different formulations, designed to investigate the effects of the following parameters:

1) Effect of ethanol (EtOH) concentration. Because gypsum solubility in an aqueous solution can be reduced by adding ethanol to the solution[11], increasing ethanol additions to water (0, 10, 30 and 50 vol%) were explored to study the influence of decreasing calcium ion concentrations, for a given DAP concentration (0.1 M).

2) Effect of DAP concentration. Increasing DAP concentrations (0.05, 0.1 and 0.5 M) were explored to study the influence of phosphate ion concentration for a given calcium ion concentration (determined by an ethanol addition of 30 vol%).

3) Effect of pH. Because higher pH is expected to favor formation of HAP instead of brushite[12], the effect of increasing the pH to 10 using ammoni-um hydroxide was investigated for increasing DAP concentrations (0.05, 0.1 and 0.2 M) and a given concentration of calcium ions (corresponding to a 30 vol% ethanol addition).

After immersion in the solutions for 24 hours, the samples were extracted and abundantly rinsed with water to remove unreacted DAP and ammonium sulfate (that is expected as a by-product[8-10]).

2.3. Characterization

The mineralogical composition of the new CaP was determined by grazing incidence X-ray diffraction (GID, Bruker D8 Discover X-Ray Diffractometer, Cu anode, incidence radiation $\theta=0.5°$, detector scan range $2\theta=3-37°$). This technique detects the composition of surface layers without interference from the substrate, thanks to the reduced penetration depth of the incoming X-rays.

The morphology of the new phases was assessed by observing the samples (after coating with carbon) using an environmental scanning electron microscope (FEI Quanta 200 FEG ESEM).

0 vol% EtOH Unidentified CaP	10 vol% EtOH Unidentified CaP	30 vol% EtOH Brushite	50 vol% EtOH β-TCP

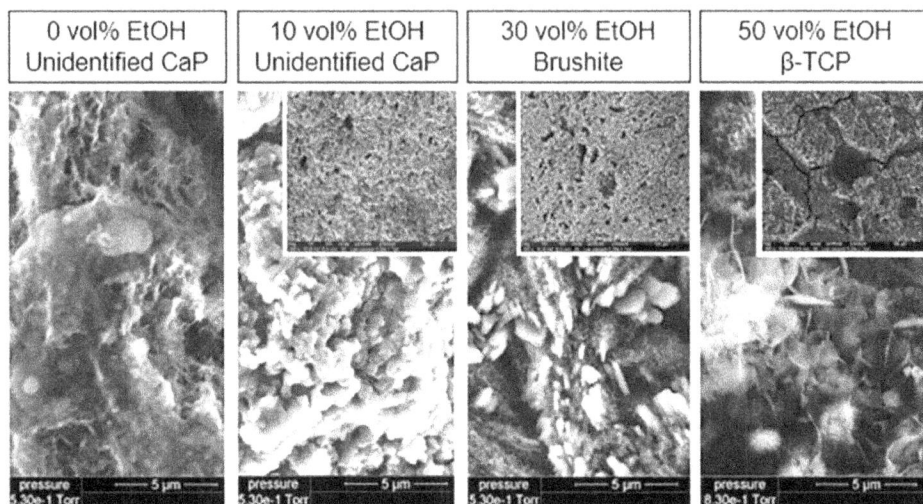

Figure 2: CaP phases formed by reaction with a 0.1 M DAP solution with increasing ethanol content at pH=8

3. Results and discussion

The composition and morphology of the new CaP phases formed by reacting gypsum with a 0.1 M DAP solution with increasing amounts of ethanol are reported in *Figure 2*. While fewer phases were formed with no or low ethanol addition, abundant new phases were found for 30 and 50 vol% ethanol additions. The new phases were identified by GID as brushite and β-TCP, respectively. While in the case of β-TCP (50 vol% ethanol addition) diffused cracking was observed, no cracks were visible in the case of brushite (30 vol% ethanol addition). Therefore, the

Increasing the pH from 8 to 10 had a strong impact on the composition of the new phases *(Figure 4)*. At DAP concentrations of 0.05 and 0.1 M, alongside brushite (the only phase formed at pH=8) also β-TCP and HAP were formed. Further increasing DAP concentration to 0.2 M (the highest concentration that did not lead to immediate precipitation at pH=10), only -TCP and HAP were found. However, in all cases where the pH was increased up to 10, diffused cracking occurred, because of excessive growth of the new CaP phases. Cracking also led to flaking and detachment from the substrate, which of course is undesired.

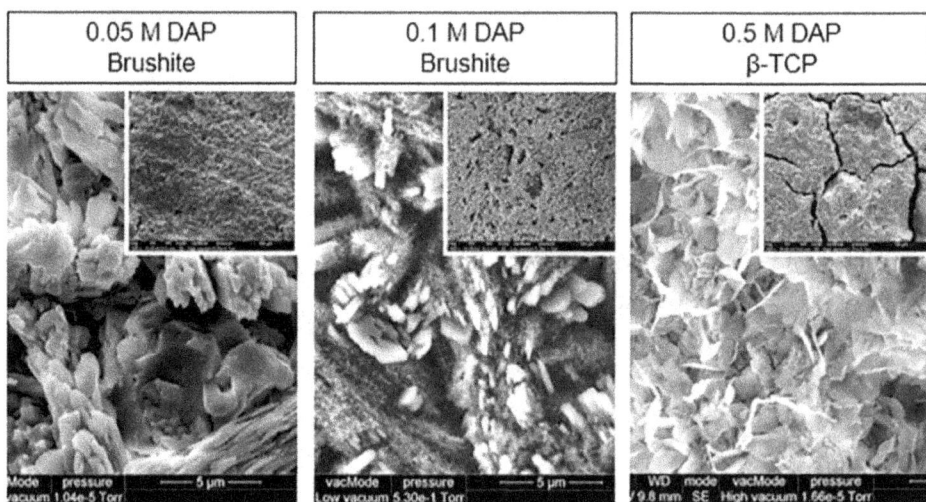

| 0.05 M DAP Brushite | 0.1 M DAP Brushite | 0.5 M DAP β-TCP |

Figure 3: CaP phases formed by reaction with a DAP solution with increasing concentration at fixed 30 vol% ethanol content at pH=8

latter ethanol concentration was selected in the prosecution of the study.

When higher and lower DAP concentrations were explored for a fixed 30 vol% ethanol addition, the new phases reported in *Figure 3* were obtained. With a DAP concentration of 0.05 M, brushite was formed, with no visible cracking. On the contrary, when the DAP concentration was increased up to 0.5 M, β-TCP formed but heavy cracking occurred.

The change in composition of the new CaP phases, passing from brushite to β-TCP, when either the DAP or the ethanol concentrations were increased might seem counterintuitive, because in both cases the Ca/P ratio was diminished. In fact, a higher Ca/P ratio in the solution would be expected to favor the formation of phases with a higher Ca/P ratio (hence, β-TCP instead of brushite). However, this was not the case and the reason

| 0.05 M DAP, pH=10 | 0.1 M DAP, pH=10 | 0.2 M DAP, pH=10 |
| Brushite + β-TCP + HAP | Brushite + β-TCP + HAP | β-TCP + HAP |

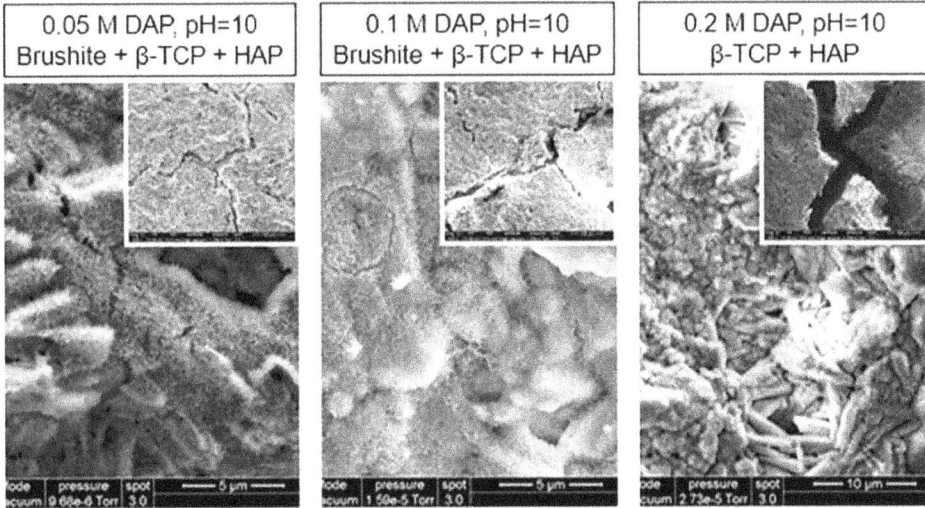

Figure 4: CaP phases formed by reaction with a DAP solution with increasing concentration at fixed 30 vol% ethanol content and pH=10

is thought to be the fact that brushite formation is kinetically favored, because brushite contains HPO_4^- ions (which are the main species originated from DAP dissociation), whereas HAP contains PO_4^{3-} ions (which are only a minor fraction). Accordingly, it has recently been proposed that precipitation of CaP phases follows the Ostwald's rule, i.e. the phase with the fastest precipitation rate (e.g. brushite) is preferentially formed, even if it is not the most stable phase (e.g. HAP)[9]. To verify this hypothesis, two solutions of $CaCl_2$ and DAP were prepared with Ca/P molar ratios of 10:6 (corresponding to that of HAP) and 1:1 (corresponding to that of brushite). By powder XRD it was found that very similar results were obtained in the two cases, i.e. formation of brushite and a minor amount of HAP, independently of the initial Ca/P ratio. The hypothesis that brushite formation is kinetically favored over that of HAP, basically independently of the Ca/P ratio in the starting solution, seems hence confirmed.

In any case, brushite is about 27 times less soluble in water than gypsum[1], hence its formation is expected to be benefi-

cial. On the contrary, even if less soluble phases such as β-TCP and HAP could be obtained, their formation was always associated with diffused cracking, which would be detrimental for the treatment success. Therefore, the formulation involving 0.1 M DAP in 30 vol% ethanol at pH 8 (leading to brushite and no cracking) appears as the most promising one, among those investigated in this study. Further tests are in progress to assess its ability to diminish the water solubility of treated gypsum. In addition to the conservation of sulfated marble, the same formulation is expected to be suitable also for the conservation of gypsum stuccoes. Further tests are in progress to assess the ability of this treatment to reduce the water solubility of gypsum stuccoes and increase their mechanical properties.

4. Conclusions

In this study, gypsum samples were treated with different aqueous solutions of DAP and the composition and morphology of the new CaP phases were investigated. The effects of ethanol addition to the DAP solution (aimed at diminishing gypsum solubility), DAP concentration, and solution pH were investigated. The results of the study indicate that it is possible to obtain phases with very low solubility (such as β-TCP and HAP) by increasing the DAP concentration, by increasing the EtOH concentration or by increasing the solution pH. However, in all these cases, the new CaP phases exhibited diffused cracking, presumably because of excessive growth of the new phases. For this reason, the formulation involving 0.1 M DAP in 30 vol% ethanol at pH 8 (leading to formation of uncracked brushite) seems like the most promising one. Indeed, as brushite is much less soluble than gypsum, a benefit in terms of reduction in dissolution in rain is expected. Treatment with the same solution with the aim of forming brushite is also expected to be beneficial in the case of gypsum stuccoes, both in terms of reduction of solubility in rain and increase in mechanical properties. Relevant experimental tests are in progress.

Acknowledgements

This project has received funding from the European Union's Horizon 2020 research and innovation programme under the Marie Sklodowska-Curie grant agreement No 655239 (HAP4MARBLE project, „Multi-functionalization of hydroxyapatite for restoration and preventive conservation of marble artworks").

References

[1] Sassoni E., Graziani G., Franzoni E., Repair of sugaring marble by ammonium phosphate: comparison with ethyl silicate and ammonium oxalate and pilot application to historic artifact, Mater Design (88) (2015) 1145-1157.

[2] Naidu S., Blair J., Scherer G. W., Acid-resistant coatings on marble, J Am Ceram Soc (99) (2016) 3421-3428.

[3] Graziani G., Sassoni E., Franzoni E., Scherer G. W., Hydroxyapatite coatings for marble protection: Optimization of calcite covering and acid resistance, Appl Surf Sci (368) (2016) 241-257.

[4] Sassoni E., Graziani G., Franzoni E., Scherer G. W., Some recent findings on marble conservation by aqueous solutions of diammonium hydrogen phosphate, MRS Advances, DOI: 10.1557/adv.2017.45.

[5] Naidu S., Scherer G. W., Nucleation, growth and evolution of calcium phosphate films on calcite, J Colloid Interf Sci (435) (2014) 128-137.

[6] Sassoni E., Naidu S., Scherer G. W., The use of hydroxyapatite as a new inorganic consolidant for damaged carbonate stones, J Cult Herit (12) (2011) 346-355.

[7] Ma X., Balonis M., Pasco H., Toumazou M., Counts D., Kakoulli I., Evaluation of hydroxyapatite effects for the consolidation of a Hellenistic-Roman rock-cut chamber tomb at Athienou-Malloura in Cyprus, Constr Build Mater (150) (2017) 333–344.

[8] Snethlage R., Gruber C., Tucic V., Wendler E., Transforming gypsum into calcium phosphate – the better way to preserve lime paint layers on natural stone? In: Mimoso JM & Delgado Rodrigues J, Stone consolidation in Cultural Heritage (2008) 1-13.

[9] Molina E., Rueda-Quero L., Benavente D., Burgos-Cara A., Ruiz-Agudo E., Cultrone G., Gypsum crust as a source of calcium for the consolidation of carbonate stones using a calcium phosphate-based

consolidant, Constr Build Mater (143) (2017) 298-311.

[10] Sassoni E., Graziani G., Scherer G.W., Franzoni E., Preliminary study on the use of ammonium phosphate for the conservation of marble-imitating gypsum-stuccoes, In: Papayianni I., Stefanidou M., Pachta V. (Eds), Proceedings of the 4th Historic Mortars Conference HMC2016, Santorini (GR), 10-12 October 2016, p. 391-398.

[11] Gomis V., Saquete M.D., García-Cano J., CaSO4 solubility in water–ethanol mixtures in the presence of sodium chloride at 25 °C. Application to a reverse osmosis process, Fluid Phase Equilibr (360) (2013) 248–252.

[12] Eliaz N., Metoki N., Calcium Phosphate Bioceramics: A Review of Their History, Structure, Properties, Coating Technologies and Biomedical Applications, Materials (10) (2017) 334.

Electrode placement during electro-desalination of NaCl contaminated sandstone – simulating treatment of carved stones

Lisbeth M. Ottosen* and Lovisa C. H. Andersson
Department of Civil Engineering, Technical University of Denmark, Lyngby, Denmark
* LO@byg.dtu.dk

Abstract

Carved stone sculptures and ornaments can be severely damaged by salt induced decay. Often the irregular surfaces are decomposed, and the artwork is lost. The present paper is an experimental investigation on the possibility for using electro-desalination for treatment of stone with irregular shape with only two electrodes. The used Gotland sandstones were contaminated by NaCl in the laboratory. Due to the relatively good homogeneity in initial salt concentration obtained in this way, interpretation of the ED process were direct. Stones with an up-side-down T-shape formed the core of the investigation. Electro-desalination experiments were made with different duration to follow the progress. Successful desalination of the whole stone piece was obtained, showing that also parts not being placed directly between the electrodes were desalinated. This is important in case of salt damaged carved stones, where the most fragile parts thus can be desalinated without physically placing electrodes on them. The Cl removal rate was higher in the areas closest to the electrodes and slowest in the part, which was not placed directly between the electrodes. This is important to incorporate in the monitoring program to decide when a desalination action is finished.

Keywords: electro-desalination, sandstone, carved stone, NaCl

1. Introduction

Salt induced decay of historic carved stone, sculptural or ornamental, is a major cause for loss of important cultural heritage. Often the carvings making up the surface pattern are lost first, and thus these constitutes the most fragile parts. Poulticing is a group of used and discussed techniques for removal of salts from stone monuments. In these methods, a poultice is attached to all outer surfaces of the monument and the idea is to transport the salts from the stone into the poultice by diffusion and/or advection. However, in case of fragile carvings, poulticing can be difficult to apply, as these parts can further decompose by the physical contact. This paper is a laboratory investigation on the use of electro-desalination (ED) to remove salt from a sandstone with a T-shaped geometry, where the electrodes are placed in poultice around the simulated carving.

ED is based on application of an electric DC potential gradient to the salt contaminated stone. In the electric field, ions in the pore solution are transported by electromigration towards the electrode of opposite polarity, and hereby the ions from the damaging salts are transported out from the stone. The electrodes are placed externally on the stone surface in electrode compartments with poultice, and the ions from the salts concentrate in the poultice during the treatment. When the poultices are removed after the desalination, the ions of the salts are removed with them.

At both metallic electrodes there are pH changes due to electrolysis:

At the anode:
$$H_2O \rightarrow 2H^+ + \frac{1}{2} O_2 (g) + 2e^- \qquad (1)$$

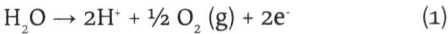

At the cathode:
$$2H_2O + 2e^- \rightarrow 2OH^- + H_2 (g) \qquad (2)$$

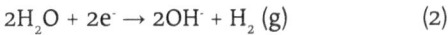

As seen from (1) and (2) pH decreases at the anode and increases at the cathode. It is necessary to neutralize the pH changes to prevent severe pH changes of the stone. The work by[1,2] underlined the importance of avoiding stone acidification, as in experiments without pH neutralization; the stones were severely damaged next to the anode. Use of a calcite rich clay poultice offers neutralization of the acidification from the anode.[3] The calcite buffers the acid and the clay gives workability, so the poultice can have optimal contact to the stone surface during the treatment.

Electro-desalination has been tested successfully in laboratory scale in different types of sandstones: Posta and Cotta sandstones[4], Gotland Sandstone[2], Nexø sandstone[5], and granite.[6] In fact no limit has been seen in stone type. Also successful removal of chlorides[4], nitrates[7] and sulphates[8] have been gained. These results were obtained with stone segments with plane surfaces. The present work focuses on the use of ED to desalinate carved stones.

Two works[9,10] have previously focused on ED of carved stones. Both these experimental works were carried out with stones, which had an uneven salt distribution, and both works confirmed that it was possible to extract salts from the elevated carvings of the stone, which were not placed directly between the electrodes. The uneven salt distribution in these stones, made interpretation of the result difficult. Further both investigations only included one experiment each, and thus the progress in desalination of the

carving needs to be clarified. The present paper is an experimental investigation, where T-shaped stone samples with a known and relatively even distribution of NaCl are desalinated from one set of electrodes.

2. Materials and methods

2.1. Stone for the experimental work

The experimental work was conducted with Gotland Sandstone, which is a stone type often carved in Denmark. The grey calcitic Gotland sandstone is composed of about 60 wt% quartz grains, cemented together by 7-10 wt% calcite, and with lower amounts of clay minerals, micas, feldspar minerals, pyrite, and glanconite.[11] The porosity is unusually large, around 15 % by volume.[11] This type of stone was chosen as it is commonly used and relatively homogeneous. T- shaped stone samples for the experiments were all cut from the same stone block, and had the same size, see *figure 1a*. To have a uniform distribution of the salt prior to the experiments, the stone samples were contaminated with the salt in the lab. The stone pieces were dried at 105°C and vacuum saturated by 80 g/l solution of NaCl in a desiccator prior to the ED experiments.

2.2. Electro-desalination experiments

Electrode compartments as seen in *figure 1b*. A frame was folded in thin plastic and jointed with tape to fit the horizontal part of the stone at each side of the attribute. The frames were filled with poultice; a mixture of kaolinite and $CaCO_3$ [3] with an initial water content of 54%. Inert platinum coated electrode meshes were placed at the top of each electrode compartment. The sandstone

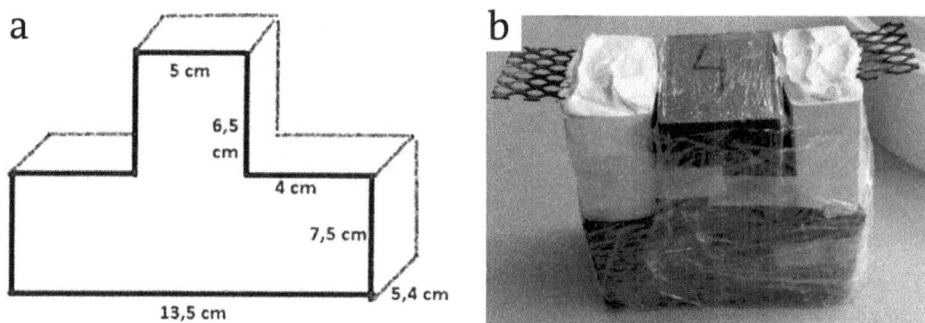

Figure 1: (a) The dimensions of the T-shaped stones and (b) The stone with electrodes.

and electrode compartments were wrapped in plastic film to hinder evaporation. A reference experiment (REF) was made with no applied current, but electrode compartments were placed at the stone exactly as in the ED experiments. The duration of the REF experiment was 7 days. Four ED experiments were made differing only in duration: 1, 4, 6 and 13 weeks (ED1, ED4, ED6 and ED13). A constant current of 10 mA was applied to the electrodes all through these experiments. Table 1 shows the dry weight of the different stones, the water content after the vacuum saturation and the duration in weeks of the experiments.

After the REF and ED experiments, the stones were segmented with hammer and chisel into 16 segments (Figure 2).

The water content in the segments was calculated as weight loss after drying at 105°C. The dried segments were grinded in a mechanical mortar. Following, 10 g powder was suspended in 25 ml distilled water and agitated for 24 h. The samples settled for 10 min and pH was measured. The samples were filtered through 0.45 µm filter. The Cl concentrations were analysed by ion chromatography (IC, Dionex DX-120).

The poultices were changed every 7 day except from ED13, where the last poultice was used in 3 weeks. The poultice was weighed, and Cl concentration, water content and pH were measured using the same methods as for the stone segments.

	Weight (g)	Initial water content (%)	Duration (weeks)
REF	1528	12.0	1 (no current)
ED1	1513	12.0	1
ED4	1536	11.6	4
ED6	1504	12.1	6
ED13	1481	11.9	13

Table 1: REF and ED experiments. Dry weight of stone, initial water content after vacuum saturation and duration of experiment

	Water content (%)	pH	Cl (mg/kg)
REF	10.4 ± 0.2	8.8 ± 0.06	4290 ± 530
ED1	9.9 ± 0.6	9.0 ± 0.9	4050 ± 1060
ED4	9.9 ± 0.3	10.8 ± 0.9	1580 ± 760
ED6	9.1 ± 1.1	11.3 ± 1.0	650 ± 500
ED13	3.0 ± 0.2	11.3 ± 0,9	21 ± 4

Table 2: Water content. pH and chloride concentration (average ± standard deviation) in the stone segments at the end of the REF and ED experiments

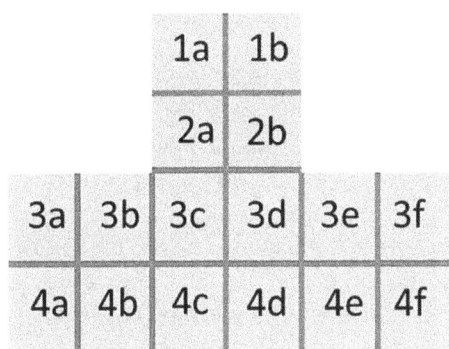

Figure 2: Segmentation and numbers after REF and ED experiments

3. Results and discussion

3.1. Overall results

Table 2 shows the overall results from the ED experiments as average values (± standard deviation) of water content, pH, Cl concentration. The water contaent was maintained at the same level when changing the poultice every week, but not if the same poultice was used for 3 weeks as was the case for the last poultice in ED13. The pH increased when the duration of ED was prolonged due to electrolysis at the cathode, which was not buffered sufficiently in the poultice. The Cl concentration decreases to a very low level in ED13.

3.2. Comparison of REF and ED1 experiments

Figure 3a and b shows respectively the resulting Cl concentrations in the different stone segments of the REF stone with poultice but without an applied electric field and in the corresponding ED1 experiment with poultice and applied electric field, both after 7 days.

The REF stone shows lower Cl concentration in the segments, which had direct physical contact with the poultice (segments 3a, 3b, 3e and 3f, *figure 2*). The Cl concentrations in the two REF poultices were 3600 and 4200 mg/kg. As the Cl concentration was non-measurable in the poultice before the experiment, the Cl was transported from stone to poultice during the 7 days of the experiment. In the same period, the water content in the poultice decreased from 53.6% to 47.3%. In addition, the average water content in the stone decreased from 12.0% *(table 1)* to 10.4% *(table 2)*. Thus, the wrapping of the stone and poultice did not fully hinder evaporation. The possible transport mechanisms for Cl from stone to poultice are diffusion and advection. Calculating the pore water concentration of Cl in the two materials based on stone concentration and water content, the pore water concentration in the stone (41 g/l in aver-

a

		4800	4500		
		4300	4100		
3600	3600	4300	4400	4000	3600
4900	5000	4600	4800	4400	4700

b

		4900	4900		
+		3800	2300	-	
4000	4000	3800	3100	2500	2100
5500	5100	5000	4700	4800	4500

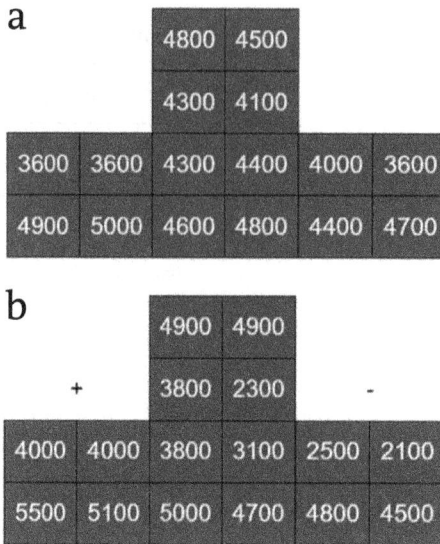

Figure 3: Chloride concentrations in the segment of the (a) REF stone and (b) and in the segment of the ED1 experiment after 7 days [mg/kg]

age) is much higher than in the poultice (8.3 g/l), and thus the concentration gradient between the two allows a continued diffusion. Whether the drying of the poultice and stone transport water from stone to poultice cannot be assessed on basis of the current data, however, this transport must be considered limited, and thus advection is not considered as being of major importance.

Due to the applied electric field, the Cl concentration pattern in the ED1 stone differs from the REF *(figure 3b)*. As Cl-electromigrates from the cathode end towards the anode, the concentration in the stone segments, which were in direct contact with the cathode poultice (3e and 3f) were lower than the concentration in 3a and 3b (in the anode end). The concentration in 3d and 2d are clearly decreased as well. The latter shows that the electric field lines distributed into the attribute already during the first week of ED. The Cl concentrations in the anode and cathode poultices were 7670 mg/kg (cathode) and 250 mg/kg (anode), which are higher and lower than the concentration in the poultices in the REF experiment, respectively.

3.3. Cl removal from stone over time during ED experiments

Figures 4a, b and c show the Cl concentrations in the different segments *(figure 2)* of the ED experiments over time. Together these figures illustrate how the desalination progresses in the stone works. The Cl concentration decreased over time in all segments and a very low level of below 30 mg Cl/kg was obtained in every segment in ED13. Thus a very important conclusion can be drawn. It is possible fully to desalinate such attribute by ED.

The removal rate differed in the different segments. *Figure 4a* show that the Cl removal was fastesr in the segments in row 2 than in row 1, i. e. the desalination occurred from the row closest to the electrodes first and later to the upper row. The same pattern is seen when comparing the desalination of row 3 *(figure 4b)* and row 4 *(4c)* when e. g. comparing the Cl concentrations after 6 weeks in the two figures. In row 3 the concentrations were less than about 500 mg/kg and in row 4, the concentrations were all higher than 500 mg/kg. That row 3 is desalinated before row 4 confirms the pattern found for ED of bricks in[3], when the electrodes were placed at the same side of the bricks. Further, it is evident that the Cl concentration decreased fastest from the end of the stone closest to the cathode *(figures 4a and b)*, e. g. 3f and 4f had.

From the Cl concentration and weight of poultice measured every week (when the poultices were changed) the accumulated Cl removal over time can be calculated. *Figure 5* shows the Cl removed into the anode poultice. The Cl removal in the four ED experiments progresses very similar and as discussed in section 3.2, the

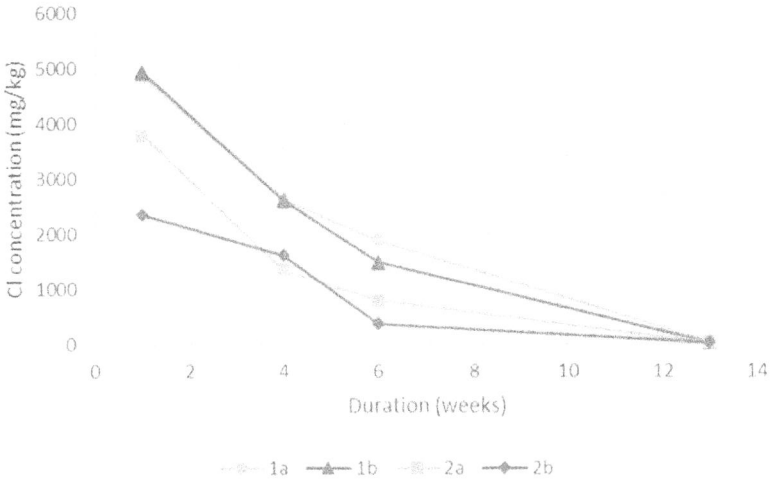

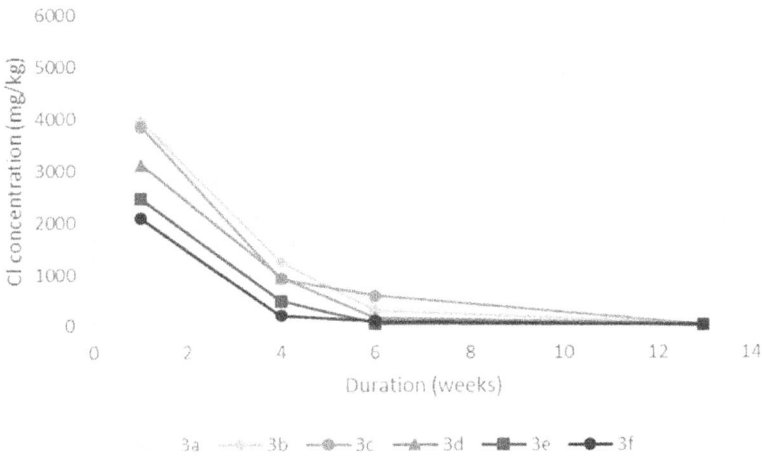

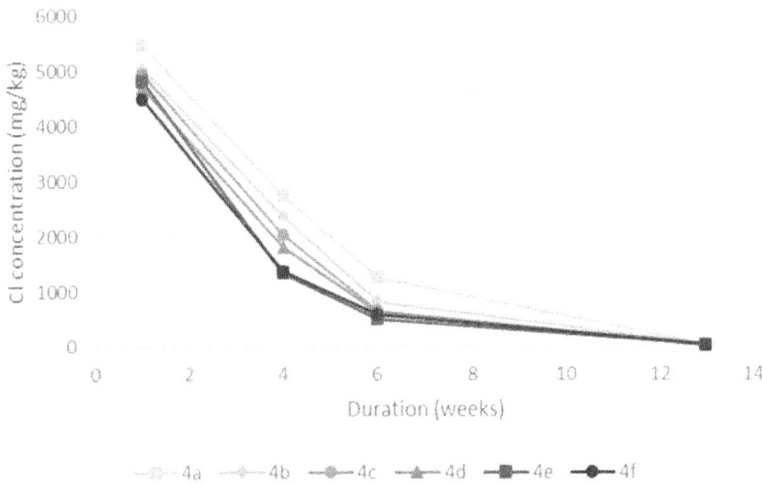

Figure 4: Cl concentration in the different stone segments as function of ED duration: (a) in the attribute, segments in row 1 and 2 (b) segments in row 3 and (c) segments in row 4.

removal into the anode poultice is about double the concentration in the REF experiment during 1 week.

During the first 3-4 weeks, the accumulated Cl removal in the ED experiments increases almost linearly with duration. Between 4 and 9 weeks, the removed amount of Cl decreases every week and after 10 weeks the removal is very low. This pattern corresponds well to the results in *figure 4*, where it is seen that after 6 weeks, the concentration is lowered significantly in every segment. However, combining the two experiments also show, that it is important to continue the ED of the carved object after the point where the Cl removal rate into the poultice decreases. Evaluating the ED only on basis of figure 5 could mislead to the conclusion that the desalination was sufficient after 6-7 weeks, but *figure 4a* shows, that at this point the Cl concentration is still rather high in segments 1a and 1b. Thus in the case of ED treatment of sculptures and ornaments, the progress of the desalination must be followed also by other means to ensure full desalination

The total amount of Cl into the anode poultice was 3300 mg Cl *(figure 5)* and for comparison, only 65 mg Cl was removed into the cathode poultice. This underlines the effect of the applied electric field.

4. Conclusions

This investigation shows that ED has potential to desalinate carved stone, as the salts can be removed from parts, which are not directly between the electrodes. In the present investigation, T-shaped stones were fully desalinated using two electrodes from an initial concentration of about 4300 mg Cl/kg to about less than 34 mg Cl/kg in all parts of the stone. The part not situated directly between the electrodes was desalinated with the slowest rate, and the Cl amount removed into the anode poultice decrea-

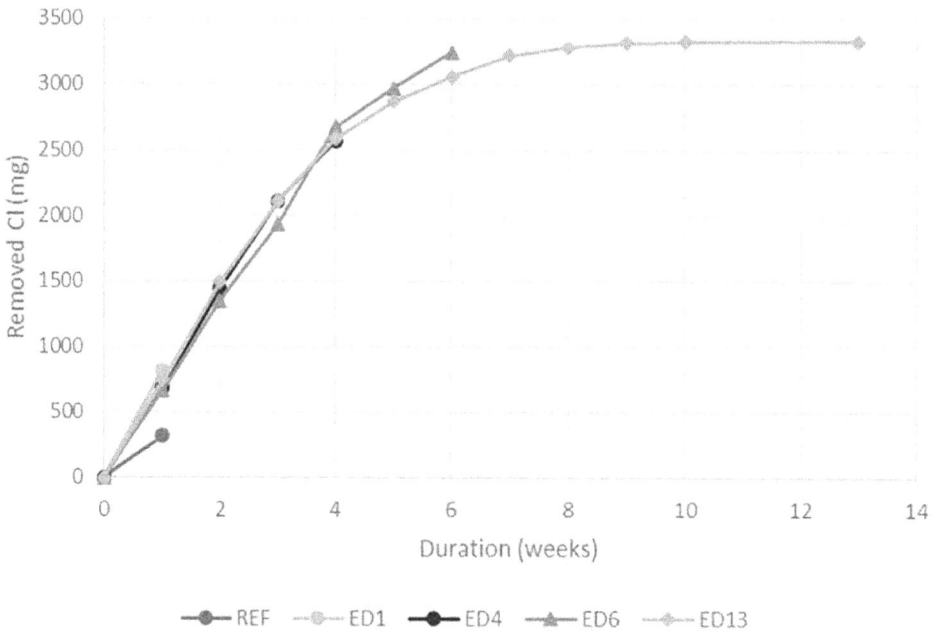

Figure 5: Accumulated Cl amount removed into the anode poultice during the REF and ED experiments.

sed to a very low overall level during while this part was desalinated. Thus it cannot be assessed whether the desalination is finished only on basis of the removal rate into the poultice at the electrodes.

References

[1] Herinckx S, Vanhellemont Y, Hendrickx R, Roels S, De Clercq H., 2011, Salt removal from stone building materials using an electric field. In: Proceedings from the international conference on salt weathering on building and stone sculptures, I. Iannou & M. Theodoridou (eds.), Cyprus 19-22 October, 357-364.

[2] SKIBSTED, G., Ottosen, L.M., Jensen, P.E., Electrochemical desalination of limestone spiked with Na2SO4 – importance of buffering anode produced acid. Paper E in Matrix changes and side effects induced by electrokinetic treatment of porous and particulate materials. PhD thesis, 2013, Technical University of Denmark.

[3] Rörig-Dalgaard, I., Preservation of masonry with electrokinetics – with focus on desalination of murals. PhD Thesis. Department of Civil Engineering, Technical University of Denmark, 2009.

[4] Ottosen, L.M., Christensen, I.V., Electrokinetic desalination of sandstones for NaCl removal - Test of different clay poultices at the electrodes. Electrochimica Acta. (86), 2012, 192– 202.

[5] Petersen, G., Ottosen, L.M., Jensen, P.E., (2010) The possibility for using electrokinetics for desalination of sandstone with low porosity. Proceedings from 8th fib International PhD Symposium in Civil Engineering. Kgs. Lyngby, Denmark, June 20-23, 455-460.

[6] Feijoo, J., Nóvoa, X.R., Rivas, T.; Mosquera, M.J., Taboada, J., Montojo, C., Carrera, F., Granite desalination using electromigration. Influence of type of granite and saline contaminant. Journal of Cultural Heritage (14), 2013, 365–376.

[7] Skibsted, G., Ottosen, L.M., Jensen, P.E., Paz-Garcia, J.M., Electrochemical desalination of bricks - Experimental and modeling. Electrochimica Acta, (181), 2015, 24-30.

[8] Ottosen, L.M., Electro-desalination of sulfate contaminated carbonaceous sandstone – risk for salt induced decay during the process. In Science and Art: A Future for Stone : Proceedings of the 13th International Congress on the Deterioration and Conservation of Stone. Eds. John J. Hughes, Torsten Howind, vol. 2, 2016, 897-904.

[9] Ottosen, L.M., Skibsted, G., Præstholm, T., Electrodesalination of sandstones with irregular shapes and uneven distribution of salts. Proceedings from SWBSS2014 3rd International Conference on Salt Weathering of Buildings and Stone Sculptures. Ed. H. De Clercq, 2014, p 405-420.

[10] Feijoo, J., Matyscák, O., Ottosen, L.M., Rivas, T., Novoa, X.R., Electrokinetic desalination of protruded areas of stone avoiding the direct contact with electrodes. Materials and Structures, (50:82), 2017, 2-15.

[11] Nord, A.G., Tronner, K., Effect of acid rain on sandstone: The Royal Palace and the Riddarholm Church, Stockholm. Water Air and Soil Pollution, (85), 1995, 2719-2724.

How *not to bother* salts while grouting

Chiara Pasian[1]*, Francesca Piqué[2], Cristiano Riminesi[3] and Albert Jornet[2]
[1] *Department of Conservation and Built Heritage, Faculty for the Built Environment, University of Malta, Msida, Malta*
[2] *Institute Materials and Constructions, University of Applied Sciences and Arts of Southern Switzerland, Lugano, Switzerland*
[3] *Institute for the Conservation and Valorization of the Cultural Heritage (ICVBC), National Research Council, Florence, Italy*
* *chiara.pasian@um.edu.mt*

Abstract

The objective of this research work was to assess if the injection of water-reduced grouts (water-ethanol-based) mobilises soluble salts to a lesser extent compared to the injection of conventional water-based grouts. Ethanol was used as a partial substitute for water in grouts preparation, being a poorer solvent for soluble salts when compared to water. Three injection grouts were evaluated; for each grout individually, the performance of mixtures prepared with 100% water was compared to the performance of the same grout prepared with water-ethanol. To assess the different performance in a salt loaded system, the grouts were tested into replicas of delaminated plaster intentionally contaminated with NaCl. The salt movement was followed by evanescent field dielectrometry using the SUSI© instrument. Grouts with reduced water content were also tested on site, at the San Vincenzo Oratory in Pazzallo, Tessin (CH), on decorated plasters afflicted by severe delamination associated with high amounts of soluble salts. Two adjoining areas, comparable in condition and in salt content, were stabilised with the same grout differing only in the nature of the suspension medium used (water-ethanol vs. water). Both in the laboratory and on site, it was experimentally verified that water-ethanol-based grouts caused a significant reduction of salts mobilisation, bothering salts to a lower extent compared to a typical water-based grouting intervention.

Keywords: injection grouting, water, alcohol, dielectrometry, SUSI©

1. Introduction and research aims

Injection grouting is an intervention aiming to stabilise and re-adhere delaminated plaster/render introducing a compatible adhesive material with bulking properties[1, 2]. Injection grouts are suspensions composed of a binder, aggregates, additives and a suspension medium, which is typically water. In lime- and hydraulic lime-based injection grouts, water is required for chemical setting as well as to improve injectability, however an excessive amount of water may lead to bleeding or segregation[3], shrinkage and thus grout failure. During grouting water is introduced in high amounts in the porous materials of the wall painting and/or historic structure (during pre-wetting and by grouting itself), potentially jeopardising water-sensitive original materials, solubilising soluble salts, and leading to their movimentation and damaging re-crystallisation. Water-reduced injection grouts were designed and studied to reduce these risks.

The objectives of this research include the evaluation of performance of designed grouts with reduced water content in relation to the solubilisation and movimentation of soluble salts in contaminated plasters. The performance of the grouts was assessed in the laboratory through evanescent field dielectrometry using SUSI© instrument* (C. Riminesi and R. Olmi, Diagnostics and monitoring

* Acronym for *Sensore per la misura di Umidità e Salinità Integrato, integrated sensor for measuring humidity and salinity.*

of moisture and salt in porous materials by evanescent field dielectrometry, 4th International Conference on Salt Weathering of Buildings and Stone Sculptures, Potsdam 20-22.9.2017), and on site, at the San Vincenzo Oratory in Pazzallo, Tessin (CH), on decorated plasters afflicted by severe problems of delamination associated with high amounts of soluble salts.

2. Materials and mixtures

Three injection grouts were evaluated: two hydraulic lime-based pre-mixed commercial grouts (PLM A CTS and LEDAN RI.STAT BASE B TecnoEdile Toscana) and one slaked lime-based grout designed in our laboratory (called in this paper grout C).[4] Grout C was formulated with the following components and proportions in parts by volume (pt/V): 1 pt/V slaked lime putty, 1.5 pt/V quartz sand (∅ < 740 µm), 1 pt/V pozzolan (pozzolana flegrea), 0.6 pt/V ammonium carbonate, and 0.1 pt/V plasticiser (Sika Viscocrete-2S, Sika AG) [4]. Ammonium carbonate, which naturally decomposes releasing carbon dioxide, was added to provide carbon dioxide to help carbonation in the deficiency of air (void in depth). The slaked lime used in this research is composed of ca. 50% $Ca(OH)_2$ – 50% water (in weight), determined gravimetrically.

The grouting mixtures were prepared using as suspension medium water (control) or water-ethanol mixed in different proportions. Alcohols are less effective solvents for ionic substances such as soluble salts when compared to water, as they have the polar hydroxyl group (–OH) and a non-polar chain. Ethanol was selected as a partial substitute for water because of its miscibility with water and because it is a poor solvent for salts.[5, 6]

The amount of suspension medium to prepare the grouts (mL suspension medium/g dry pre-mixed grout or g slaked lime) used for this research is significant-ly lower compared to the water amount suggested in the technical data sheets of PLM A and LEDAN RI.STAT BASE B. In the suspension medium to prepare the grouts, the amount of water was progressively decreased. The three suspension media selected were: 100% water ('control' representing the traditional grouts preparation), 50% water: 50% ethanol, 25% water: 75% ethanol.[4] PLM A and LEDAN RI.STAT BASE B required the addition of a filler (quartz sand, inert and non-porous) when mixed with water-ethanol to obtain a cohesive grout with no shrinkage.[4]

Set grout performance and working properties of the water-reduced grouts were tested in the laboratory following international standards and tests specifically designed for injection grouts.[7] The grouts mixed with water-ethanol showed adequate shrinkage, porosity, cohesion, and adhesion in replicas of delaminated plaster, as well as good injectability, and therefore they are potentially suitable for implementation on site.[4]

3. Soluble salts movement – Assessment in the laboratory

In order to assess the mobilisation and solubilisation of salts while grouting, replicas of delaminated plaster intentionally contaminated with NaCl were prepared. The effects of the injection into such replicas of water-ethanol grouts were compared with those of water-based grouts.

3.1. Materials and methods

As ethanol is a poorer solvent for soluble salts when compared to water[5, 6], the injection of water-ethanol-based grouts should solubilise salts in the delaminated plaster to a lesser extent compared to the injection of water-based grouts. This was verified in the laboratory with SUSI©* system (acronym for Sensore per la misura di Umidità e Salinità Integrato, integrated sensor for measuring humidity and salinity). SUSI© is an instrument based on Evanescent Field Dielectrometry (EFD), which is a technique deriving from dielectric spectroscopy.[8] SUSI© works in the microwave range (1-1.5 GHz), it is portable and allows real-time non-invasive measurements of sub-surface moisture content (MC) and the detection of soluble salts (SI, i. e. salinity index). The values of MC and SI can be separately and independently measured by the instrument. The moisture content (MC) of porous materials such as plaster is measured due to the dielectric contrast between water and the host material (plaster).[9] The water content which can be measured ranges between 0 and 20% (water volume fraction); the range of salinity index varies between 1 and 10 from no-salt to saturated salt conditions.[10] SI is a measure of the salt ions present in solution in the host material.

Replicas of delaminated plaster were prepared (in stratigraphy: brick, coarse plaster composed of slaked lime : sand 1:3 pt/V, fine plaster composed of slaked lime : sand 1:2 pt/V). The delamination (void) was produced between the coarse and the fine plaster, using the following procedure: the coarse plaster was applied on a brick tile; on this plaster, in the middle of the tile, small cylinders of dry-ice (solid CO_2) were stacked together in the shape of a truncated pyramid; the second plaster layer (fine plaster) was applied on the dry-ice cylinders. As the dry-ice sublimes, an empty void between the two plaster layers is created.[4] The fine plaster was contaminated with 0.5% NaCl (0.5% of the weight of the sand used to prepare the plaster and 0.7% of the total weight of the plaster).

The grouts chosen for the experiment were the grouts with the least amount of water (25% water : 75% ethanol) vs. the controls (100% water).

3.2. Calibration – Determination and correction of ethanol contribution

SUSI© typically works in presence of water to determine MC and SI. In this research the authors wanted to assess the solubilisation of salts in presence of a water-ethanol grout, therefore a new calibration of the instrument was needed.

Samples (10 cm x 10 cm and 2 cm thick) made of the same plaster of the replicas (fine plaster, slaked lime : sand 1:2 pt/V) were prepared. 9 plaster samples were prepared with no soluble salts and 9 analogous samples were prepared with plaster contaminated with 0.5% NaCl (0.5% of the weight of the sand used to prepare the plaster and 0.7% of the total weight of the plaster). The samples were immersed in three different solutions: 100% water, 50% water : 50% ethanol and 25% water : 75% ethanol. In each solution 3 non-contaminated samples and 3 contaminated samples were immersed. The samples were kept lifted from the bottom of the container with the solution through glass rods; the samples were immersed in the solution up to the height of 1 cm of their total thickness.

For every sample, MC and SI were measure at t0 (dry samples); after the samples were placed in the solutions, for each sample MC and SI were monitored at 30 seconds interval until complete saturation. The time for complete imbibition was recorded for each sample. A calibration of the instrument taking into consideration the contribution of ethanol, with and without soluble salts, was performed.

3.3. Injection of water-reduced grouts into contaminated replicas

Six replicas of delaminated plaster *(see Section 3.1)* were prepared.[4] The following grouts were injected, one for each replica:

* Pat. CNR n. 1626 – FI 2004A000187.

1) PLM100-0 (100% water) vs. PLM25-75 (25% water : 75% ethanol)

2) LEDAN100-0 (100% water) vs. LE-DAN25-75 (25% water : 75% ethanol)

3) C100-0 (100% water) vs. C25-75 (25% water : 75% ethanol)

Before the injection, measurements with SUSI© of MC and SI were performed on the dry replicas. The replicas were placed in a vertical position to simulate the injection into a wall. Before injecting the grout, replicas were pre-wetted. Pre-wetting is performed to clear the cavity and to wet the internal surfaces of the void, with the aim of reducing the absorption of the liquid contained in the grout mix and reduce grout shrinkage.[4] The pre-wetting liquid employed for each replica was the one used to prepare each grout respectively (100% water for grouts PLM100-0, LEDAN100-0 and C100-0; 25%

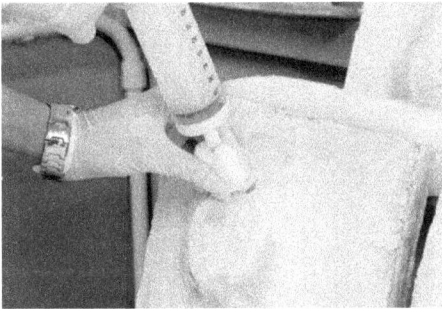

Figure 1: Injection of grout PLM25-75 into the replica of delaminated plaster

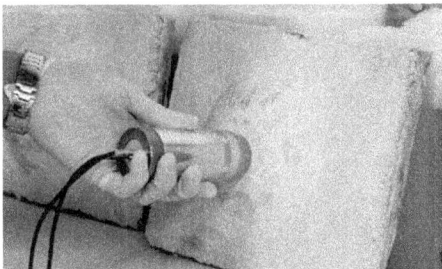

Figure 2: SUSI© measurements on a replica after 48 hours from the injection

water : 75% ethanol solution for grouts PLM25-75, LEDAN25-75 and C25-75). Measurements of MC and SI were performed with SUSI© immediately after pre-wetting. Injection of the grout was then performed, one grout in each replica *(Fig. 1)*.

After the injection, measurements of MC and SI were performed *(Fig. 2)*: immediately after the injection; at 10 minutes interval within the first hour, after 2, 3 and 4 hours, after 24, 48 and 72 hours. In this way, variations of MC and SI could be followed over time for each replica.

3.4. Discussion

Graphs in *Fig. 3a-f* show the variation of MC and SI over time for each replica. For each grout typology (respectively PLM, LEDAN and grout C) it is visible that MC is much higher for grouts prepared with 100% water (controls) compared to grouts prepared with water-ethanol *(Fig. 3a-c-e)*. This was predictable, as MC indicates the water volume fraction in the host material (plaster).

More interesting are the results concerning SI *(Fig. 3b-d-f)*; for all grouts the mixture prepared with 25% water : 75% ethanol (red in the graphs) has a much lower SI compared to the control prepared just with water (black in the graphs). This because a 25% water : 75% ethanol solution solubilises salts to a lesser extent compared to water.[5, 6] Data regarding SI are presented in *Table 1*:

Grout	Range of SI over time
PLM100-0	2.0 – 8.5
PLM25-75	2.2 – 4.0
LEDAN100-0	2.3 – 6.5
LEDAN25-75	1.0 – 3.8
C100-0	2.3 – 5.5
C25-75	0.0 – 1.8

Table 1: Range of salinity index (SI) over time of the replicas

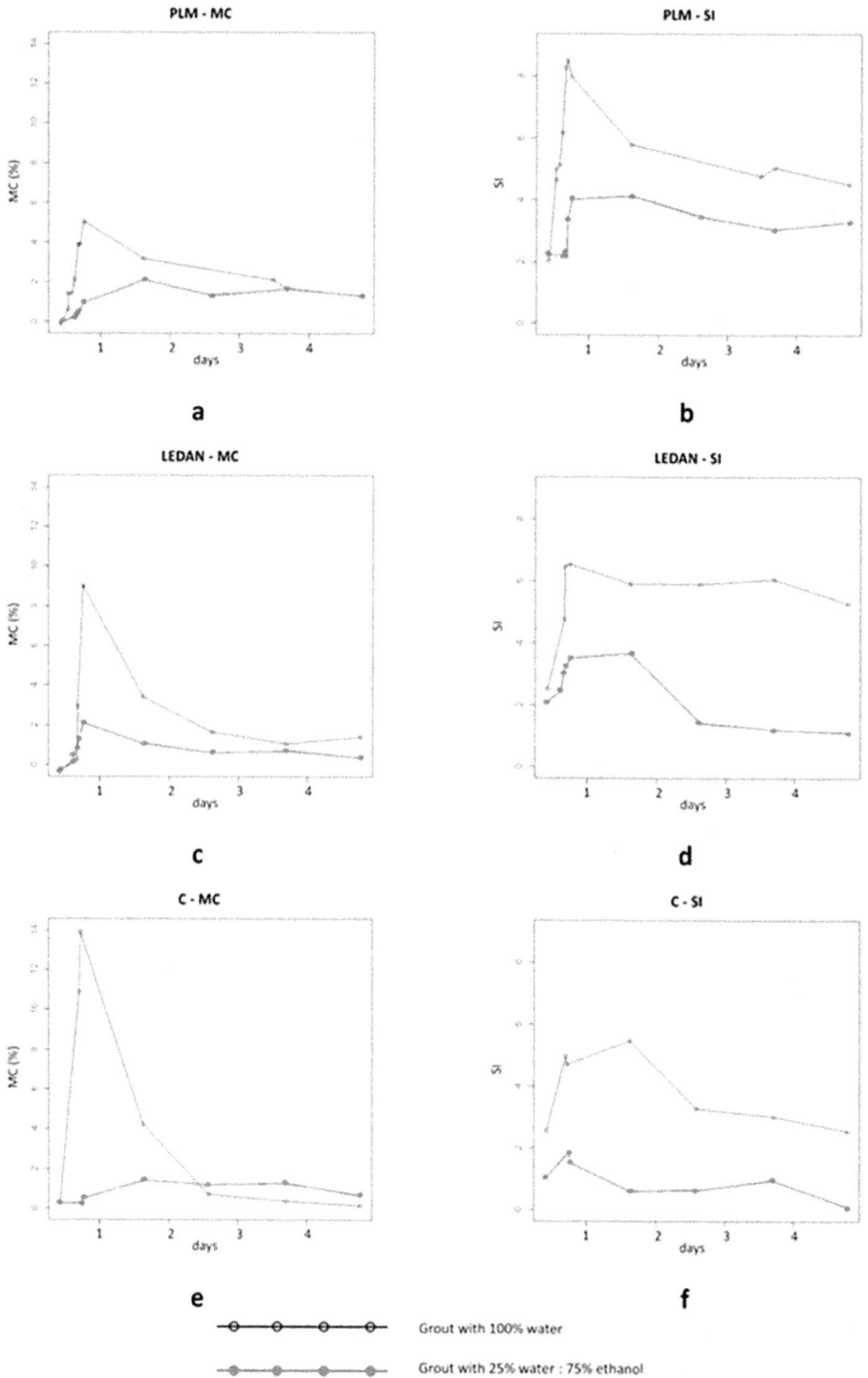

Figure 3: Graphs showing MC (3a PLM, 3c LEDAN, 3e grout C) and SI (3b PLM, 3d LEDAN, 3f grout C) over time, from t0 (dry replica) to 72 hours after the injection. The first point of measure on the graph is performed on the dry replica, the second point immediately after pre-wetting, the third point immediately after the grout injection; measures are then taken at 10 minutes interval within the first hour, after 2, 3 and 4 hours, after 24, 48 and 72 hours

The maximum value of SI after the injection of PLM25-75 is less than half compared to the maximum value after the injection of the control (PLM100-0). The maximum value of SI after the injection of LEDAN25-75 is about 2/5 lower compared to the maximum value after the injection of the control (LEDAN100-0). Values referred to grout C25-75 are particularly low (maximum value of SI 1.8): the maximum value measured for C25-75 is about 1/3 of the maximum value measured for the control (C100-0).

To increase wetting and reduce the overall provision of water, pre-wetting is often performed by conservators with a solution of water and alcohol[4], typically followed by the injection of a water-based grout.

The graphs reporting the measures performed with SUSI© show that the injection of grouts prepared with 25% water : 75% ethanol (after pre-wetting with the same solvent mixture) solubilises salts in the replicas to a lower extent compared to the injection of the same grouts prepared with 100% water.

4. On site application of grouts with reduced water content

Grouts with reduced water content were tested on site, at the San Vincenzo Oratory in Pazzallo, Tessin (CH). The Oratory, decorated with wall paintings and stuccoes and dated to the XVIII c., was abandoned in the last decades and it is affected by severe problems of delamination associated with high amounts of soluble salts. The north wall of the Oratory was the most affected by problems deriving from soluble salts and high humidity (RH constantly around 80% with 19-21°C T in the months during which the research was carried out; environmental monitoring by A. Kueng, SUPSI). The north wall is adjacent to a building which used to be a cowshed and which was thus a vehicle for nitrates.

The pilaster located on the north wall of the Oratory presented delamination between the brick support and the coarse plaster *(Fig. 4)* affecting an area of approximately 30x80 cm and 3 cm thick. Diffuse salt efflorescence was present both inside the delamination void and on the plaster surface; the salts present were identified as manly composed of nitrokalite (KNO_3 – equilibrium relative humidity 94.6% at 20°C [11]) (A. Kueng, SUPSI, internal report).

This area was considered suitable for testing and comparing the effects of reduced water grouts with those of regular grouts.

Figure 4: Delaminated pilaster in San Vincenzo Oratory, north wall

4.1. Objectives, materials and preliminary testing

The objective of the application on site was to verify that also in a real case the injection of a grout prepared with water-ethanol mobilises salts to a lesser extent compared to the injection of a grout prepared just with water.

A piece of plaster, already detached inside the delamination, was extracted; its porosity was measured in the laboratory (standard SIA 262/1:2003 [12]) as well as its compressive strength (standard UNI EN 1015/11 [13]). The porosity and the compres-

40 cm above the ground. The salts distribution in a wall is extremely heterogeneous both on the surface and in depth. In particular, in the case of capillary rising damp, ions are distributed in the wall in different concentrations also according to the height from the ground [11]; therefore it is preferable to compare two areas at the same height from the ground.

Before performing the intervention, crystallised salts were removed from the surface (area of 600 cm2) with a dry brush and weighed (0.26 g). Salt efflorescence was evenly distributed on the sur-

	n (%) tot. por.	UE (%) capill. por.	LP (%) air pores	σc (N/mm2) compr. strength
Original plaster	35.2	28.5	6.7	3.24
PLM25-75	46.4	39.7	6.7	3.57

Table 2: Porosity and compressive strength of the original plaster and grout PLM25-75

sive strength of the original plaster were compared with those of the grouts with reduced water content, tested in the laboratory with the same standard procedures. The injection grout with the compressive strength most similar to that of the original plaster was PLM prepared with 25% water : 75% ethanol (PLM25-75) *(see Table 2)*. PLM25-75 has a higher porosity compared to that the original plaster *(see Table 2)*; this is considered good because the porosity of the injection grout should be similar or higher to that of the original material [1].

4.2. Intervention

The lowest part of the delaminated pilaster, an area of 30 cm width x 20 cm height (delamination 3 cm thick) was chosen for the test. This area was about

face. Salts were also removed with a dry brush inside the delamination aided by a mini vacuum cleaner, which allowed removing both salts and dust inside the void.

The selected area was divided vertically by placing in the middle of the delamination a piece of foam rubber (3 cm x 20 cm, 3 cm thick as the delamination itself) to create a barrier. In this way the two grouts (PLM100-0 and PLM25-75) could be injected in the two separate chambers created in the delamination area at the same height *(Fig. 5)*.

The intervention was followed with a thermocamera (Agema 570 Thermovision PRO). The two different areas were monitored before the intervention, during the pre-wetting (performed for both areas with a solution 25% water : 75% ethanol), during the injection respectively of PLM100-0 and PLM25-75, and af-

Figure 6a: Thermoimage after the injection of PLM100-0

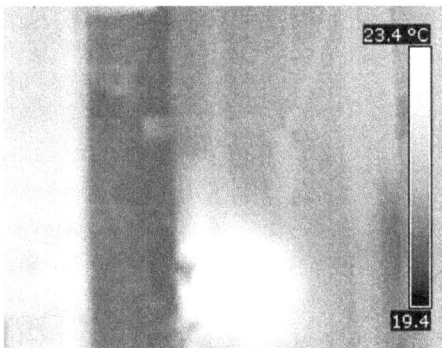

Figure 6a: Thermoimage after the injection of PLM100-0

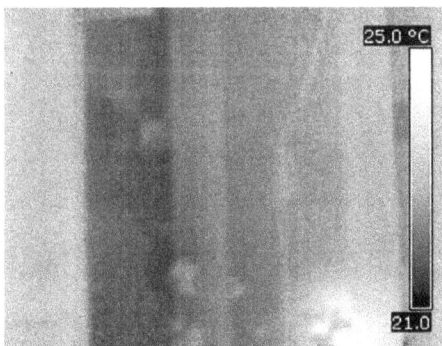

Figure 6b: Thermoimage after the injection of PLM25-75)

ter the injection over time to follow the grouts setting (every 5 minutes for the first hour; 2, 3, 4 hours after the injection; 24 and 48 hours after the injection). In order to better assess the positioning of the grouts in the delamination *(Fig. 6)*, the grouts were both slightly heated in a water bath prior to injection. In Fig. 6a PLM100-0 is more clearly visible compared to PLM25-75 *(Fig. 6b)* because it was possible to heat it longer (PLM25-75 is ethanol-based: it could not be heated long because its suspension medium is more volatile than water). It was verified that the positioning of two grouts injected in the delamination was kept separate by the foam rubber, as planned.

4.3. Assessment

After two months, the salt crystallisation was assessed on the surface of the pilaster where the intervention was performed. Salt efflorescence was removed with a dry brush in the two separate but equivalent areas (where grout PLM100-0 and where grout PLM25-75 were injected) and separately weighed. In the area where the water-ethanol-based grout PLM25-75 was injected the salt crystallisation was about 40% less (in weight, 0.12 g) compared to the area where the water-based grout PLM100-0 was injected (0.21 g).

5. Final discussion and conclusions

Injection grouts with reduced water content were evaluated, comparing the water-ethanol-based grout (25% water : 75% ethanol) with the correspondent control (100% water). In the present research their performance in terms of salts mobilisation was assessed:

- in the laboratory with SUSI© on replicas of delaminated plaster contaminated with NaCl

- on site in a limited area of intervention affected by delamination and severe soluble salts problems

In the laboratory, all the replicas injected with the water-ethanol-based grout showed a much lower salinity index (SI) compared to replicas injected with the correspondent water-based grout (control). This verified because a lower amount of salts was brought in solution by the water-ethanol grout suspension medium compared to the 100% water grout suspension medium. The data of SI measured with SUSI© prove that water-ethanol-based grouts solubilise salts to a much lesser extent compared to water-based grouts.

In the limited area treated on site, the injection of a water-ethanol-based grout led to a much lower salts re-crystallisation on the plaster surface compared to the injection of the water-based control.

Water-reduced grouts (25% water : 75% ethanol) proved to be suitable for site implementation with the advantage of bothering salts to a lower extent compared to a typical water-based stabilisation intervention.

Acknowledgements

We would like to thank Dr Christian Paglia, director of the Institute Materials and Constructions (SUPSI), allowing the conduction of the tests at the IMC. Many thanks to Andreas Kueng (IMC, SUPSI), who performed the salts analyses and the environmental monitoring in San Vincenzo Oratory. Thanks are also due to Prof Tiziano Teruzzi (IMC, SUPSI) for his help in the use of the thermocamera.

References

[1] I. Griffin, Pozzolanas as Additives for Grouts: An Investigation of Their Working Properties and Performance Characteristics, Studies in Conservation 49 (1) (2004) 23-34.

[2] S. Rickerby, L. Shekede, Fan Zaixuan, Tang Wei, Qiao Hai, Yang Jinjian, F. Piqué, Development and Testing of the Grouting and Soluble-Salts Reduction Treatments of Cave 85 Wall Paintings, in: Conservation of Ancient Sites on the Silk Road. Proceedings of the Second International Conference on the Conservation of Grotto Sites, Mogao Grottoes, Dunhuang, People's Republic of China, June 28-July 3, 2004, Ed. N. Agnew. Getty Conservation Institute, Los Angeles, 2010, pp. 471-479.

[3] B. Biçer Şimşir, I. Griffin, B. Palazzo-Bertholon, L. Rainer, Lime-based injection grouts for the conservation of architectural surfaces, Reviews in Conservation 10 (2009) 3-17.

[4] C. Pasian, F. Piqué, A. Jornet, Non-structural injection grouts with reduced water content: Changes induced by the partial substitution of water with alcohol, Studies in Conservation 62:1 (2017) 43-54.

[5] S.P. Pinho, E.A. Macedo, Solubility of NaCl, NaBr, and KCl in water, methanol, ethanol, and their mixed solvents, Journal of Chemical & Engineering Data 50 (2050) 29-32.

[6] A. Kolker, J. de Pablo, Thermodynamic Modelling of the Solubility of Salts in Mixed Aqueous-Organic Solvents, Industrial & Engineering Chemistry Research 35 (1996) 228-233.

[7] B. Biçer-Şimşir & L. Rainer, Evaluation of Lime-Based Hydraulic Injection Grouts for the Conservation of Architectural Surfaces – A Manual of Laboratory and Field Test Methods, Los Angeles: The Getty Conservation Institute, 2013.

[8] R. Olmi, M. Bini, A. Ignesti, S. Priori, C. Riminesi, & A. Felici, Diagnostics and monitoring of frescoes using evanescent-field dielectrometry, Measurement Science and Technology, 17:8 (2006) 2281.

[9] R. Olmi & C. Riminesi, Study of water mass transfer dynamics in frescoes by dielectric spectroscopy, Il Nuovo cimento della Società Italiana di Fisica C 31:3 (2008) 389.

[10] N.Proietti, D. Capitani, V. Di Tullio, R. Olmi, S. Priori, C. Riminesi, A. Sansonetti, F. Tasso & E. Rosina, MOdihMA at Sforza Castle in Milano: Innovative Techniques for Moisture Detection in Historical Masonry. In Built Heritage: Monitoring Conservation Management 2015, pp. 187-197, Springer International Publishing.

[11] K. Zehnder & A. Arnold, Monitoring wall paintings affected by soluble salts, in: The Conservation of Wall Paintings, ed. S. Cather, The Getty Conservation Institute, Marina del Rey, 1991, 103-108.

[12] SIA 262/1 2003, Perméabilité à l'eau, Construction en Béton–Spécifications complémentaires, Société suisse des ingénieurs et des architectes, Zurich.

[13] UNI EN 1015/11 1999, Methods of Test for Mortar for Masonry–Part 11, Determination of Flexural and Compressive Strength of Hardened Mortar, Milano.

Moisture transport during poultice application

Christoph Franzen[1]*, Martin Aulitzky[2] and Stephan Pfefferkorn[2]
[1] *Institut für Diagnostik und Konservierung an Denkmalen in Sachsen und Sachsen-Anhalt e. V., Dresden, Germany*
[2] *Fakultät Bauingenieurwesen/Architektur, Hochschule für Technik und Wirtschaft Dresden, Germany*
*franzen@idk-denkmal.de

Abstract

Water movement is the key parameter in poultice application for desalination. Water is the eluent and transport medium for the salts to be extracted from the porous substrate. While from the hypothetical point of view the dynamical processes of water ingress, dissolution of salts, transportation and precipitation seems to be well known, measured data on the water quantities at specific stages of the process are scarce. Especially, time approximations about the change in main movement direction of the water are unidentified. By experimental data we give indications to the total amount of water used with different poultice compositions and the development of the water ratio during the process until equilibrium dryness is reached. In conclusion we suggest increasing the ratio of water and probably the poultice layer thickness from cycle to cycle to increase the effective cross section stepwise.

Keywords: desalination, water movement, salt transport, poultice material

1. Introduction

Desalination by poultice applications is a widely applied technique in restoration.[1] The general idea is to apply temporarily an additional layer on the salt contaminated porous material. That poultice layer brings water to and into the substrate material. Salts dissolve in the water. With time water evaporates at the surface, which is the poultice surface. The water in the porous material moves to the poultice where it evaporates. The dissolved salt ions are transported in the poultice. When the water evaporates the salts precipitate. The salts are eliminated from the object by removal of the poultice. - Thus there are three moisture pathways during poultice application to be regarded which happen partially in parallel and change their magnitude during the process. There is the ingress of water from the poultice into the substrate and in the substrate, and then there is the reflow from the substrate to the poultice and finally the evaporation from the poultice. Evaporation, also to be regarded as a driving force, acts from the very beginning, although the first situation to be regarded is the water ingress from the wet poultice into the dry substrate due to concentration balancing and capillary suction. Effects of gravitational flow will not be discussed here, but should be taken into consideration much more in future.

Of course in practical execution of that simple and effective idea some more effects do happen, sometimes limiting the efficiency of the application. First to mention is that the procedure acts on the surface of salt contaminated porous material. Thus it is to be assumed, that it does not effect to salts sitting deeper in the material. What could be a drawback could also be regarded as advantage. In the very most cases the salts are concentrated near to the surface, thus poultices act where the problem is and so not affect other parts. Nevertheless the question occurs to which depth that argument is valid. Especially in cases where no or minor evaporation happens and the poultices are wet all the time of application[2] the cross section where the effects of diffusion is effective is not clear. However, we focus here on drying poultices and address their limitations by formulating the requirements: The poultice material should be free of salts. The same applies to the water to prepare the poultice. Thus the use of demineralised water is recommended. The ready to use wet poultice shall be workable on site. The material is either applied by trowel or spittle, or is sputtered, than in terms of viscosity it has to fulfil the requirements of the spraying device. The poultice has to adhere to the surface to the object. The adherence has to stand all the time of application, as lost of contact does terminate the function. On the other hand the poultice should be removable from the surface easily and free of residues. The poultice has to bring the optimal amount of water to the system, enough to get salts in solution but on the same time not flush the salts deeper into the material. Finally the poultice has to catch the salts. To reach all these demands nearly all poultices for desalination nowadays are multi component systems. There is material to carry the water, clays to secure adherence and workability, and matrix components for stability, minimising the shrinking and providing a pore structure.[1, 3]

In the study presented here we show the time resolved moisture development in three different substrate materials treated with five different poultice types.

2. Materials and methods

2.2. Building material to desalinate

Three different important building materials were chosen for the experimental work. Two varieties of natural sandstone material of Elbe Sandstone: Postaer Sandstone and Cottaer Sandstone and one type of brick was used for the tests. In *table 1* some data of those porous building materials is presented, detailed material description is given by.[4]

2.3. Poultice material

Five commercially available poultice materials from four trade specific producers were chosen: RE and RA are from building material companies, RI, for fine grained materials, and RII, for coarse grained, from a producer specialises on restoration material and material A is pure cellulose (BC200). The material types were characterised optically, by XRD and sieve measurements, the workability of the pasted material was noted, as well as the development during application. Furthermore on prisms of the hardened poultice strength measurements were executed.

Material/sample code	Postaer Sandstone/P	Cottaer Sandstone/C	Brick/Z
Total pore volume Vol. (%)	21	22	-
W-Value (kgm2h$^{-1/2}$)	9.12	5.61	4.94

Table 1: Three important building materials were chosen for experiments

2.4. Methods, sampling procedure

To win insight in the state of water movement at eight pre-defined time stages dummies were destructively sampled. Consequently for each of the five poultice materials eight dummies from three substrate materials were prepared. The natural sandstone was cut into blocks of 10x15x10 cm, two bricks pieces were glued together to win comparable surfaces for the experiments of 11.3x15 cm. The block sides were sealed, the surfaces contaminated with NaCl-salt spray. All dummies were marked with a 50 µl drop of ink.[5] Samples were taken in increasing time intervals after 4 hours, 7 hours, 1 day, 2 days, 3 days, 6 days, 2 weeks, and 3 weeks. At each sampling stage the state of the system was visually inspected, the poultice removed for visual inspection, moisture and salt determination, and the substrate block gravimetrically inspected, drill powder sampled in 12 stages of 1 cm depth and analysed for moisture content[6] and salt content by electrical conductivity. Finally the block was chopped in stone masons fashion for further visual inspection of the ink drop.

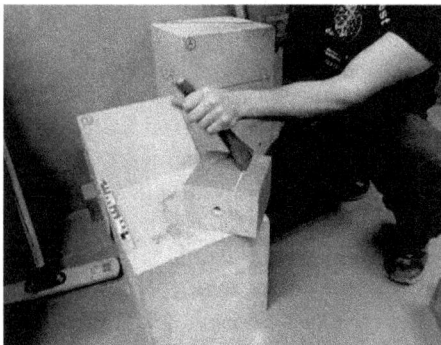

Figure 1: Chopping of the block

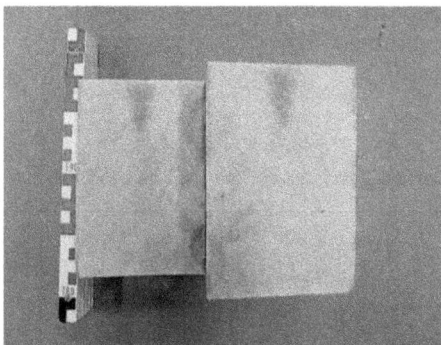

Figure 2: Ingress of ink drop

3. Results and Discussion

3.1. Poultice material

Generally looking quite similar with naked eye you could notice in dry raw poultice material RE sand, clay and cellulose, RA contents lesser sand but foam glass spheres, R I looks like pure powder and R II appears as powder with some sand. Table 2 shows significant material data of the poultices, from mass related water content (Wc) as recommended by the producers and the resulting mass of water provided to the wall (Wp). The strength data was determined on samples comparable to the end stage of the desalination action.

Obviously there are major differences in the poultice performances due to divers material compositions. *Figure 3* shows their size fractions with standard sieves. When it comes to the creation of specific pore sizes for effective desalination as recommended by[7] we suggest to define adapted mesh combinations for the future. Of course the cellulose content in all mixtures hampers the sieving. For displaying the results (diagramms in *Figure 3*) giving the sieve fraction in mass could be misleading due to different densities e.g. sand and foam glass. Reference to volume lead to other difficulties.

Figure 4 gives the XRD patterns of the four smallest mesh fractions, where the clay is to be expected. Here significant

poultice/ code	Wc [-]	Wp [Lm⁻²]	shrink [mm/m]	Comp. Strength	Flex. Strength	Emod$_{dyn}$
RE	0,322	7.0	21.1	1.63	0.79	2.95
RA	0,600	7.7	21.6	0.68	0.41	0.34
R I	0,475	8,7	72.0	nd	1.65	3.13
R II	0,475	6,6	18.7	1.16	0.50	1.03
A	4,500	14.8	52.1	nd	0.06	nd

Table 2: Five commercially available poultice materials

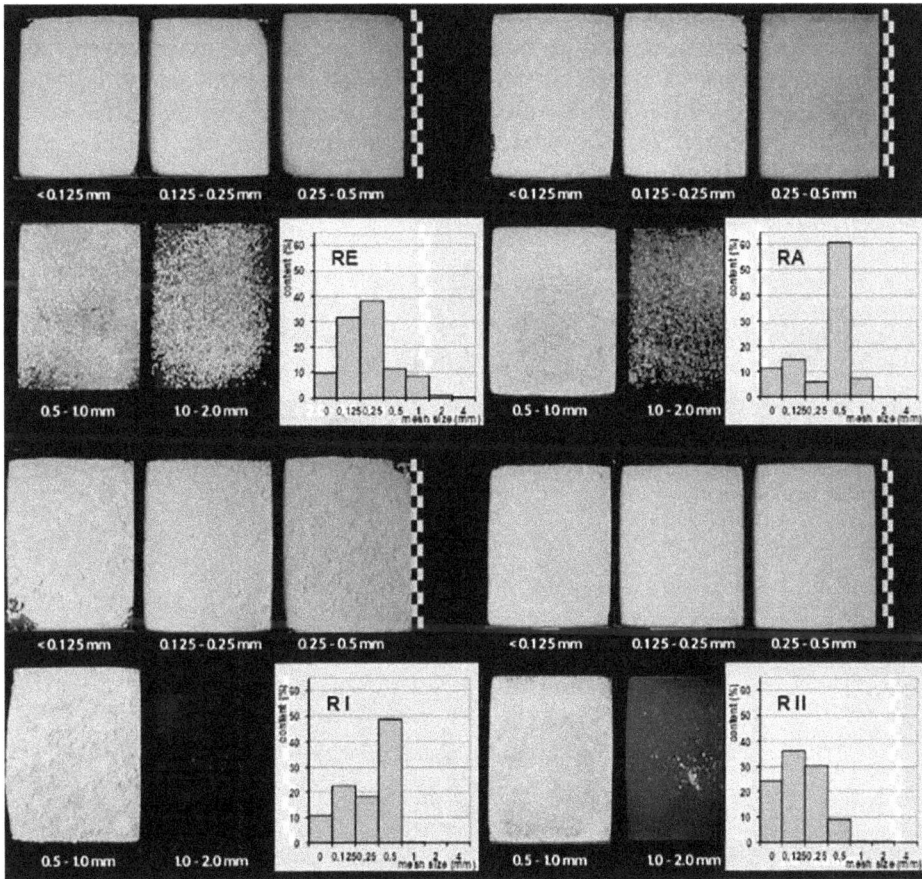

Figure 3: Size fractionated poultice materials (A – cellulose not displayed)

material differences become obvious as sometimes bentonite, sometimes kaolinite is used in the industrial produced material mixtures. Thus as the clays manipulate the workablity, effect the work of later cleaning und influence the final strength we could not yet decide if it is the type of clay or the amount or the total sieve line, which leads with RI to those problematic results of high shrinkage, which got also visible through the experiments.

3.2. Water distribution vs. time

The customer has to find out, which poultice material works best in his subject on the given substrate. Wendler[8] suggested a classifiing parameter to evaluate the effectivity of different poultice mixtures by the time the water movement is changed. As to be seen in *Figure 5* (first column) it is evaluated when the wetted substrate begins to loose weight. That

happens in most cases after 4h or 7h, sometimes 24h. Despite the lowest W-value laboratory determined for brick the water uptake in the very first (4) hours is always higher than into the sandstones. The data provides the willingness of the poultice to release water to the system: A>>RA≥RE>>RI≥RII. In that very first 4 hours the sandstones under all poultices take always about 20-30% of the water equal to the minimum of what the brick does take. There the pore system of the substrate is the limiting factor of water uptake. Taking into account the experimental procedure we conclude, that the graviational effects of water flow in those partially supersaturated water sytems have to be taken into account much more.

When the poultice gets dryer than the subtrate the general water movement is returned at least mathematically. Nevertheless in parallel the water front in the substrate moves on. That is to be seen in *Figure 5* 2nd column, where the water con-

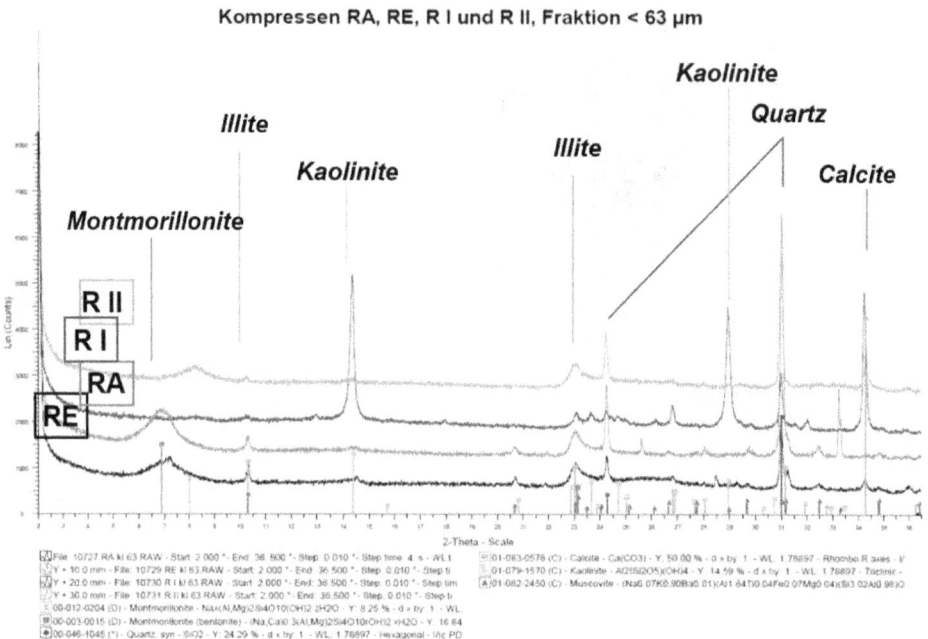

Figure 4: XRD pattern of four poultice materials, smallest mesh fraction

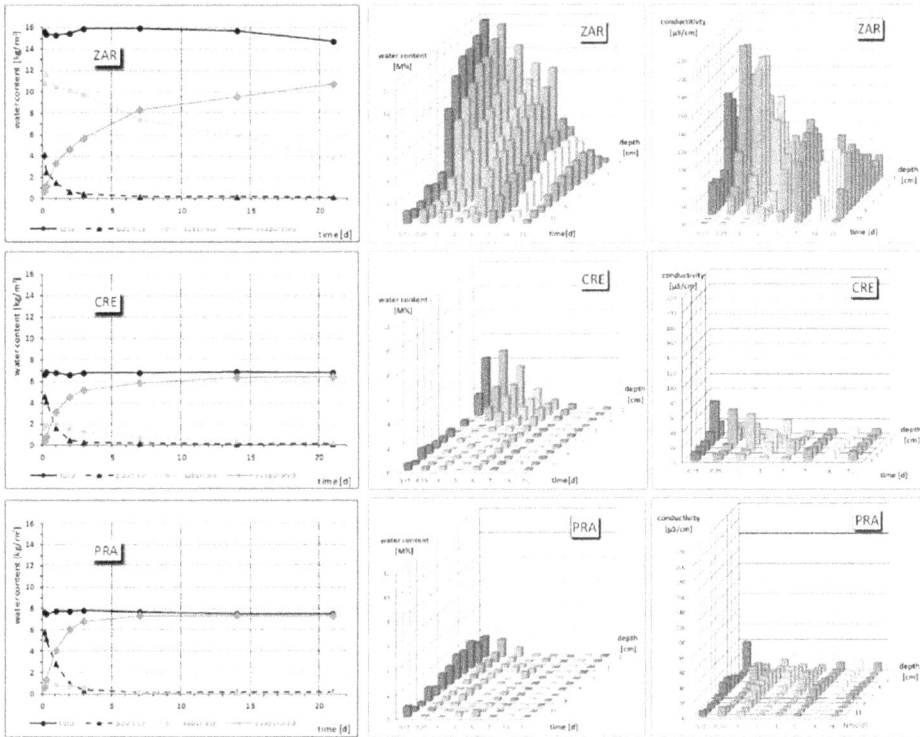

Figure 5: 1ˢᵗ column: water contents in experiment: total, poultice, substrate and mount evaporated vs time; 2ⁿᵈ: depth resolved water content; 3ʳᵈ: conductivity results

tent in the substrate is shown in all depth at all investigated time stages. In brick material most poulitces affect a depth about 4 to 5 cm. In Postaer sandstone RE an RA reach about 7 cm while R I and R II on the other hand affect 4 cm which partially can be explained by a problematic performance during the application.

In spite of the lower W-value the Cottaer sandstone it tends to take up more water than the Postaer Sandstone. It is important to note, that Cottaer sandstone in many cases has not completely dried out after 21 days. Pure cellulose carrys about double of the water into the system compared to the poultice mixtures. With such a cellulose treatment Postaer sandstone can be wetted totally down to 12 cm depth.

3.3. Water related effects

The results of the conductivity measurements are heterogenious. Most probably the concentration of contamination was too low, nevertheless trends are to be seen. In the first hours in all cases the salt peak from the very top is flushed into about 3 cm depth. After 1 day it broadens and, of course, decreases. A low, decreased restitic contamination in all the profile where water took access to is visible until the very end of investigation. Under poultice material which looses eager contact during the application, most probably due to high shrinkige and low adherence to the substrate, a top peak and a tail into the materials is visible.

Also the ink drop test has limited interpretation potential. The ink was practically invisible in the brick, showed a cone in Postaer sandstone. In Cottaer sandstone it was hardly visible but tended to be vanished on the top and was most intensive in about 2 cm depth. Also in the poultice material the ink was solely visible in the pure cellulose material AC. However, we recommend to use much more than 50 µl ink for contamination for possible future experiments.

3.3. Water related effects

probably the concentration of contamination was too low, nevertheless trends are to be seen. In the first hours in all cases the salt peak from the very top is flushed into about 3 cm depth. After 1 day it broadens and, of course, decreases. A low, decreased restitic contamination in all the profile where water took access to is visible until the very end of investigation. Under poultice material which looses eager contact during the application, most probably due to high shrinkige and low adherence to the substrate, a top peak and a tail into the materials is visible.

Also the ink drop test has limited interpretation potential. The ink was practically invisible in the brick, showed a cone in Postaer sandstone. In Cottaer sandstone it was hardly visible but tended to be vanished on the top and was most intensive in about 2 cm depth. Also in the poultice material the ink was solely visible in the pure cellulose material AC. However, we recommend to use much more than 50 µl ink for contamination for possible future experiments.

4. Conclusion

In the study five different poultice materials were characterised and tested on three substrate materials. The commercial produceres use different material combinations to offer workable poultices. Refering to the amount of water provided by the pultice for the desalination the materials have minor differences, while pure cellulose is able to really flood a system. The performance of the poultices already when simple material data like shrinkige measurement is taken into account, is for some of the comercially available products problematic. The advantages and disadvantages of specific clays taken for the mixtures are not well understood. Here much more investigation on one and but also information, what kind is implemented, is needed.

The study shows that the moisturing of the substrate happens within the first hours of poultice application. Depth of several centimeters are reached, a distribution of salts from the top into the stone at that stage is unavoidable. Water depth penetration seems solely partially predicable and connected to the W-value of the substrate. Gravitation effects and teh water release properties of the poultice are effects much more to be regarded. Thus for successively progressive maintaining deeper zones it is adviseable to start with thinner (less water) poultices in the beginning and increase the layer thickness for cycle to cycle to increase the effective cross section stepwise.

References

[1] HERITAGE, A. et al., Current use of poultices in the conservation of monuments. In: A. HERITAGE, A. HERITAGE & F. ZEZZA, (eds). Desalination of Historic Buildings, Stone and Wall Paintings. London: Archetype Publications Ltd, 2013.

[2] WTA Guideline 3-13-01/E Non-destructive desalination of natural stones and other porous materials with poultices. WTA Publications, Munich, 2005.

3 WTA Guideline 3-13-18/E, Salt reduction of porous building material with poultices. WTA Publications, Munich, in preparation, 2018.

4 Grunert S, Der Elbsandstein: Vorkommen, Verwendung, Eigenschaften Elbe Sandstone: deposits, use, properties-GeologicaSaxonica. J Cent Eur Geol 52(53), 2007, 3–22.

5 Lenz R, Systemoptimierte Salzminderungskompressen, Material – Wirkung Tagungsbeitrag Retrospektive & Perspektive(n) in der Wandmalereirestaurierung Werkstattgespräch 2016, in prep., 2017.

6 WTA, 2002, WTA-Merkblatt 2-11-02/D: Messung der Feuchte von mineralischen Baustoffen. München: Fraunhofer IRB Verlag, 2006.

7 Franzen, C., Analytische Begleitung von Salzreduzierungsmaßnahmen, in: Praxisorientierte Forschung in der Denkmalpflege – 10 Jahre IDK-, Hrsg.: Institut für Diagnostik und Konservierung an Denkmalen in Sachsen und Sachsen-Anhalt e.V., 2006, 31 – 40.

7 Pel, L., Sawdy, A. & Voronina, V., Physical principles and efficiency of salt extraction by poulticing. Journal of Cultural Heritage 11(1), 2010, p 59-67.

8 Wendler E., Möglichkeiten und Grenzen einer Salzminderung durch Kompressen: Materialoptimierung und aktuelle Fallstudien an Naturstein, Ziegel, Putz und Wandmalerei, Natursteinsanierung Stuttgart, 2007, S. 29-38.

The application of hydroxyapatite-based treatments to salt-bearing porous limestones: A study on sodium sulphate-contaminated Lecce Stone

Gabriela Graziani[1, 2]*, Enrico Sassoni[1], George W. Scherer[3] and Elisa Franzoni[1]

[1] Department of Civil, Chemical, Environmental and Materials Engineering (DICAM), University of Bologna, Italy

[2] Laboratory of NanoBiotechnology (NaBi), Rizzoli Orthopaedic Institute, Bologna, Italy

[3] Department of Civil and Environmental Engineering (CEE), Princeton University, USA

*gabriela.graziani2@unibo.it

Abstract

Salt weathering is among the most severe phenomena affecting porous limestone, often leading to the loss of precious material from historical sculptures and building façades. Together with exerting a weathering action, salts can also hamper the success of consolidation treatments, by interfering with consolidants' penetration and/or curing reactions. On site, application of consolidants on non-contaminated stone is rarely feasible, especially in the case of porous limestones; for this reason, testing application of consolidants on salt laden stone is of particular relevance to guarantee their successful application in the field.

In this paper, two hydroxyapatite (HAP)-based treatments have been applied to Lecce Stone, a very porous organogenic limestone, highly susceptible to salt weathering. The two treatments differ for DAP concentration, application procedure and presence/absence of ethanol in the formulation. Prior to treating, specimens were subjected to salt crystallization cycles in a sodium sulphate solution, to cause salt contamination and induce weathering. They were then desalinated, purposely leaving a percentage of salts in the stone (SO_4^{2-} after desalination ~0.15-0.3 wt%). Phase formation and distribution as a result of different contamination levels were investigated, and the efficacy of the treatments in comparison to ethyl silicate was determined, as it is currently the most used consolidant for this lithotype. Finally, the durability of the most promising treatment to further salt crystallization was evaluated, again in comparison with ethyl silicate.

Promising results were obtained, as the presence of sodium sulphate did not prevent the HAP- based consolidants from uniformly distributing in the stone, sealing cracks and forming insoluble phosphate phases. Interestingly, the application on non-desalinated stone revealed that the nature and the amount of the phases that form as a result of the treatment are dependent on the amount of salts in the stone. Finally, most promising HAP-based formulation was found to enhance the stone's durability towards further salt crystallization, also compared to ethyl silicate.

Keywords: salt weathering, cultural heritage, calcium phosphates, limestone

1. Introduction

Salt weathering is among the main causes of stone deterioration, especially for porous limestone, that can undergo severe damage, such as flaking, scaling and pulverization, all leading to the loss of significant amounts of material and of architectural/sculptural details. Together with causing weathering, salts inside the stone may interfere with consolidants' penetration and hardening reactions, thus possibly having a negative impact on the treatments' outcome: for this reason, desalination procedures are normally performed prior to consolidation. However, even after desalination some salts may remain inside the stone, interfering with the success of consolidation. Therefore, the effects of the application of consolidants on salt-bearing stones needs to be investigated.

Treatments based on hydroxyapatite (HAP) (produced by treating the stone with an aqueous solution of diammonium hydrogen phosphate, DAP) have been proposed for limestone consolidation, exhibiting remarkable efficacy, compatibility and durability.[1, 2] However, tests carried out so far mainly focused on uncontaminated samples, which is rarely the case on site. Salt-related issues could arise for real specimens because not only HAP but also metastable calcium phosphate (CaP) phases and, possibly, phosphate salts could form, depending on the substrate contamination[3]; moreover, HAP might easily incorporate foreign ions, possibly altering its crystallinity and solubility.[4, 5]

In the literature, the presence of foreign ions has been found to promote HAP growth, but also to cause the formation of soluble phases, depending on substrate composition, nature of contaminants and surface roughness.[6] For this reason, the investigation of application of a hydroxyapatite-based treatment on salt-laden stone is of particular interest.

In this study, we studied the effects of applying HAP-based treatments on salt-contaminated Lecce stone, a porous organogenic limestone characteristic of Baroque Architecture of southern Italy. This stone contains significant amounts of silica, aluminum, phosphorus and magnesium, which could also have an impact on the formation of hydroxyapatite

or soluble phases.[4, 5] Stone samples were contaminated by sodium sulphate, which is one of the most widespread and noxious among the salts that can be found on site, then were treated with two different HAP-based formulations, previously developed by the authors for limestone and marble consolidation.[1, 2, 3, 6] The effects of the treatments on phase formation were analyzed and compared to those obtained by ethyl silicate, currently the most used product for the consolidation of limestone and, in particular, of Lecce Stone.

2. Materials and Methods

2.1. Materials

Tests were carried out on Lecce Stone (LS), which has non-uniform microstructure and mechanical properties, even within a single stone slab. Cores 5 cm in height and 2 cm in diameter were used for all the tests, except for the evaluation of the durability after consolidation, for which 5x5x5 cm^3 cubes were used, given the low resistance of the stone to salt weathering. All cores and cubes were cut from one quarry slab.

Samples were contaminated by cycles of sodium sulphate crystallization (samples "SALT"), performed as follows:

1) Wetting phase (24 h): samples were immersed to a depth of about 1 cm in a 14 wt% solution of sodium sulphate decahydrate ($Na_2SO_4 \cdot 10H_2O$) in deionized water;

2) Drying phase (22 h): samples were dried in ventilated oven at 50°C;

3) Cooling (1h) and measuring of weight and dynamic elastic modulus (total: 2 h).

Two cycles were performed on the cores, 5 on the cubes. The number and duration of the cycles were selected based on preliminary tests, with the aim of contaminating the stone with a suitable amount of salts, without causing rupture, and conserving a morphology suitable for mechanical testing. Salt content as a result of the cycles was measured by ion chromatography (IC).

To reproduce the conditions on site, where stone is normally desalinated before the application of the consolidant (even if a fraction of salts usually remains in the stone), part of the cores and all the cubes were partially desalinated after contamination. This also allowed evaluation of the impact of different levels of salt contamination on phase formation (samples "DESALT"). Desalination was performed by applying a deionized water poultice (water:cellulouse pulp ratio 1:5, thickness ~1 cm). The poultice was kept sealed for 24 hours to allow solubilization of the salts, then left to dry in contact with the samples to allow advection of the salts towards the external surface. Residual salt content was evaluated by IC performed on the ground stone after salt extraction and filtering. The investigation of phase formation and of mechanical properties after treatment was also carried out on uncontaminated, uniformly weathered samples. For this purpose, a batch of stone cores (samples "HEAT") was artificially weathered by heating at 400°C for 1h, according to a procedure previously developed by the authors for promoting cracks formation in the stone.[1,2]

2.2. Treating procedure

Two HAP-based treatments were tested that were previously developed for the consolidation and protection of limestone and marble, respectively:

1) Treatment "3M": A 3 M DAP solution is applied by brushing (10 strokes) followed by the application of a limewater poultice (1.7 g/l Ca(OH)2, limewater:poultice ratio 1:6). Limewater poultice is applied with the aim of providing further calcium ions to boost HAP formation and to remove possible soluble phosphate remaining in the stone after treatment;

2) Treatment "0.1M": A 0.1 M DAP solution also containing 0.1 mM CaCl2 and 10 vol% ethanol is applied, followed by the application of a 0.1 M DAP + 0.1 mM CaCl2 solution without the addition of ethanol. Both solutions are applied by brushing (10 strokes)[7]. This procedure was developed for the consolidation of marble, for which it was found that ethanol promotes HAP formation (and thus better efficacy) with lower amounts of precursors.

Ethyl silicate (ES) was also tested for comparison's sake, as it is currently the most used product for consolidation of Lecce Stone. For ES treatment a commercial formulation (Estel 1000, CTS, Italy, composed of 75 wt% ethyl silicate and 25 wt% white spirit D40, also containing 1 wt% dibutyltin dilaurate as a catalyst) was applied by brushing (10 brush strokes). After treatment, samples were left to cure for 1 month in room conditions prior to testing. A deionized water poultice was applied to ethyl silicate-treated cubes before performing durability cycles (water:poultice ratio 5:1, poultice kept on the samples for 48 h), to remove possible hydrophobicity linked to residual ethoxy groups, that could interfere with the uptake of saline solution.[8] The poultice was not applied to ES-treated cores to avoid desalination and because they did not need to be subjected to any characterization test involving absorption of water or saline solution.

Stone cores were treated on the whole external surface, while cubes were treated on one face perpendicular to the bedding planes.

2.4. Evaluation of the treatments' effects

Phase formation was examined by FT-IR. For a better understanding of the effect of the presence of salts on the HAP based treatments, FT-IR was performed on heated (HEAT), desalinated (DESALT) and non-desalinated (SALT) samples, to highlight the impact of different levels of salt contamination. FT-IR was also performed on ES-treated samples to verify whether the presence of salts interferes with ES curing.

Phase morphology and distribution were evaluated by SEM/EDS, to understand whether the presence of salts prevents consolidants from penetrating deeply into the stone and to evaluate the treatments' capability of sealing cracks.

Mechanical properties after treatments were evaluated in terms of dynamic elastic modulus (5 samples were taken for each condition). Dynamic elastic modulus was selected as it is non destructive and allows testing of the same samples before and after salt contamination and treatment, which mitigates the impact of the high heterogeneity of the stone on the evaluation of the treatments' performance. To have a preliminary evaluation of the efficacy of the treatments, dynamic elastic modulus was also measured on samples artificially weathered by heating (instead of salt weathering), so as to determine their effects on uniformly weathered and uncontaminated stone.

For the durability tests, after desalination and treatment, 4 stone cubes for each condition were subjected to 10 additional salt crystallization cycles, performed exactly as during the contamination phase. The effects of the cycles were evaluated by measuring weight loss and dynamic elastic modulus after each cycle and after 10 cycles and desalination.

3. Results and Discussion

Morphology and composition of Lecce Stone was investigated by SEM/EDS *(Figure 1)*. Untreated stone consists of a cement

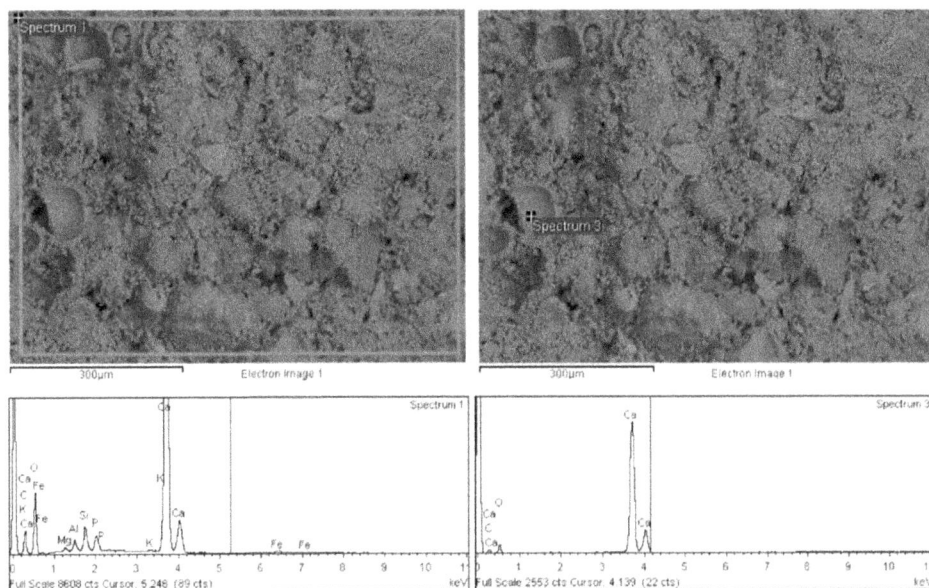

Figure 1: SEM images and EDS spectra of untreated Lecce Stone

mainly constituted by calcite with traces of phosphorous, magnesium, aluminium, iron and silicon, that envelops fossil shells only composed of calcium carbonate *(Figure 1)*. To have a preliminary evaluation of the treatments' efficacy prior to application on contaminated stone, they were first applied to samples artificially weathered by heating, that also exhibit more uniform weathering *(Table 1)*. Significant efficacy is obtained for the application on heated stone, especially for treatment 3M that causes higher increases in Ed compared to ethyl silicate. Remarkably, both ES and 3M fully restore

about 0.15-0.30 wt% of salts remain in the stone, depending on the specimen and the initial contamination, their residual presence being confirmed by FT-IR and EDS spectra *(Figure 2 and 3)*. Interestingly, salt content after the application of treatment 3M significantly decreases (average $SO_4^=$ content = 0.0244 wt%), indicating that the application of a limewater poultice allows for some desalination of the stone, which can be considered an advantage of this treatment. Salt content in 0.1M and ES samples, instead, remains substantially unaltered in all the examined areas of the samples.

Ed [GPa]					
	Before heating	After heating ("HEAT")	ΔEd	After heating and consolidation	ΔEd
3M	10.5±0.6	7.7±0.4	-27%	12.5±0.8	+62%
0.1M	10.6±1.2	7.8±0.6	-26%	9.1±0.7	+17%
ES	12.8±0.5	9.3±0.2	-27%	13.0±1.1	+40%

Table 2: Dynamic elastic modulus on samples artificially weathered by heating. Values for stone before heating are different between the different batches of samples because of the heterogeneity of the stone.

the initial dynamic elastic modulus of unweathered stone. A lower increase is obtained by treatment 0.1M, which however can still be considered promising, particularly in view of the much lower amounts of precursors used.

Lecce Stone exhibits a very low resistance to salt weathering, significantly variable between different slabs and even different areas of the same slab. For this reason, two salt crystallization cycles were selected to achieve salt contamination without causing rupture in the stone. As a result of the cycles, about 1.43 wt% sulphates (expressed as $SO_4^=$) are deposited inside the stone, the total amount being quite variable between different specimens. After desalination,

Phase formation after treatments is reported in *Figure 2*. FT-IR spectra confirm the presence of residual sulphate in desalinated stone. No formation of soluble calcium phosphates phases nor phosphate salts or other soluble compounds is detected for any of the treated specimens. For sample 3M the application on non-desalinated ("SALT"), compared to desalinated stone ("DESALT"), leads to a higher formation of reaction products. However, given the position of the bands, it seems that octacalcium phosphate (OCP) is the phase that forms in the presence of higher amounts of salt (samples "SALT"), while HAP forms in its absence and for lower levels of contamination (samples "HEAT" and "DESALT").[9]

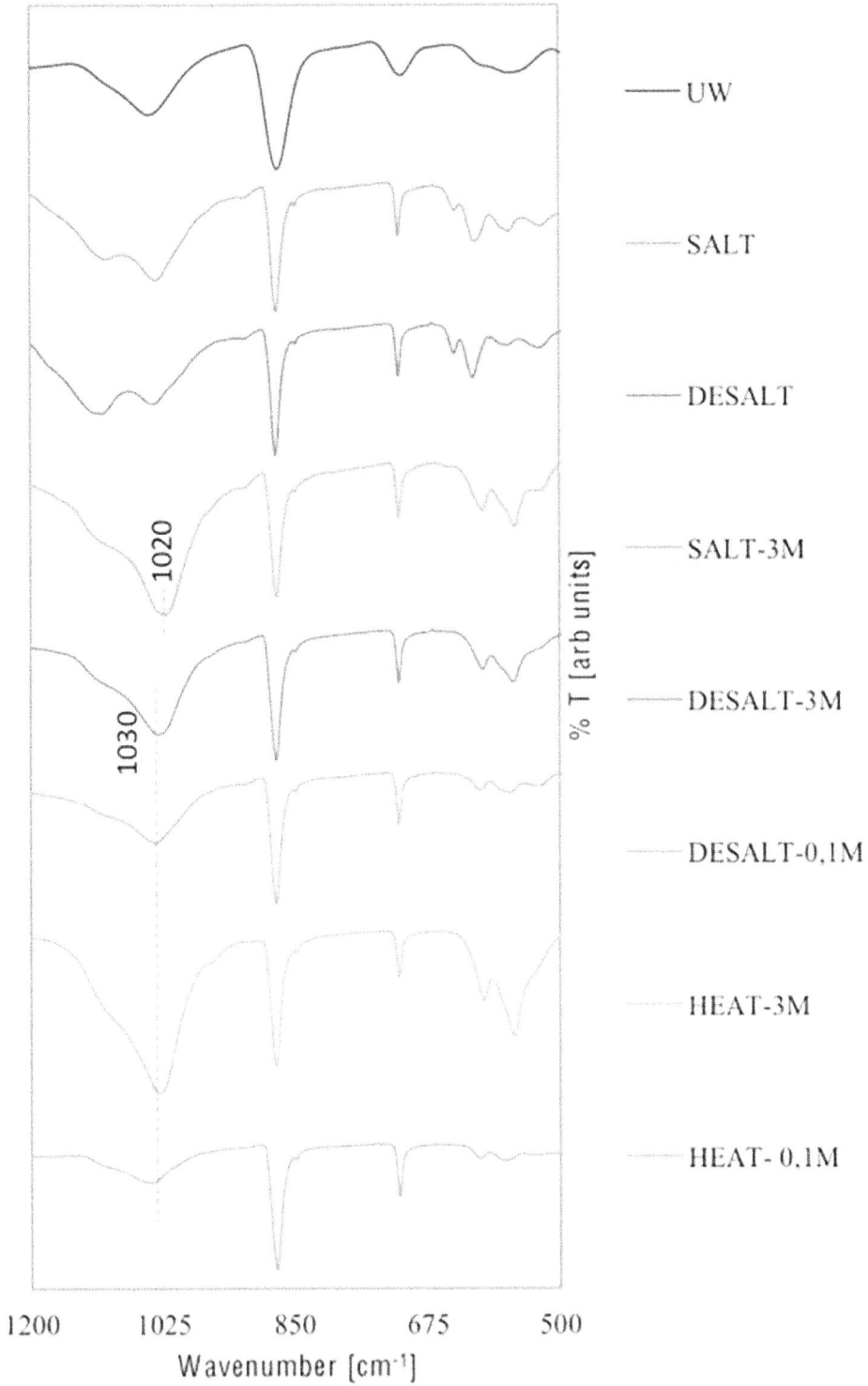

Figure 2: FT-IR spectra on untreated and treated specimens, for different levels of salt contamination

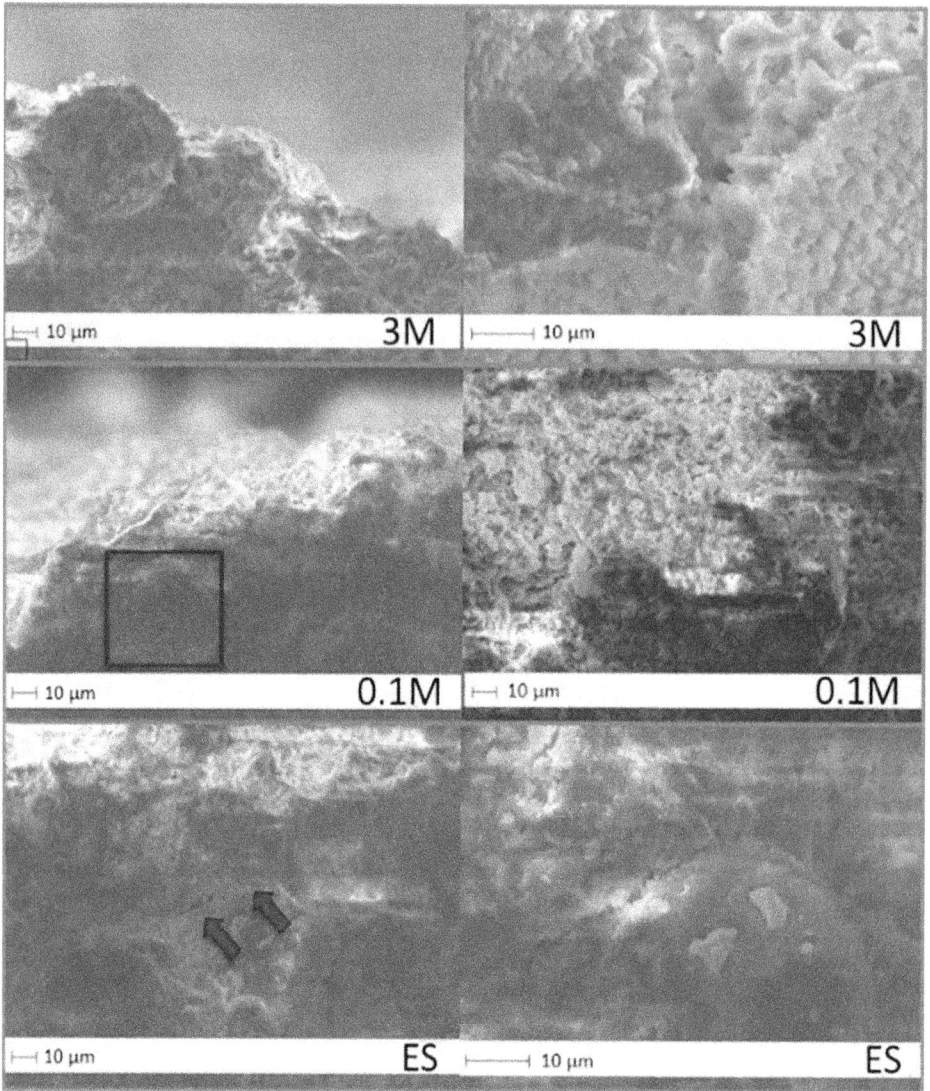

Figure 3: SEM images of treated samples' cross sections. In figure, the red square indicates significant formation of phosphate phases in correspondence of one shell, the arrows indicate a crack that remains unselaed after ethyl silicate treatment.

The formation of higher amounts of calcium phosphates in the presence of higher amounts of salts could be due to increased surface area and roughness (increase in surface roughness being caused by salt damaging the stone), similar to what was found in [10] for ammonium oxalate treated samples, where more calcium oxalate was found to form on non-desalinated samples because of the higher specific surface and the different surface texture of the samples. Conversely, the formation of OCP instead of HAP is possibly due to the fact that different ions (including Na) can exert opposite impact on phase formation, and on crystallinity and solubility of reaction products, so not only the ions themselves but also the relative amount of each ion can have an effect on phase formation[4, 5].

In general, the effect of salts could be due to the fact that their presence and the damage they cause affect surface roughness, specific surface and the presence of sites available for nucleation. This indicates that the level of salt contamination has an impact on phase formation; however, despite OCP being a metastable phase (eventually converting in HAP), it is far less soluble than calcite, hence its formation is not considered a drawback of the treatment. The absence of any band corresponding to sodium sulphate, in agreement with IC, confirms that salts were removed by the application of the limewater poultice during the 3M treatment. Treatment 0.1M leads to the formation of HAP (see the presence of the phosphate stretching bands at 600 and 560 cm^{-1}, and the shift in the main band with the appearance of a shoulder at $\approx$1100 cm^{-1}), but a significantly lower amount of phosphate phases forms compared to 3M, in contrast to what was reported for marble.[7] However, because the stone itself contains phosphate fractions, the lower phase formation complicates the exact discrimination between HAP and other calcium phosphate phases. Data for SALT-0.1M treatment are not reported here, as no significant differences can be assessed between the spectra for SALT and DESALT samples. Sample ES exhibits a broad band in the 1040-1100 cm^{-1} area, probably resulting from the overlapping of the bands at 1037 cm^{-1} (Lecce Stone), 1080 cm^{-1} (Si-O-Si of hydrolyzed ES)[9] and 1104 cm^{-1} (sodium sulphate). No bands at 1170 cm^{-1} are present, indicative of Si-O-C.[11]

Phase morphology and distribution in treated samples was evaluated by SEM/EDS. In sample 3M, phosphate phases were detected both on the surface and in depth in the samples, indicating that the presence of salts does not prevent a uniform penetration of the consolidant. Interestingly, phosphate formation is higher adjacent to and inside the shells, which exhibit no ion-contamination, suggesting that HAP nucleation could be inhibited by ions contained in the stone matrix (probably magnesium, which is known to be an HAP growth inhibitor [4, 5]). For Lecce stone, therefore, the presence of trace elements in the stone seems to have a stronger impact on HAP formation than sodium sulphate has.

Consistently, also in sample 0.1M, phosphate formation is favoured on the shells, indicating that, even with ethanol in the solution phase, formation is lower in ion-doped areas of the stone. The formation of calcium phosphates is generally lower compared to 3M and it is quite non-uniform (in some areas P is not detected by EDS). However, because the stone itself contains phosphorous in variable concentrations, it is not possible to detect the formation of phosphate phases unless a large amount forms. The treatment, however, is effective in sealing cracks, also in depth in the sample, again indicating that the presence of salts does not prevent the penetration of the consolidant. The ability of the HAP-based consolidants to seal cracks is reflected in their ability to fully restore the dynamic elastic modulus of untreated stone *(see Table 2)*. The lower phase formation in 0.1M treatment compared to 3M, however, combined with the lower increase of dynamic elastic modulus, indicates that the formulation needs further optimization for being applied to Lecce Stone, and that a higher concentration of DAP should be used, given the high porosity of the stone (experiments are in progress). However, because HAP is found to form and seal cracks also in depth in the sample, it seems that ethanol has a positive impact on HAP formation. SEM images on ethyl silicate treated samples indicate abundant phase formation, but also reveal the presence of unsealed cracks that were not noticed in HAP-based treatments. For these samples, EDS analysis did not provide meaningful information, as a high and very variable

Ed [GPa]					
	Before salt weathering cycles	After 2 cycles and desalination	ΔEd	After consolidation	ΔEd
3M	16.5±0.3	16.2±0.6	-2%	17.8±0.3	+10%
0.1M	10.9±0.3	10.8±0.2	≈	11.4±0.3	+6%
ES	11.1±0.3	10.7±0.6	-4%	13.9±0.4	+30%

Table 2: Dynamic elastic modulus of samples subjected to salt crystallization cycles

Ed [GPa]			
	Before salt weathering cycles	After 10 cycles and desalination	ΔEd
Untreated	18.1±0.4	14.2±0.3	-22%
3M	18.1±0.4	16.0±1.1	-12%
ES	17.4±0.3	13.6±0.7	-22%

Table 3: Durability of treated samples in terms of dynamic elastic modulus

amount of silica is contained in the stone itself.

The efficacy of the treatments was evaluated by dynamic elastic modulus *(see Table 2)*. Weathering resulted mainly in the pulverization and consequent detachment of the external layers of the samples, but did not cause significant decrease in the elastic modulus of the non-detached parts. None of the HAP-based treatments provides dramatic increases in dynamic elastic modulus compared to the values after weathering. The increase is higher for samples treated by ethyl silicate, which, however, exhibit the highest amount of residual salts in the pores, which could cause some pore occlusion resulting in increases in Ed not ascribable to a consolidating effect. However, the values after treatment exceed those of the unweathered specimens, thus indicating that low increases in dynamic elastic modulus are registered because the stone was not severely damaged by weathering cycles, which were aimed at investigating the effects of salt contamination rather than causing significant damage to the stone. However, because phosphate formation was found to occur also deep in the samples *(Figure 3)* and because only HAP and OCP and no other CaP phases were detected by FT-IR *(Figure 2)*, it can be concluded that no pore occlusion occurred, which would prevent the consolidant from penetrating deeply into the stone, and no formation of metastable phases took place as a result of salts contamination, which were the main concerns regarding the application on salt weathered stone. This implies that higher increases in mechanical properties can be obtained for more severely damaged samples (experiments are in progress).

To better investigate the effect of the 3M treatment compared to ES, their capability of increasing stone durability towards further weathering cycles was evaluated *(results in Table 3)*. Untreated specimens were also considered for comparison's sake. As can be seen in Table 3, 3M samples experience the lowest decrease in dynamic elastic modulus after the cycles, thus indicating that this treatment has the best durability. This also indicates that, despite apparent low efficacy obtained for the cores (caused by limited weathering level of the specimens), still this treatment is promising as it enhances the stone durability towards salt weathering better than ethyl silicate does.

4. Conclusions

The application of two HAP-based treatments was tested on Lecce Stone contaminated by sodium sulphate. The following conclusions can be derived:

1) The presence of salt in the stone does not prevent penetration of the HAP-based consolidants nor cause formation of phosphate salts instead of calcium phosphates;

2) Depending on the amount of salts contained in the stone, either HAP or OCP are obtained, both suitable for consolidation of limestone;

3) Increasing levels of sodium sulphate affect the nature of the phases that form, but promote the formation of higher amounts of calcium phosphates. Instead, other ions contained in the stone (possibly magnesium) affect the nucleation of HAP, leading to a nonuniform coverage of the substrate, and to the preferential formation of HAP in uncontaminated shells.

4) Comparing the 0.1M and 3M treatments, the latter exhibits better efficacy, as it results in more calcium

phosphate formation. The 0.1M treatment, however, still guarantees sealing of cracks even in the depth of the sample, which suggests that optimization of this treatment is worth pursuing;

5) Treatment 3M is promising as it increases resistance of the stone to further salt weathering, whereas ethyl silicate treatment does not.

5. References

[1] Sassoni E., Graziani G., Franzoni E., An innovative phosphate-based consolidant for limestone. Part 1: Effectiveness and compatibility in comparison with ethyl silicate, Construction and Building Materials (102), (2016), 918-930.

[2] Sassoni E., Graziani G., Franzoni E., An innovative phosphate-based consolidant for limestone. Part 2. Construction and Building Materials (102), (2016), 931-942.

[3] Sassoni E., Naidu S., Scherer G. W., The use of hydroxyapatite as a new inorganic consolidant for damaged carbonate stones, Journal of Cultural Heritage, (12), (2011), 346-355.

[4] Supova M., Substituted hydroxyapatites for biomedical applications: A review, Ceramics International, (41), (2015), 9203-9231.

[5] Boanini E., Gazzano M., Bigi A., Ionic substitutions in calcium phosphates synthesized at low temperature, Acta Biomaterialia (6), (2010), 1882-1894.

[6] Sassoni E., Graziani G., Franzoni E., Repair of sugaring marble by ammonium

phosphate: comparison with ethyl silicate and ammonium oxalate and pilot application to historic artifact, Materials and Design 88 (2015) 1145-1157.

[7] Sassoni E., Graziani G., Franzoni E., Scherer G.W., Some Recent Findings On Marble Conservation By Aqueous Solutions Of Diammonium Hydrogen Phosphate, MRS Advances, DOI: 10.1557/adv.2017.45.

[8] Franzoni, E.; Graziani, G.; Sassoni, E., TEOS-based treatments for stone consolidation: acceleration of hydrolysis–condensation reactions by poulticing, Journal of Sol-gel Science and Technology, (74), (2015), 398 – 405.

[9] Tao J., FTIR and Raman Studies of Structure and Bonding in Mineral and Organic–Mineral Composites, Methods in Enzymology,(532) (2013), 533 – 556.

[10] Fort R., Alvarez de Buergo M., Gomez-Heras M., Vazquez Calco C., Heritage, weathering and conservation, Proceeedings of the International conference on heritage, weathering and conservation, 21-24 June 2006, Madrid, Spain.

[11] Rubio F., Rubio J., Oteo J.L., A FT-IR study of the hydrolysis of tetraethylortosilicate [ES], Spectroscopy Letters, (3), (1998), 199–219.

Salt reduction

Evaluation of desalination and restoration methods applied in Petra (Jordan)

Wanja Wedekind* and Helge Fischer
Applied Conservation Science (ACS), Goettingen/Berlin, Germany
* wwedekind@gmx.de

Abstract

One of the major causes for the deterioration of historical monuments made up of sandstones can be attributed to the circulation and evaporation of salt laden fluids percolating through the rock material, an aspect that has been either overlooked or neglected in previous restoration attempts, often with damaging consequences. Particularly the lack of consideration for the importance of a quantitative analysis of the salt content of the rock, has led to adverse effects in subsequent restoration attempts.

In the course of restoration of antique tombs no. 825 and 826 in the world heritage site of Petra the role of salt in the weathering process was fully taken into account and the restoration process executed accordingly.

With regard to the desalination two different methods were applied: the poultice method and the sprinkling method. The degree of desalination was in both cases intermittently measured until the concentration of the rock was evaluated to be low enough to initiate the restoration process with application of the selected restoration mortar. Its condition was then probed by long term successive inspections.

This study discusses the results and limits of both procedures and the techniques and methods applied in the course of the restoration of the two monuments.

Keywords: Petra, sandstone, salt weathering, desalination procedures, restoration mortar, restoration procedures, control of success

1. Introduction

The ancient city of Petra is situated south of the Dead Sea in the Kingdom of Jordan. It has gained international recognition when it was declared a UNESCO World Heritage Site in 1985 for its unique, approximately 650 relatively well preserved facades carved out of the sandstone bedrock. This magnificent monumental rock architecture, however, is in danger of severe and accelerating disintegration due to various forms weathering and neglect.

In order to halt or slow down the observable decay of these monumental structures and preserve them for future generations, a project was devised and executed aiming at establishing a Jordanian institution capable of independently handling all aspects of the restoration effort on a continuous and permanent basis. The result was the establishment of an institute that is now known as the "Conservation and Restoration Center in Petra (CARCIP)". The project was managed by one of the authors of this paper, funded by the German government and executed within the framework of the "German International Cooperation Agency (GIZ)" and the "Department of Antiquities (DOA) of the Jordanian government in the years 1993 to 2002. From its start the project furthermore secured the support of the "Bavarian State Conservation Office (BLfD). It was for most of its duration also under the patronage of His Roy-

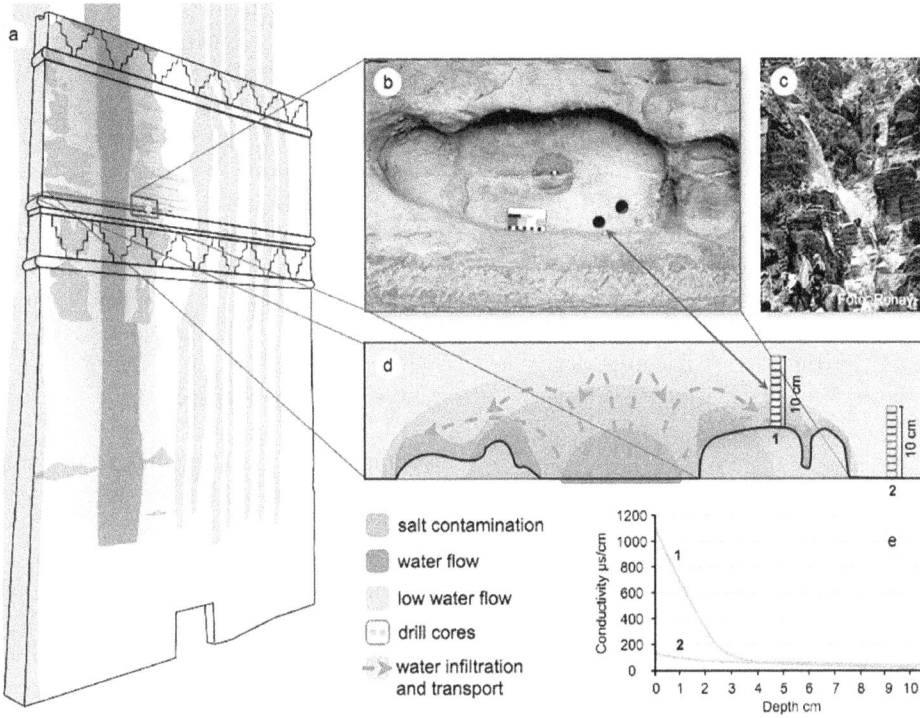

Figure 1: a) Distribution of alveolar weathering and traces where the rainwater drains at façade no. 826. b) Tafoni 3 with sampling point of the crust and drill core holes. c) Cascade-like water stains after a rainfall. d) Idealized cross section of the sampling area. e) Electrical conductivity at depth in the drill core samples.

al Highness Prince Hassan bin Talal, the then Crown Prince of the kingdom.

Scientific investigation and research continued intermittently from the year 2000 to 2016 by the University of Goettingen (GZG), the University of Applied Arts and Sciences of Hildesheim (HAWK), and the private institution Applied Conservation Science (ACS). This work focused primarily on the role of salt (NaCl) in the weathering process, its impact on the behaviour of restoration mortars, as well as the development of efficient and safe desalination procedures.

Particularly on the basis of these continuing activities, new methods in the conservation effort, as well as new restoration materials were devised and implemented by the authors and presented in this paper.

2. Environmental conditions

2.1. Climate

Petra is located in a semiarid climatic zone with a mean annual rainfall of 190 mm.[1] Heavy rainfalls may occur during the winter months, whereas the winters are cold and wet. Due to the vicinity to the Mediterranean Sea lying some hundred kilometres to the west and winds blowing primarily from that direction a high load of soluble elements are contained in the rainwater.[2]

2.2. Rock material

The tomb facades of Petra were chiselled out of the reddish sandstones of the Cambrian Umm Ishrin Formation. In detail colours vary from reddish-brown to yellowish-brown, gray or white. The brown colour is attributed to the deposition of limonite ($FeO.OH.nH_2O$) in the pores during formation of the sandstones, while the reddish colours stem from the precipitation of hematite (Fe_2O_3) flakes in the pores. The different lithological layers of the sandstone are medium- to fine-grained and show porosities ranging from 4.2 % to 20.6 %. Average values are 13 %.[3] In contrast, the limonite rich varieties show a low porosity of 7%.[4]

3. Weathering processes and agents

3.1. Weathering in arid and semi-arid environments

Areas in arid or semi-arid climatic zones like Petra show an increasing amount of salt deposition.[5] Therefore, weathering due to salts plays a crucial role.

Weathering phenomena observed in the Cambrian sandstones in Petra are mainly caused by salt crystallization on the retreating surface of the rock, due to evaporation of percolating saline water reaching the surface. This process constitutes the most important factor for the destruction of built and rock-cut architecture in Petra. [6, 7, 8, 9, 10, 11]

The resulting formation of tafoni comprise locally limited, as well as completely deteriorated areas, their size ranging from more centimetres to of several meters in diameter *(Figure 1b)*. This kind of extreme deterioration is generally surrounded by mostly undamaged rock.

3.2. Weathering processes

Comparative examinations of the sandstone monuments of Petra have shown that two varieties of tafoni development can be distinguished. Damaged spots evolve either as horizontal or vertical tafoni *(Figure 2)*.

Horizontal tafoni are usually found above the microporous limonite layers which in turn are frequently observed parallel to sedimentation layers. This is to be expected due to the blocking and accumulation of seeping aquifer water above the less permeable limonite layers.

Vertical tafoni develop left and right of areas affected by rainwater flowing down

alveolar and tafoni weathering
extensive weathering
limonite vains
salt crusts

Monument no. 825   Monument no. 826

alveolar and tafoni weathering
extensive weathering
limonite vains

Figure 2: The treated monuments. a) Mapping by the CARCIP team. b) Monument no. 825 and 826 and c) mapping of monument no. 826.

the surface during heavy rainfalls. *(Figures 2c, 6b)*

4. Description of the tombs no. 825 and 826

The tombs no. 825 and 826 are located beside each other along the southern tip of the so-called "Kings Wall" *(Figure 2 b)*.

Tomb no. 825 is on the left and is also known as the tomb of the 14 graves *(Figure 2 b)*. It was evidently built after the construction of the water pipes in the second century within the rock. It is 19 meters high and 9 meters wide.

Tomb no. 826 has a double tin frieze *(Figure 2b)*. It is of the Assyrian type and 21 m tall. The maximum width of the pedestal part is 12 meters, whereas the top part is 10 meters wide. The monument's actual ground level is invisible because of submersion in rubble.

4.1. The monuments in relation to encountered weathering processes

Tomb no. 825 is affected by seeping aquifer water moving through the massive rock, whereas tomb no. 826 is affected by run off water. Therefore, at tomb 825 horizontal oriented alveoles are created, whereas at monument no. 826 mainly vertical ones are observed *(Figure 2)*.

5. Evaluation of restoration methods and materials

5.1. The "Silica sol repair mortar"

The mortar chosen for restoration on monument 825 was a silica-sol mortar, based on good experiences made by the BLfD on several sites under restoration in Bavaria, where it was applied on sandstones. After promising results of its application in a Nabatean quarry the project

decided to start its use on monument no. 825.

There, however, surprisingly it soon turned out to respond in a rather unexpected and unfavourable way. The material started to crumble, sometimes already a short time after application. Careful evaluation of the problem led to the conclusion that the rapid disintegration of the mortar was due to salt contamination of the rock onto which the mortar was applied.

This can also easily be explained by the fact, that Silica-sols are colloidal solutions of SiO_2 in water that are negatively charged.[12] The negative charges are balanced and stabilized with a defined amount of positively charged ions, namely Na+. After application of the mortar, the Na-ions migrate from the rock into the mortar leading to a destabilization of the silica-sol binder. *(Figure 9b)*. Sometimes even salt efflorescence appeared on the surface of the freshly applied mortar. In consequence, collapsing of the mortar took place *(Figure 9 a)*. It thus turned out that salt reduction was of prime importance in any further attempt to apply the chosen mortar.

5.2. Salt extraction by application of poultices

The negative results experienced with the repair mortar on monument 825 stressed the necessity to devise a method to effectively desalinate the sandstone before its renewed application poultices. One method to reduce the salt load was to chisel off strongly salted areas.[13] The method chosen, however, was salt extraction by application of poultices. In first trials, clay based poultices were used. These turned out to be unsuitable as they clinged to the suface and upon removal would even sometimes destroy delicate Nabatean chisel marks. They were finally replaced by poultices developed by the

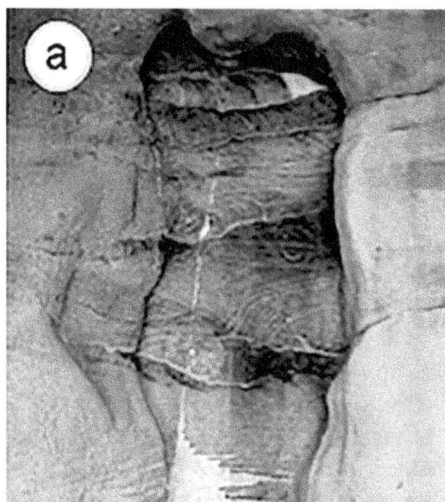

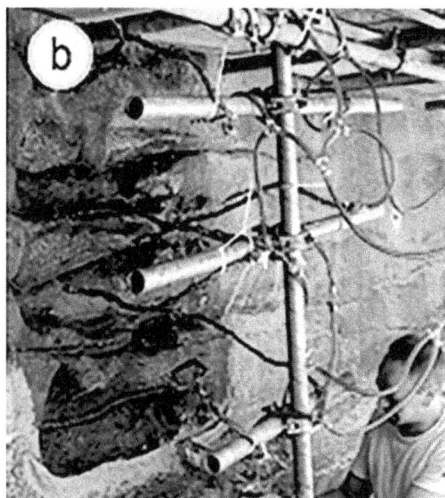

Figure 3: Procedure of desalination by the sprinkling method. a) Tafoni no. 9 at monument no. 826. b) The installed sprinkling device, sprinkling and collection of the eluate. c) Measurement of the collected eluate. d) Application of the poultice and e) sampling of the poultice.

conservator Egon Kaiser and one of the authors. These consisted of a wet mixture of cellulose and washed sand in a proportion of 1:5, and in general, showed a good workability as well as promising results in the beginning. The moist poultices were applied in square patches on the salt contaminated areas and were removed after drying *(see Figure 4d)*.

However, even after the application of several cycles of poultice application salt reduction turned out to still be insufficient, as the silica sol repair mortar would not bind as expected in some alveolar areas *(Figure 9b)*.

Obviously the efficiency of the desalination process needed to be further improved. In order to find a proper solution, a systematic quantitative analysis of the progress of desalination after each individual cycle of poultice application was undertaken in defined sample areas. Thus samples from the poultices were taken after each cycle, dissolved in distilled water and their relative salt content measured by electrical conductivity *(Figure 6a)*. But even after up to 17 cycles of poultice application an adequate improvement of the behaviour of the silica sol mortar could not be observed. The results thus indicated that the chosen method of salt extraction was unfeasible, not only with respect of its efficiency but also for the sheer enormity of the restoration effort, considering the size of the monuments. The poultice application method thus might work well on smaller, isolated objects, that are cut off from the water cycle encountered in objects that exposed to geological processes.

Nevertheless, the restoration of monument 825 was finalized and considered a success, keeping in mind that on large restored areas of its surface the salt load was rather small, no damage was done to the monument and that all in all the final result was satisfactory, though at a very high price that stood in no relation to the success. At least it can be said that

it was a valuable learning experience. It should also be stressed in this connection that the main focus of the project was to train local Jordanian staff in its ability to manage and master a wide range of techniques employed in the conservation and restoration effort, in accordance with the highest international standards, including aspects such as planning, documentation, assessment procedures, site preparation, the use of heavy and specialized equipment, as well as the mastering of very specialized skills like surveying, core drilling or the use of analytical tools.

Towards the end of the project, however, the authors concluded that a new method for the efficient desalination of the monuments in Petra had to be developed and that the approach chosen so far and still supported by the project advisor [14]. Though possible alternative methods had already been suggested by the authors, these had been outright rejected by the project advisors.

5.3. The sprinkling method

After finalizing restoration monument 825, work was shifted to the adjacent monument no. 826, which showed rather different types of degradation and thus presented new challenges. Like before the different weathering forms were mapped, quantified and documented with other relevant observations. Additionally electrical conductivity measurements were carried out on selected surface areas using a portable measuring device (type: protimeter). Thereupon a number of samples were taken from defined areas in selected alveolae. First samples of crust material, then samples of poultices applied to the very same areas *(Figure 4e)*. Finally two drill cores were taken from these areas and also investigated *(Figures 2a, d, e)*. Different ions within the stone material of the drill cores were identified in different sections by ion-chromatography.

The totally new approach in the restoration effort here, however, rests in the introduction desalination method devised and implemented by the authors. This approach was inspired by previous observations on the development of vertical tafoni only adjacent to the areas that were marked by run off water, a process that had also been observed on some rare occasions of heavy rainfall. This clearly indicated that rainwater washes out salt accumulations on and under the rock surface. This observation then led to the development of a method, that mimicked the natural process of desalination. Thus a device was put into operation that would sprinkle water under controlled conditions onto the surface of the salt affected surfaces. This procedure was then labelled the sprinkling method.

By this method water is sprayed onto the stone wall surface through fine nozzles *(Figure 3b)*. At the start of this procedure water is predominantly absorbed by the porous stone surface through capillary forces. Water absorption is dependent upon the transport properties of the rock and these are essentially controlled by the pore space properties, like porosity and pore radii distribution and are a time-dependent process.[15]

The excess water not absorbed by the stone runs off the treatment area and is collected at the bottom in 1l containers *(Figure 3c)*. Electrical conductivity of the eluate of each sample container is then measured thus giving an indication of the amount and concentration of the dissolved substances *(Figure 3c)*. After the sprinkling a poultice of the same composition as used in previous applications was applied onto the treated area *(Figure 3d)* in order to detect if additional salt would be extracted that way. A sample taken therefrom *(Figure 3e)* was dissolved and tested by electrical conductivity,

indicating no significant further salt extraction.

The spray pressure applied to the stone during the sprinkling process is very low. In addition, this is a particularly suitable method in situations where the prevailing climate is very warm due to the rapid evaporation rates there. Thus in Petra the water soaked up by the sandstones normally evaporates within a few days. After each subsequent washing cycle the amount of salt extracted from the surface was, as expected, observed to continuously decrease. Eventually salt concentrations decrease to a value, where mortars can be safely applied.

Measurements on the drill cores taken before and after desalination give a quantitative indication of the amount of salt dissolved from the rock.

It should be mentioned in this connection, however, that in spite of the fact, that salt contamination is efficiently reduced in areas treated by this method, side effects could be observed in some instances in the immediate vicinity to the not treated adjacent areas. Obviously some of the salty solutions would migrate towards the fringe of the treated are, a side effect that was actually observed. Though the effects are minor, we are aware that they are of concern. It must be stated here, however, that such effects cannot be precluded in situations, where the treated rock is part of a huge geological entity.

It may be added here, that similar side effects were also encountered at the fringes of poultices that were used in the early stages of the project for salt extraction.

6. Results

It was realized already at an early stage of application of the sprinkling method, that it was far more efficient, economical and less time consuming than the me-

Tafoni	Salt content of the crust (g)	Salt content in the first two poultices (g)	Salt content in the eluate by sprinkling (g)	Number of collected litres
1	0.33	0.43	-	-
2	0.04	19.5	100.4	68
3	0.25	13.3	91.3	105
4	0.36	0.4	-	-
5	0.47	17	10.7	73
6	2.81	25	145.6	220
7	0.89	20.5	114.1	145
8	0.37	18.3	396.7	349
9	0.17	20.7	133.7	145

Table 1: Evaluation of the near-surface salt contamination and desalination at monument no. 826

Figure 4: a) Ongoing damage progress of restored areas treated with silica-sol mortars during a period of 15 years on monument 825. b) Sample of a dry-slaked lime mortar just after application in 2006 and c) after 10 years.

thods previously employed.

In order to demonstrate the effectiveness of desalination by use of the sprinkling method, data collected from monument 826 are shown as an example in *Table 1*.

The table indicates that the highest degree of desalination by the sprinkling method was encountered in tafoni 8, the lowest one in tafoni 5. In total around 1000 g of solvent material was extracted from all tafoni areas by sprinkling. Tafoni 1 and 4 only show a very low salt content within the first two poultices, and therefore were not treated by sprinkling. However, from tafoni 2, 3, 6, 7 and 9 an amount of solvent material of around 100 g to 150 g were extracted.

While five years after restoration only a few damages in the restored areas could be observed on both monuments, around 10 years later almost all mortar applications were affected by crumbling on monument 825 *(Figure 4a)*.

As can be seen on monument no. 826, only very slight signs of degradation could be observed 10 years after the conservation treatment. *(Figure 4a)*.

Interestingly, a lime mortar newly devised by one of the authors also exhibits no signs of alteration after 10 years *(Figures 4b and c)*, indicating that there is also room for the development of alternative the repair mortars. Results of this potentially also far more economic effort will be presented in a separate paper.

7. Summary and conclusions

The sprinkling method for desalinating natural stone in Petra proved to be a tool that can effectively and easily reduce the salt load of areas affected by salt contamination, while the quantity of salt extracted can be continuously monitored at the same time. Quantitative analysis of the degree of desalination also provides better insights into the weathering processes of the individual buildings.

Taking into account the particular conditions encountered in Petra, where e.g. the extended presence of a scaffold structure in front of a monument can be an issue, the use of the more efficient and faster sprinkling method for desalination constitutes an additional advantage. Fifteen years after the application of silica-sol-mortar in areas treated this way, the mortars can still be seen to exhibit almost no sign of degradation, in contrast to the observations made on monument 825 treated with the previous method as shown in *Figure 4*.

A faster degradation of the repair mortar is clearly visible on monument no. 825, as compared to monument 826.

References

[1] Wedekind, W, Jordan - Petra. In Truscott M, Petzet, M, Ziesemer J, Heritage at Risk – ICOMOS World Report 2004/2005 on Monuments and Sites in Danger, Munich, 2005.

[2] Kuehlenthal M, Fischer H (2000) Petra, Die Restaurierung der Grabfassaden, The Restoration of the Rockcut Tomb Facades, Munich, 2000.

[1] Gaudi A, Viles H, Salt Weathering Hazards. John Wiley & Sons, 1997.

[2] Al-Khashman, O.A., Jaradat, A.Q., Salameh, E. Five-year monitoring study of chemical characteristics of wet atmospheric precipitation in the southern region of Jordan. Environ Monit Assess (185), 2013, 5715–5727

[3] Heinrichs K, Diagnose der Verwitterungsschäden an den Felsmonumenten der antiken Stadt Petra / Jordanien, Aa-

chener Geowissenschaftliche Beiträge (Value 4) Wissenschaftsverlag Mainz, Aachen, 2005.

[4] Wedekind W, Weathering and Conservation of monuments constructed from tuff and sandstone in different environment, PhD-thesis, Goettingen, 2016.

[5] Wellmann HW, Wilson AT, Salt weathering, neglected geological erosive agent in coastal and arid environments. Nature (205), 1965, 1097-1098.

[6] Wedekind W, Ruedrich J, Salt-weathering, conservation techniques and strategies to protect the rock cut facades in Petra/Jordan. In: Fort R., Álvarez de Buergo M., Gomez-Heras M. & Vazquez-Calvo C. (eds.). Heritage, Weathering and Conservation. Taylor & Francis, London, 2006, 261-268.

[7] Abd El-Hady M. The deterioration of Nubian sandstone blocks in the ptolemaic temples in Upper Egypt. Proceedings of the 9th International Congress on Deterioration and Conservation of Stone. Venice, June 19-24. Elsevier, 2000, volume 2, 783 – 792.

[8] Bala'awi, F, Wind speed and salt simulation tests: Towards a more comprehensive approach. In: Salt weathering on buildings and stone sculptures (SWBSS): proceedings from the international conference, The National museum, Copenhagen, Denmark, 22-24 October 2008, 41-50.

[9] Eklund S, (2008) Stone weathering in the monastic building complex on Mountain of St. Aaron in Petra, Jordan. Master of Arts Thesis. University of Helsinki.

[10] Gomez-Heras M, Lopez-Arce P, Bala'awi F, Vazquez-Calvo C, Fort R, Ishakat F, Alvarez de Buergo M, Allawneh F, Characterisation of salt combinations found at the 'Silk Tomb'(Petra, Jordan) and their possible source. In: Ioannou, I., Theodoridou, M. Salt Weathering on Buildings and Stone Sculptures (SWBSS), Limassol, 2011, 81-88.

[11] Heinrichs K, Diagnosis of weathering damage on rock-cut monuments in Petra, Jordan. Environmental Geology, volume 56, issue 3, 2008, 643-675.

[12] Snethlage R, Testing and Evaluation of Stone Repair Materials in Petra. In: Kuehlenthal Mn Fischer H (eds.) Petra – The Restoration of the Rockcut Tomb Facades, Arbeitsheft Bayer Landesdenkmalamt f Denkmalpflege, Munich, 2000, 184-190.

[13] Kuelenthal M, Guidelines and Procedures for the Restoration of the Monuments in Petra. In: Kuehlenthal M, Fischer H (eds.) Petra – The Restoration of the Rockcut Tomb Facades, Arbeitsheft Bayer Landesdenkmalamt f Denkmalpflege, Munich, 2000, 87-88.

[14] Kuehlenthal M, Bavarian State Conservation Office, Assesment and restoration concept of Monument 826, intern project report, CARCIP, 2000.

[15] Wittmann FH, Feuchtigkeitstransport in porösen Werkstoffen des Bauwesens. In: Goretzki L. (ed.): Verfahren zum Entsalzen von Naturstein, Mauerwerk und Putz. Feiburg, 1996, 6-16.

Salt extraction by poulticing in the archaeological site of Coudenberg

Sebastiaan Godts*, Roald Hayen and Hilde De Clercq
Royal Institute for Cultural Heritage (KIK-IRPA), Brussels, Belgium
*sebastiaan.godts@kikirpa.be

Abstract

A common method for the extraction of salts is the use of poultice materials. Despite extensive scientific research, the salt extraction process in practice often remains empiric. This paper aims to further the understanding of a salt extraction by poulticing in practical experiments. Three types of poultice materials are tested on brick masonry in 12 different locations in the subterranean archaeological site of Coudenberg in Brussels. The tested poultice materials have different compositions including kaolin clay, sand and/or cellulose fibres, theoretically favouring advection and/or diffusion. It is shown that for an accurate interpretation of the results it is crucial to evaluate the salt content in the poultice and to a depth of at least 20 cm in the different materials of the substrate, while considering the different densities and surface area of each material. Furthermore, it is shown that the presence of gypsum overshadows the results, hence the exclusion of equimolair contents of Ca^{2+} and SO_4^{2-} is essential for the evaluation. Finally, the results of these experiments reveal that regardless of the different procedures, such as, the poultice type, pre-wetting or application time, salts are being redistributed into the wall rather than extracted.

Keywords: salt extraction, desalination, poultice, historic masonry, archaeological site

1. Introduction

The archaeological site of Coudenberg, the last material evidence of the former Palace of Brussels, is situated at the heart of Belgium's capital underneath the Royal Square and its 19th century buildings. The project is the culmination of twenty-five years of archaeological work, which have progressively enabled different areas of the site to become accessible to the public. Not all the materials of the site have been preserved in the same condition and the project faces considerable challenges in both conservation and presentation considering the relatively recent exposure from burial.[1] Major water infiltrations and salt damage are now visible in a large part of the site.

The presence of salts in the subterranean site is essentially caused by water and ion transport that occurred over an extended period of time. During this time salts accumulated unevenly at the evaporation front, near the surface at certain heights and preferentially in the materials with the smallest pores. Damage phenomena are naturally explained by the accumulation of salts and the occurrence of crystallisation cycles. These cycles are promoted by changes in relative humidity (RH). Daily RH changes typically influence the first centimetres and diminish rapidly in depth of the substrate, with crystallisation cycles occurring less frequently further in the substrate from daily, weekly, monthly to seasonally. The amount of cycles that influence the first centimetres can easily run up to more than a 100 cycles per year.[2] Hence, visible salt damage is seen at certain heights or in specific regions, with mainly the first centimetre of a substrate being affected. Lowering the salt content near the surface and diminishing daily changes in RH both in intensity and recurrence will mitigate salt damage, contributing to a sustainable preservation and an appropriate presentation for visitors.

While the removal of salts from movable stone objects by means of the water bath method is quite well known, in-situ salt extraction of brick masonry by means of poultices is often carried out on the basis of general assumptions, and in most cases without control of the efficiency or the effect in the depth of the substrate. This paper goes deeper into the practice of salt extraction on brick masonry using three different poultice types, focusing on the influence of some of the execution parameters on the salt extraction efficiency.

2. Materials and methods

2.1. Sampling, salt analysis and efficiency evaluation

Before and after each poultice application samples are taken from the mortar and brick at different depths. Samples are obtained by powder drilling (Ø8 mm) in intermediate steps, 0-1, 1-3, 3-5, 5-10, 10-15 and 15-20 cm, to obtain an in-depth moisture and salt profile. For each sample, the actual and the hygroscopic moisture contents are determined gravimetrically. The quantity of anions (Cl^-, NO_3^-, and SO_4^{2-}) and cations (Na^+, K^+, Ca^{2+}, and Mg^{2+}) of the filtered aqueous extract is measured by ion chromatography (IC, Metrohm). The theoretical amount of carbonates (CO_3^{2-}) is determined from the excess of analysed cations. The results are presented as kg•m^{-2} (per sampled depth), based on the specific density (drilled powder) of the dry mortar/brick.

The density, the porosity and the pore size distribution of the dry poultice materials, brick and mortar are determined by mercury intrusion porosimetry (MIP, Autopore IV, Micromeritics). From the results the theoretical properties of the poultice, that is, the advective or diffusive properties are determined.

The evaluation of the poultice efficiency and the salt extraction/migration processes is based on a comparison of the evolution of the salt content in the depth of the masonry and in the poultice, before and after salt extraction treatment. The total amount of ions, including or excluding equimolar amounts of Ca^{2+} and SO_4^{2-} (reflecting the gypsum content), in kg.m^{-2} (per sample depth) are considered during the evaluation. The salt content at a certain sampling depth in the masonry (brick and mortar) is determined as follows:

$$\varphi_{brick} \cdot x_{brick} \cdot d \cdot A_{brick} + \varphi_{mortar} \cdot x_{mortar} \cdot d \cdot A_{mortar}$$

(I)

With φ_{brick} and φ_{mortar}, respectively the mass density (specific density) of the brick and mortar (in kg•m^{-3}), x_{brick} and x_{mortar}, respectively the total content of soluble ions in each material, excluding equimolar amounts of Ca^{2+} and SO_4^{2-} (in wt%), d, the sampling depth (in meter) and Abrick and Amortar, the proportional surface area (in m²/m² wall surface) of respectively brick and mortar. The total amount of salts in the sampled volume is obtained by adding the amount of salts found at every sampling depth for both brick and mortar. The result is expressed as the amount of salts per unit of surface area (kg•m^{-2}). The comparison of the salt content in the substrate after treatment with the initial salt content indicates the amount of salts which have disappeared out of the sampled volume and which either have been absorbed by the poultice or either migrated away within the substrate to a greater depth or outside of

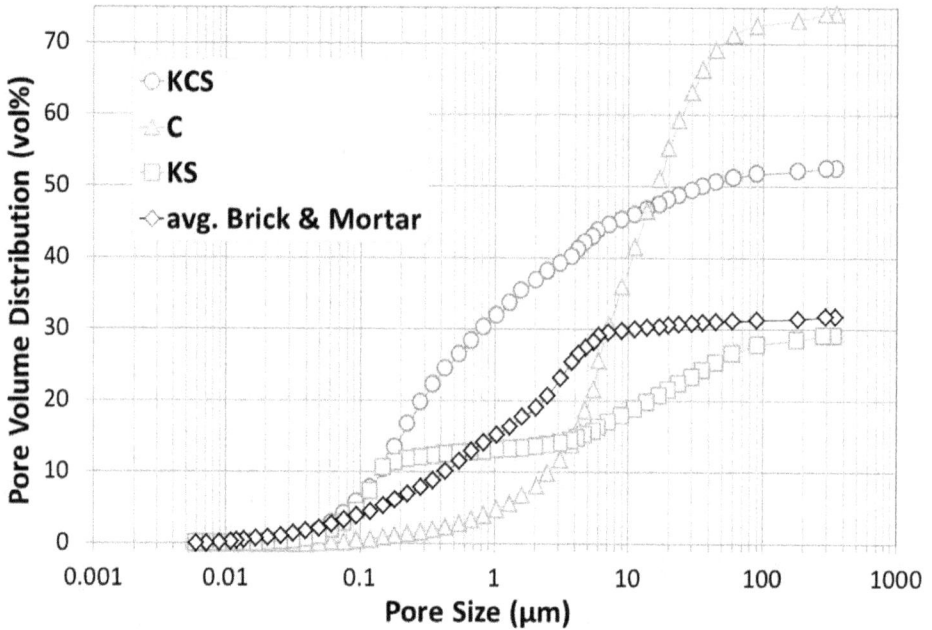

Figure 1: Cumulative pore volume distribution (vol%) of the dried poultice materials KCS (o), C (Δ) and KS (□) compared to the avg. pore volume distribution of the brick and mortar from 7 locations (◊).

the sampled area. As only the salts absorbed by the poultice are effectively extracted, the amount of salts disappeared out of the sampled volume is compared to the amount of salts detected within the poultice (kg·m⁻²), determined as follows:

$$\varphi_{poultice} \cdot x_{poultice} \cdot t$$

(II)

With $\varphi_{poultice}$ the mass density (bulk density) of the poultice (in kg·m⁻³), $x_{poultice}$ the total content of soluble ions in the poultice, excluding equimolar amounts of Ca^{2+} and SO_4^{2-} (in wt%) and t the total thickness of the poultice (0.01 m). The efficiency of the salt extraction process is determined as the amount of salts that have been absorbed by the poultice in comparison to the initial salt content in the masonry, expressed in %.

2.2. Poultice materials and application procedures

The following poultice materials have been tested:

Poultice KCS (Saltpull type 1, Rewah) is a poultice based on the findings described by Lubelli & van Hees [3, 4] and was selected for the purpose of this project as it has enough small pores compared to the substrate to theoretically favour an advective flow towards the poultice *(Figure 1)*. The poultice contains cellulose fibres (Arbocel PZ8 1,4 mm), kaolin clay (IMERYS quality China Clay Speswhite) and calibrated sand (Sibelco CEN196 sand fraction 0,5-1 mm) (1:2:1 by weight). The poultice is mixed with water (61 wt% H_2O for 39 wt% of dry poultice material) and contains 0,2% biocide.

Poultice C (Cellulose Saltpull, Rewah) is a commonly and traditionally used poultice in the restoration practice and was selected as it has enough large pores compared to the substrate to theoretically favour a diffusive flow towards the poultice *(Figure 1)*. The poultice contains cellulose fibres (Arbocel PZ8 1,4 mm). The poultice is mixed with water (78 wt% H2O for 22 wt% of dry poultice material) and contains 0,2% biocide.

Poultice KS is based on a poultice described and recommended by Bourgès & Vergès Belmin.[5,6] This poultice has enough small and large pores compared to the substrate to theoretically favour both an advective and a diffusive flow towards the poultice *(Figure 1)*. The poultice contains kaolin clay (IMERYS quality China Clay Speswhite) and calibrated sand (Sibelco CEN196 sand fraction 0,5-1 mm), 1:5 by weight (0.8:1 by volume). The poultice is mixed with water (20 wt% H_2O for 80 wt% of dry poultice material).

The three different poultices (KCS, C and KS) are tested on brick masonry in a total of 12 different locations (1-12). Locations 1, 6, 10 and 11 are free standing walls suffering from rising damp. Locations 2, 3, 7 and 8 are located on vertical earth retaining walls, while locations 4, 5, 9 and 12 are situated on vaults that are presumably in contact with earth. The main salts, excluding gypsum, found within the walls are identified as sodium chloride (NaCl) in locations 1, 2, 4, 5, 6, 8, 9, 10 and 12, together with calcium nitrate ($Ca(NO_3)_2$) in locations 4, 5, 8 and 9. Sodium sulfate ($NaSO_4$) is found in locations 3, 7 and 11, presumably in combination with smaller amounts of sodium carbonate (Na_2CO_3), as theoretically derived from the excess of sodium ions in these locations.

Poultices KCS, C and KS are applied to the salt contaminated masonry in respectively 5 (1-5), 4 (6-9) and 3 locations (10-12), and removed after respectively 11, 19 and 47 days. Before each poultice application samples are taken from the substrate (as described earlier). Before application of poultices KCS and C the

masonry is pre-wetted with a known amount (0.5 2 or 4 l·m-²) of demineralized water by spraying. Pre-wetting is carried out with intermediate steps to prevent runoff and allow maximal absorption. In the case of the application of poultice KS, the masonry is not pre-wetted. All poultices (KCS, C and KS) are applied to the masonry with a spatula in a thickness of approximately 1 cm on a surface of approximately 1 m². Poultice C is immediately covered with a water/vapour impermeable foil after application. The different poultice application conditions are based on experiences and traditions found in literature and discussions with practitioners. At the end of the experiments samples are taken from the poultice material and from the substrate as described earlier. For the purpose of this paper the results of the experiments are averaged per poultice type.

3. Results

In total 3 poultice types were tested in 12 locations with the results of each experiment showing considerable variations, making the interpretation rather challenging. However, several important results can be described looking at the individual test results (not shown) and at the averaged results per poultice type. Although tests were carried out with different amounts of water for pre-wetting the surface, no significant influence could be derived. Exceptionally, an irregular migration of salts was recorded to unknown locations in the substrate when pre-wetting the surface. Similar exceptions were derived when the substrate showed high moisture contents before the experiments; water added during the pre-wetting procedure or from the poultice itself provoked in such a case an oversaturation of the bedding mortar which allowed water and salts to migrate from the mortar towards the brick. Where the moisture content was lower before the experiments this phenomenon was naturally seen reversed as the mortar in all locations has smaller pores than the brick, and hence tends to absorb rather than expel its pore water (and salts).

3.1. Poultice KCS

The average results of the five experiments with poultice KCS are shown in *figure 2*. The amount of salts, excluding gypsum ($CaSO_4$), that have disappeared from the masonry (0-20 cm) after the poultice application is 4% of the total salt content before application, whereas in the case the presence of gypsum is included an increase of the salt content with 14% is observed. The salt content, either excluding or including gypsum, that has migrated towards the poultice amounts to only 6 and 4%, respectively, of the total salt content before application. Hen-

ce, as compared to the total amount of salts in the substrate, the poultice treatment is not so efficient, certainly in areas where salts might migrate back to the surface. When considering the first cm of the substrate (0-1 cm) the results are easily misinterpreted as an efficiency of 54 and 35% is obtained. When looking at the results, excluding gypsum, a general decrease of the salt content in the wall is seen up to a depth of 10 cm. Some of these salts will have migrated to the poultice, however, the rest most probably diffused into the depth of the substrate as an increase of the salt content is observed from 15 cm onwards.

3.2. Poultice C

The average results of the four experiments with poultice C are shown in *figure 3*. The amount of salts, either excluding or including gypsum, that has disappe-

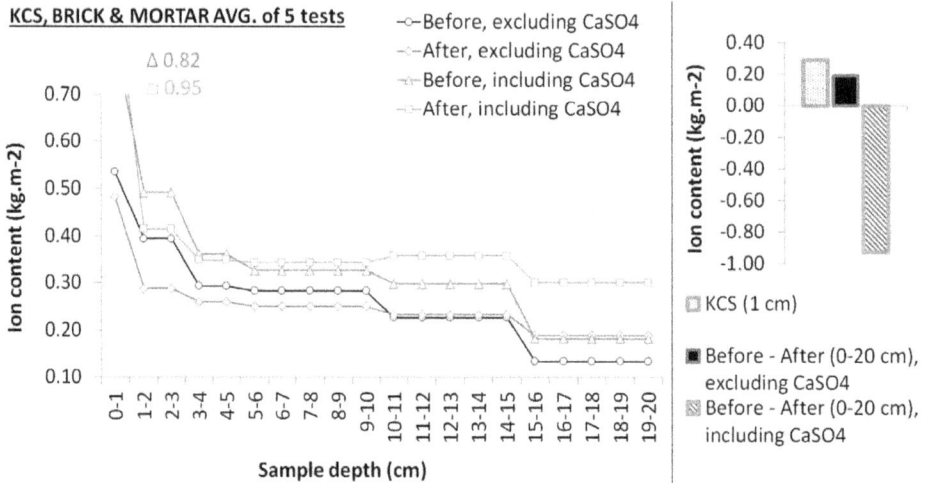

Figure 2: Poultice KCS, avg. of 5 locations (1-5). Left: Results of the avg. ion concentration (kg·m⁻², per cm), detected in the masonry, while taking the respective surface areas of the brick and mortar into account: before (o, black line) and after (◊, blue line) excl. CaSO₄, and again before (Δ, red line) and after (□, green line) incl. CaSO₄. The vertical bars on the right show respectively the avg. amount of salts deposited in the poultice (dotted and orange column, kg·m⁻²) and the avg. amount of salts, excl. CaSO₄, that have disappeared from the substrate (full and blue column, kg·m⁻², total of sampled depth 0-20 cm) and idem, incl. CaSO₄ (striped and green column). For the latter the negative value indicates an average increase of the salt content over the sampled volume.

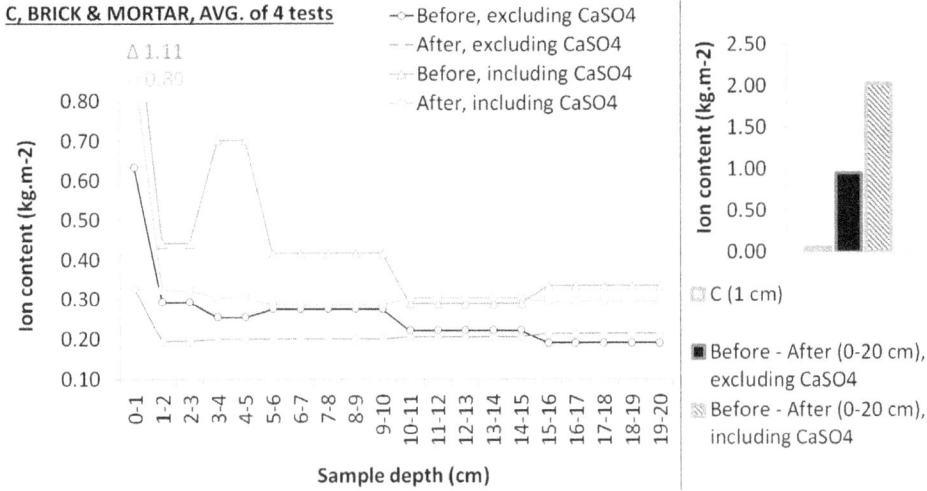

Figure 3: Poultice C, avg. of 4 locations (6-9). Left: Results of the avg. ion concentration (kg·m⁻², per cm), detected in the masonry, while taking the respective surface areas of the brick and mortar into account: before (○, black line) and after (◊, blue line) excl. CaSO₄ and again before (△, red line) and after (□, green line) incl. CaSO₄. The vertical bars on the right show respectively the avg. amount of salts deposited in the poultice (dotted and orange outlined column, kg·m⁻², total thickness of the poultice: 1 cm) and the avg. amount of salts, excl. CaSO₄, that have disappeared from the substrate (full and blue outlined column, kg·m⁻², total of sampled depth 0-20 cm) and idem, incl. CaSO₄ (striped and green outlined column, kg·m⁻², total of sampled depth 0-20 cm).

ared from the masonry (0-20 cm) after the poultice application is respectively 18 and 24%, while the amount of salts that migrated towards the poultice is in both cases only 1% of the total salt content before application. When considering the first cm of the substrate (0-1 cm) the results would indicate an efficiency of 8 and 5%. Although the efficiency concerning the extraction is slightly lower compared to poultice KCS, the salt content in the substrate, excluding gypsum, has decreased significantly further from the surface up to a depth of 15 cm. While practically none of these salts have migrated to the poultice, approximately 17% of them have disappeared, and therefor most likely migrated to lesser salt contaminated areas and into the depth (beyond 20 cm) of the substrate.

3.3. Poultice KS

The average results of the three experiments with poultice KS are shown in *figure 4*. The amount of salts, either excluding or including gypsum, that has disappeared from the masonry (0-20 cm) after the poultice application is respectively 14 and 27%, while the amount of salts that migrated towards the poultice is respectively only 4 and 3% of the total salt content before application. Taking into account the first cm of the substrate (0-1 cm) the calculated efficiency is instead 73 and 25%. A general decrease of the salt content, excluding gypsum, is seen up to a depth of 20 cm of which only a small amount have migrated to the poultice.

4. Conclusions

The results and interpretation of the extraction process show important differences when considering the gypsum content and different sampling depths in the substrate. An inaccurate interpretation is highly possible when including gypsum in the calculations as it is often

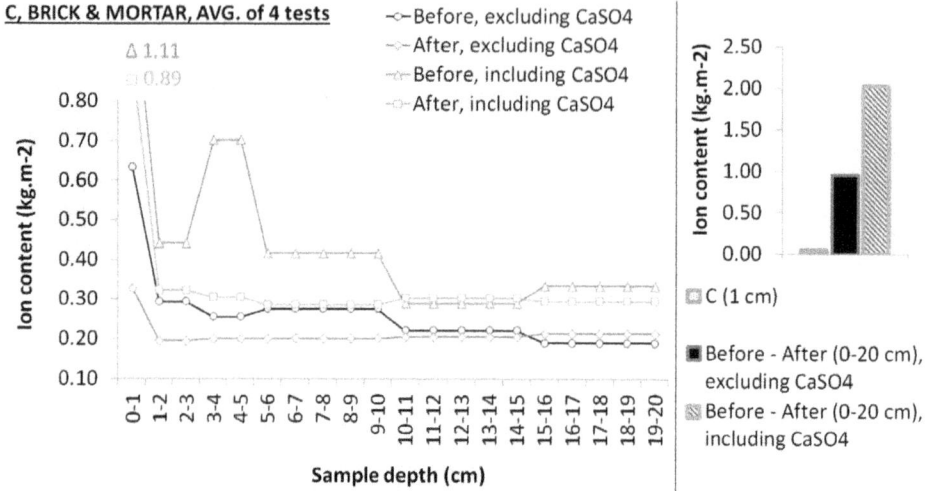

Figure 4: Poultice KS, avg. of 3 locations (10-12). Left: Results of the avg. ion concentration (kg·m⁻³, per cm), detected in the masonry, while taking the respective surface areas of the brick and mortar into account: before (○, black line) and after (◇, blue line) excl. CaSO₄ and again before (△, red line) and after (□, green line) incl. CaSO₄. The vertical bars on the right show respectively the avg. amount of salts deposited in the poultice (dotted and orange column, kg·m⁻²) and the avg. amount of salts, excl. CaSO₄, that have disappeared from the substrate (full and blue column, kg·m⁻², total of sampled depth 0-20 cm) and idem, incl. CaSO₄ (striped and green column).

unevenly distributed in the substrate and generally occurs in much higher concentrations as compared to other salts. As shown in the results these aspects, in combination with its low solubility (~2.0-2.5 g/l at 25 °C), overshadow the overall extraction process. When comparing the salt content in the poultice to the total salt content in the substrate (from 0-20 cm depth), the average results of the poultice applications indicate a rather low efficiency for the tested poultices, KCS, C and KS, with respectively an efficiency of only 6, 1 and 4% of salts extracted. When considering only the first centimetre of the substrate, extraction efficiency values of 54, 8 and 73% are obtained, which evidence the ease of misinterpretation of the salt extraction efficiency. Although the extraction of salts at the surface is of utmost importance, as it is the area mainly influenced by daily RH cycles and, hence, most likely subjected to severe damage, the significant influence of the poulticing proce-

dure on the migration processes further into the depth of the substrate forces an evaluation of the salt content up to greater depths. A general rule applies in that salts tend to migrate to further depths and to areas with lower salt contents.

The overall reduction of the salt content in the substrate to a depth of 20 cm is 4 and 14% for the poultices KCS and KS (favouring advection). Although the extraction of salts seems the least effective for poultice C, favouring diffusion, it is clear that the salt content in the sampled depth to 20 cm has decreased with 18%. Although hardly any salts are extracted, the results of this poultice show that the salt content is sufficiently low to prevent further salt damage at the surface, at least for the time being.

Surprisingly poultice C is the preferred poultice in this case, the decision is essentially based on practical concerns and its repeatability. As long as the subterranean site is subject to water infiltrations it can be assumed that any treatment will have a limited lifespan. Besides the ease of use of poultice C and a significant difference in cost concerning the raw materials, both poultices KCS and KS leave a white haze behind on the substrate, caused by the kaolin clay, which raises concerns regarding the aesthetical changes and loss of material during the subsequent cleaning of the surface, increasingly so when considering a repeated treatment.

Int Conference on Salt Weathering on Buildings and Stone Sculpture, ed. J.S. Albertsen, Copenhagen 2008.

[6] Bourgès, A. and Vergès Belmin, V., personal communication during the 13th International Congress on the Deterioration and Conservation of Stone, Glasgow 2016.

References

[1] Williams, T., Coudenberg, Brussels, Conservation and Management of Archaeological Sites (CMAS), (16-1), (2014), 1-3. http://dx.doi.org/10.1179/135050331 4Z.00000000074.

[2] Godts, S., Hayen, R and De Clercq, H., Investigating salt decay of stone materials related to the environment, a case study in the St. James church in Liège, Belgium, Studies in Conservation, (2016) DOI: 10.1080/00393630.2016.1236997.

[3] Lubelli, B., van Hees, R.P.J. and De Clercq, H, Fine tuning of desalination poultices, try-outs in practice, in proceedings of the 2nd Int Conference on Salt Weathering on Buildings and Stone Sculptures, eds. Ioannou, I. & Theodoridou, M., Limassol 2011, 381-388.

[4] Lubelli, B. and van Hees, R.P.J., Desalination of masonry structures: Fine tuning of pore size distribution of poultices to substrate properties, Journal of Cultural Heritage, (11), (2011), 10-18.

[5] Bourgès, A. and Vergès Belmin, V., Comparison and optimization of five desalination systems on inner walls of Saint Philibert Church in Dijon, France. 1st

Desalination of Cotta type Elbe sandstone with adapted poultices: Optimization of poultice mixtures

Julia Maitschke[1] and Heiner Siedel[2]*
[1] *University of Applied Sciences Potsdam, Dept. of Conservation and Restoration, Potsdam, Germany*
[2] *TU Dresden, Institute of Geotechnical Engineering, Chair of Applied Geology, Dresden, Germany*
*Heiner.Siedel@tu-dresden.de

Abstract

Cotta type Elbe sandstone has been frequently used as construction material and ornamental stone on buildings over centuries. Desalination of this sandstone (often performed with cellulose poultices in restoration practice) is ineffective in many cases due to its high amount of fine pores.

A mixture of cellulose, kaolin and sand (1:2:1 by weight, CKS_121) with a high portion of fine pores was applied for poultice desalination on artificially salt-loaded specimens of Cotta type sandstone. Moreover, another mixture of the same components (CKS_128 with cellulose-kaolin-sand 1:2:8 by weight) and two poultices that are frequently used in restoration practice (ready-made Rajasil (RAJ) poultice and Arbocel® CC1000/BWW40 (ARB) mixture) were applied for comparison. The pore size distributions and the structures of all poultices were characterized by MIP. The results showed highest amounts of extracted salt for the poultices with clay and higher portions of sand (CKS_128, RAJ). The CKS_121 poultice, although fitting best with regard to pore size distribution, shows high shrinkage due to the high amount of kaolin, which leads to loss of adhesion to the substrate during the desalination process. The results clearly demonstrate that in desalination practice of stones with fine pores compromises have to be found between the competitive parameters of "ideal" pore structure and low shrinkage of the poultice, both influenced by clay contents in the mixture.

Keywords: salt reduction, desalination, Cotta type Elbe sandstone, poultice mixtures

1. Introduction

Salt reduction or "desalination" with poultices is a well-known and widely used technique in conservation of natural stone buildings and sculptures.[1] Recent investigations into the transport phenomena of salt and water led to better general understanding of the interaction of poultice and substrate properties during the desalination process.[2-3] As a result, in current discussions of the poultice materials main attention is drawn to the role of an appropriate pore size distribution of the poultice with respect to the substrate material to be desalinated. Analyses of the pore size distribution of the substrate by mercury intrusion porosimetry (MIP) can give useful hints towards the pore size properties needed for the poultice to achieve an efficient desalination. Generally, the wet poultice should provide a portion of pores coarser then those in the substrate to offer enough water to the stone material in the wetting phase on one hand. On the other hand another portion of pores in the poultice finer then those in the stone substrate is needed to extract the water with dissolved salts from the stone by advection in the drying phase.[2] Appropriate structures can be designed in mixtures of cellulose, clay and sand by varying the proportion of the respective components.[3] These principles established in

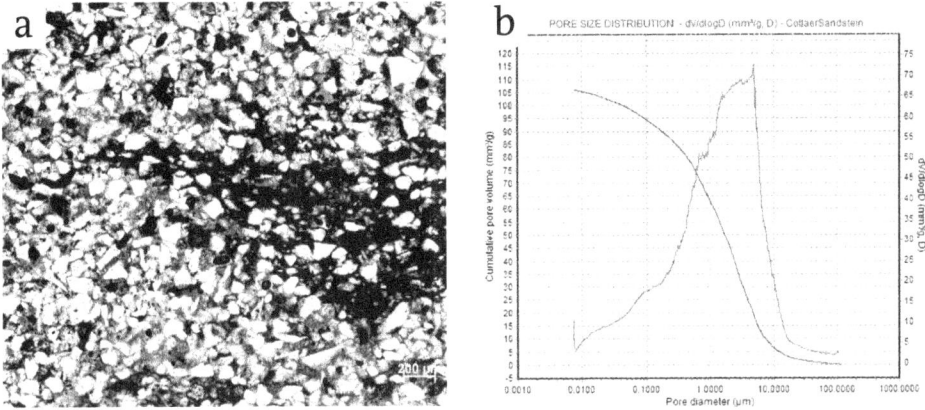

Figure 1: Thin section image of Cotta type sandstone (Nicols II) with pores coloured by blue dye (a) and typical pore size distribution obtained by MIP (b)

the laboratory have already been transferred to conservation practice[4] or tried out in the laboratory for local important building stones. [5]

From a practitioner's point of view properties of the poultice material like adhesion to the substrate, shrinkage while drying, and the removal of the poultice from the stone surface without remnants after drying are of importance beside efficiency. Higher clay contents in the poultice do not only provide the appropriate portion of fine pores needed for efficient desalination of stones with high contents of smaller pores but also change the properties of the poultice towards higher shrinkage and therefore worse adhesion to the substrate. Moreover, very small clay particles tend to adhere to the stone surface and in the pore structure of the substrate after drying and removal of the poultice layer. To match both the problems of salt transport from the stone into the poultice and the practical workability of poultices for desalination, compromises with respect to the clay content are needed in most cases. The study presents investigations carried out to optimize a poultice for desalination of a fine-pored, widely used building and sculptural stone. Moreover the efficiency of poultices

often used by restorers in practice is compared to the one designed with respect to pore size distribution of the Cotta type sandstone.

2. Materials and methods

2.1. Cotta type Elbe sandstone and test specimens

Cotta type sandstone is a variety of the Cretaceous Elbe sandstone which has been used for building purposes in East Germany and especially in Saxony over centuries. The sandstone is a fine-grained, clay-bearing quartz arenite with kaolinite and illite contents. The average diameter of grains is 0.12 mm. The sandstone structure in thin section and its pore size distribution (total porosity ca. 20 vol.%) is shown in *Figure 1*. As can be seen from *Figure 1b*, the majority of pores is concentrated between 0.1 and 10 μm. Since the material is vulnerable to salt attack, desalination is a standard procedure in the course of most of the restoration measures planned for historic objects made of Cotta type sandstone.

To investigate the efficiency of desalination, stone specimens of unweathered

a

20 cm

15 cm   10 cm

b

c

total salt content [wt.%]

0,7
0,6
0,5
0,4
0,3
0,2
0,1
0

0-1   1-2   2-3

profile depth [cm]

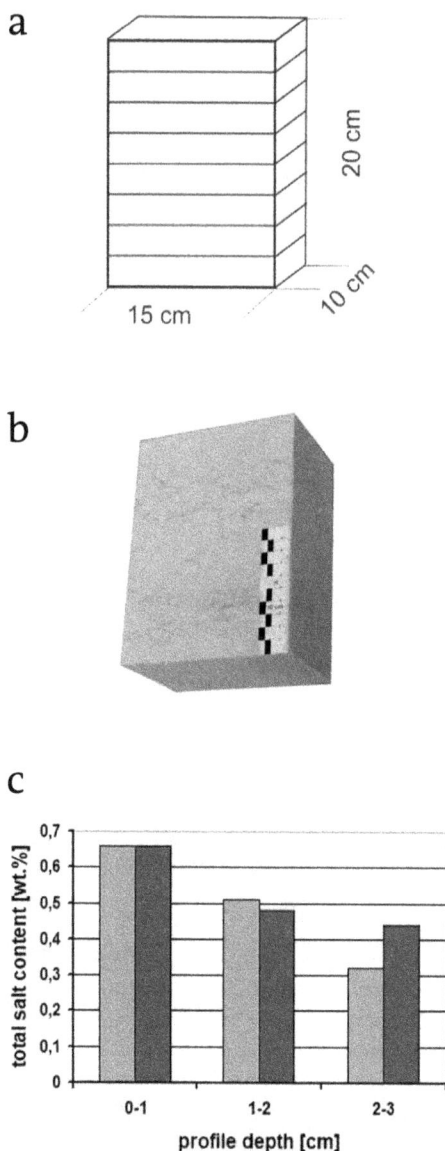

Figure 2: Dimension of the stone specimens used for desalination experiments (a, b) and analytical control of the salt distribution on two random samples (c)

Cotta type sandstone in the dimension of 15x20x10 cm with the largest areas cut perpendicular to the bedding were prepared *(Figure 2a, b)*. After sealing the specimens at the 4 smaller sides with epoxy resin, they were loaded with 3% sodium sulphate solution by capillary suction parallel to bedding over 7 days. After

sealing also the bottom side with epoxy resin, the specimens were dried at room conditions over 14 days. Profile analyses of the total salt content on random samples showed a comparable salt distribution within the test specimens, similar to that found in monuments *(Figure 2c)*.

2.2. Poultice materials

According to the pore size distribution of the Cotta type sandstone *(Figure 1b)* and the results presented in[3] a cellulose-kaolin-sand (CKS) 1:2:1 mixture (by weight) with pore sizes in the same range was selected for the starting experiments. Kaolin and sand came from local deposits (Caminau and Ottendorf-Okrilla, respectively) but had grain size distributions very similar to those used in [3] *(cf. Figure 3)*. The chemical composition of the Caminau kaolin *(Table 3)* differs somewhat from the ideal kaolinite (46.5% SiO_2 and 39.5% Al_2O_3) but is similar to that of the Polwhithe™ C material (SiO_2 47%, Al_2O_3 37%) used in.[3] The difference to ideal kaolinite in case of Caminau kaolin might be explained by the presence of small quantities of illite/mica, as indicated by 2.4% K_2O. The mineralogical composition of Caminau washed kaolin is described by contents of 80-90% kaolinite and 10-20% illte/mica, but no swelling minerals like montmorillonite or mixed layers [6]. For comparison, two poultices that are often used in restoration practice were involved in the tests: a pure cellulose poultice (mixture of Arbocel BC 1000 and BWV 40 1:2.1, [5]) and a ready-made poultice currently provided on the market (produced by Rajasil; mixture of Arbocel®, Ca bentonite, Poraver® and quartz sand) were applied. Different poultice mixtures were characterized with respect to their pore size distri bution as well as to their workability and shrinkage behavior. The latter was determined at the beginning of the experiments by

Material	Sand	Kaolin	Cellulose
Origin	Ottendorf-Okrilla (Euroquarz GmbH)	Caminau (Kaminauer Kaolinwerk GmbH)	Arbocel® BWW 40 (Kremer)
Chemical omposition	SiO_2 98% Al_2O_3 1,15%	SiO_2 46.6% Al_2O_3 35.9% K_2O 2.4%	Cellulose 99.5%

Table 1: Material parameters of the poultice components in own mixtures (source of data: data sheets of the producers)

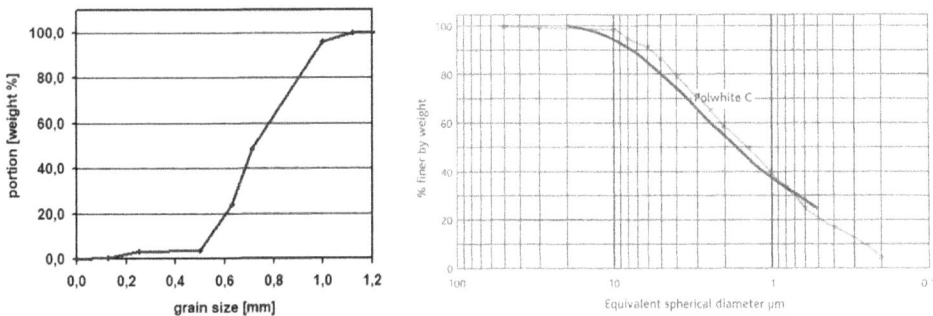

Figure 3: Grain size distribution of the used sand (a) and kaolin (b) components (red line = kaoline Caminau, black line kaoline Polwhite¨ C for comparance)

application of 1 cm thick poultices on water saturated specimens (15x20x3 cm) of Cotta type sandstone. Since the shrinkage was remarkable high for the CKS_121 mixture, a mixture CKS_128 of the same components, but with higher amounts of sand was additionally investigated.

The selected different poultice systems *(Table 2)* were applied to the artificially salt-loaded Cotta type sandstone specimens described in 2.1 in the following way: First, all stone surfaces were pre-wetted with 25 ml of demineralised water. Afterwards, the wet poultices were applied with a thickness of 1 cm. Each type of poultice was applied to 3 sandstone specimens. The specimens treated this way were then stored in the workshop (in upright position) for 3 weeks at 18 °C/42% r. H. Observations of visual changes (efflorescence, loss of adherence, cracks) during the drying process were registered. After 3 weeks, the poultices were removed from the sandstone surface for analyses of the extracted salt amounts.

2.3. Analytical methods

MIP measurements for characterization of the pore size distribution of dry poultices were carried out with Porotec Pascal 140 and Pascal 440 systems.

For the assessment of the quantity of salt moved from the stone substrates into the poultice after desalination, parts of the dry poultices (area 13x18 cm, thickness 1 cm each) were put in plastic bottles, and 1 litre of demineralised water was added. The bottles were moved in a shaking apparatus for 12 hours and afterwards placed in the laboratory 30-80 hours for total sedimentation of the solids at the bottom of the bottle. 500 ml of the clear solution were then pipetted from every bottle and filtered. The

Material	Mixture (weight portions)	Water – dry solid ratio (by weight)	Dry weight of poultice (per 100 cm², 1 cm thick) (g)	Shrinkage after drying, length (%)/width (%)
Arbocel® (cellulose) poultice (ARB)[5]	BC 1000 : BWW 40 = 1 : 2.1	4.5*	20.5	1.3/3.0
Rajasil poultice (RAJ)	Ready-made mixture of Ca bentonite, cellulose, sand and Poraver®	0.6**	87.5	0/0
CKS_121 (own mixture)	Cellulose (Arbocel® BWW 40) : kaolin : sand (0.5 - 1 mm) = 1:2:1	0.8***	86.3	4.0/6.0
CKS_128 (own mixture)	Cellulose (Arbocel® BWW 40): kaolin : sand (0.5 - 1 mm) = 1:2:8	0.39***	126.1	0/0

*taken from [6] **as recommended by the producer ***comparable to [3]

Table 2: Material parameters of the applied poultice mixtures

Material	Median pore diameter (µm)	Main range of pore diameters (µm)	Maximum of pore diameter distribution (µm)	Total pore volume (%) (from MIP)
ARB	22.05	5 - 30	22	84.67
RAJ	12.38	1 - 60	50	59.29
CKS_121	1.25	0.1 - 5	1.8	51.75
CKS_128	6.74	0.4 - 30	17	41.91
Cotta stone	1.51	0.1 - 6	5	21.99

Table 3: Data obtained from MIP measurements for different poultice materials

solvent was evaporated in a dry box at 60 °C. The respective evaporate was weighed, and the salt amount extracted per cm2 could be calculated. Moreover, two profiles of drill powder for salt analysis (from the upper and the lower part) were taken from each sandstone specimen before and after desalination, respectively. The samples of drill powder were taken stepwise from the surface to depths of 0-1, 1-2 and 2-3 cm. 2.5 g of each drill powder sample were eluted with 50 ml of demineralised water for 24 hours. Afterwards, the electrical conductivity of the solution was measured with a Hanna Instruments conductivity meter HI 991300.

3. Results

3.1. Characterization of poultice properties

MIP investigations gave insight into the structure and pore size distribution of the applied poultice mixtures. The results are displayed in *Table 3 and Figure 4*. As can be seen from *Table 3*, the distribution of pores in the CKS_121 poultice fits best to the pore size distribution of the Cotta type sandstone with respect to the theoretical optimum transport conditions.[2, 3] The pore size range of the Arbocel® poultice lies above that one of the Cotta type stone; i.e. the wet poul-

tice is expected to provide much water to the substrate but not to actively extract salt solution from the stone by advection in dryer state. The Rajasil ready-made poultice and the CKS_128 mixture have both pores in the range of the Cotta type sandstone and above.

3.2. Observations during the poultice treatment

After the application *(Figure 5a, b)* the poultices were controlled visually every day during the entire desalination cycle of three weeks. Under the specimens with the Arbocel® poultice, which was

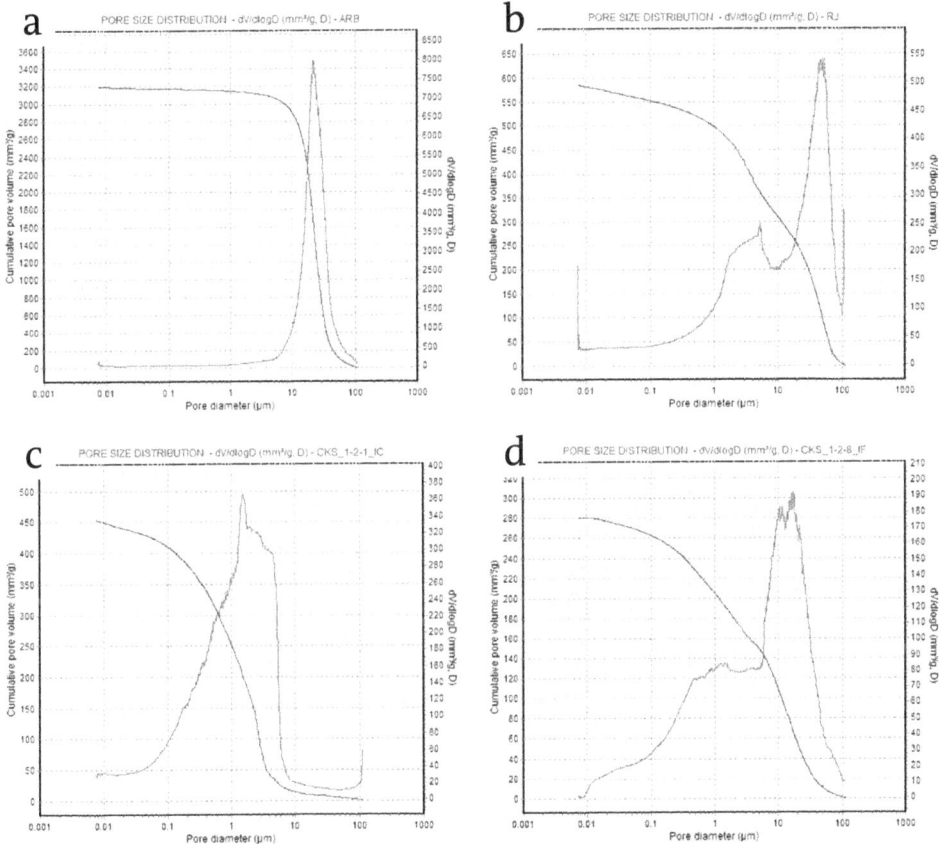

Figure 4: MIP diagrams of the applied poultices: a) pure Arbocel® poultice, b) Rajasil ready-made mixture, c) cellulose-kaolinite-sand 1:2:1 mixture (CKS_121), d) cellulose-kaolinite-sand 1:2:8 mixture (CKS_128)

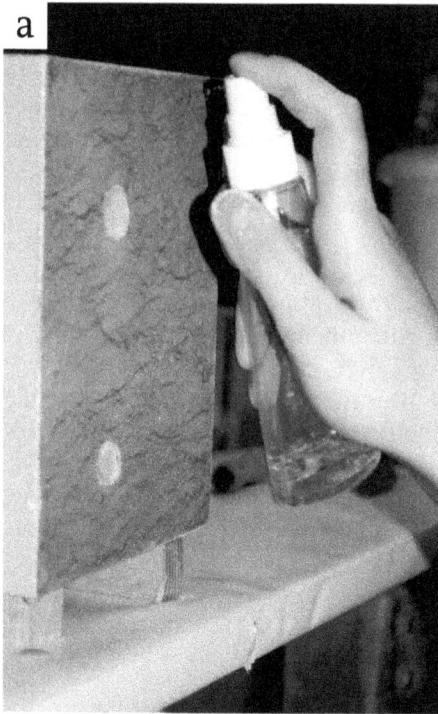

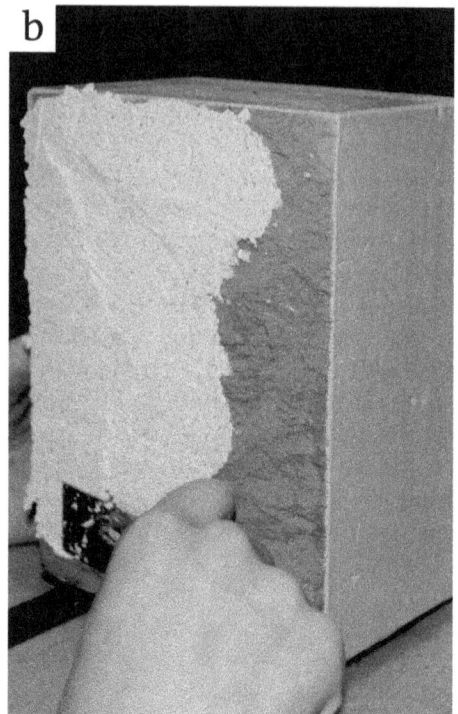

Figure 5: a) Pre-wetting of the stone surface, b) application of the CKS_121 poultice on the pre-wetted stone specimen, c) loss of adhesion on the CKS_121 mixtures one week after application

still very wet after one day, some droplets of water were found that had come out during the first day after the application. In contrast, the CKS_128 mixture already showed first small cracks on the surface after one day. On the third day, the upper parts of the CKS_121 poultice started to visibly loose adhesion to the substrate. On the fourth day, rims of efflorescing salts appeared around the poultices, apart from the CKS_121 and Arbocel® mixtures (which were the mixtures with the highest water contents, cf. Table 1). After one week, the upper parts of all CKS_121 poultices had lost adhesion; one of them had a distance up to 9 mm to the stone surface *(Figure 5c)*. Moreover, they showed single cracks with a width > 1.8 mm. Cracks were also found on the surface of the CKS_128 poultices, but no loss of adhesion to the substrate was detected here. The Rajasil ready-made mixture showed only few small cracks and still a good adhesion to the substrate. The Arbocel® poultice mixtures started to loose adhesion in the upper parts and showed salt efflorescence behind the poultice in these areas. Two weeks after the application, the Arbocel® poultices had lost contact in the upper parts (distance between poultice and substrate up to 2 mm). The CKS_121 mixture showed now visible shrinking and loss of adhesion also at the bottom of the stone specimens. Nearly no further changes could be observed on the other poultices, apart from RAJ, were in some places efflorescences on the rim lead to minimal loss of poultice material.

3.3. Salt contents in the poultices after desalination

As can be seen in *Figure 6*, different amounts of salts were extracted from the stone specimens by different poultice mixtures under the same drying conditions. The highest average value for all three specimens investigated for each mixture (79.6 g/m²) as well as the highest absolute value for a single specimen (108.1 g/m²) was obtained from the Rajasil ready-made mixture experiment. It is closely followed by the CKS_128 mixture with 74.6 g/m² salt extracted on average, with only small differences between the three specimens (72.6-76.1 g/m²). The CKS_121 mixture extracted only 61.0 g/m² salt on average, whereas the worst result (47.0 g/m² extracted salt on average) was found for the Arbocel® mixture. Although the latter could be expected from theoretical considerations of salt and moisture transport in the poultice and the sandstone substrate (cf. 3.1 and [2, 3]), the low extraction rate for the CKS_121 mixture is astonishing with regard to the appropriate pore size distribution. However, this might be explained by its early loss of adhesion to the stone substrate (see 3.2).

3.4. Efficiency of desalination

The efficiency of desalination was calculated for the different profile depths of

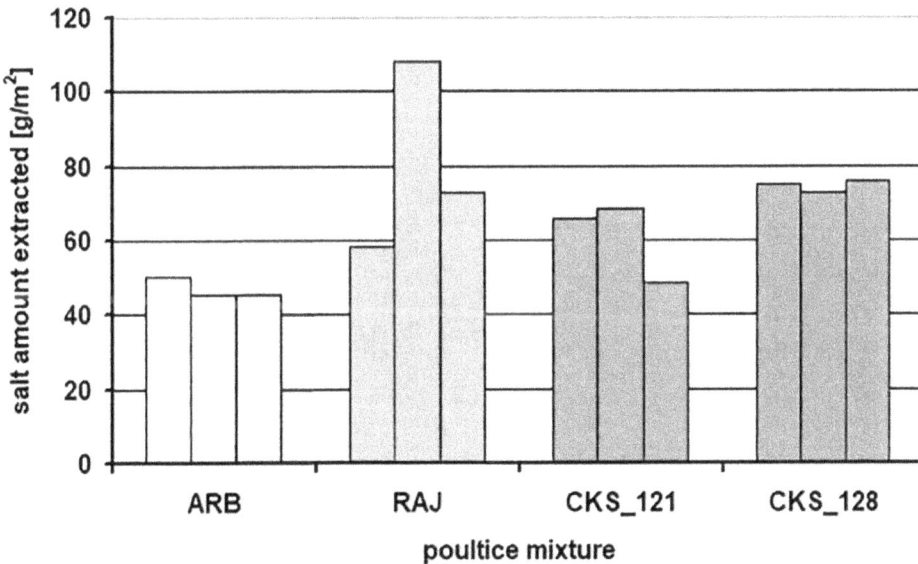

Figure 6: Salt amounts extracted from the salt-loaded Cotta type sandstone specimens with different poultice mixtures during a 3 week's desalination cycle

the sandstone specimens in the stone before and after desalination by using the electric conductivities measured after extraction of salts from the drill powders and the following equation:

$$\text{Efficiency } [\%] = (ec_b - ec_a) \times 100 / ec_b \quad (1)$$

with ec_b = electric conductivity [μS/cm] before and ec_a = electric conductivity [μS/cm] after the poultice cycle. As can be seen from *Figure 7*, the Rajasil ready-made mixture and the CKS_128 mixture show reductions of salt content in nearly all sections of the investigated profiles. Efficiency is between 10 and 40% (RAJ) and between 5 and 35% (CKS_128) in the majority of the investigated profile sections. In contrast, the Arbocel® mixture and the CKS_121 mixture show a re-markable increase in salt content (negative efficiency) in the outermost profile section (0-1 cm). The efficiency in deeper sections (1-2, 2-3 cm) is comparable to this of RAJ and CKS_128 (between 5 and 35% for ARB and between 5 and 40% for CKS_121).

4. Discussion and conclusions

According to theoretical considerations on transport of salts from the stone substrate to the poultice by advection[2, 3], an appropriate cellulose-kaolin-sand poultice mixture with high kaolin content (cellulose-kaolin-sand = 1:2:1 by weight) was designed for Cotta type Elbe sandstone. However, this poultice showed high shrinkage due to the high clay content

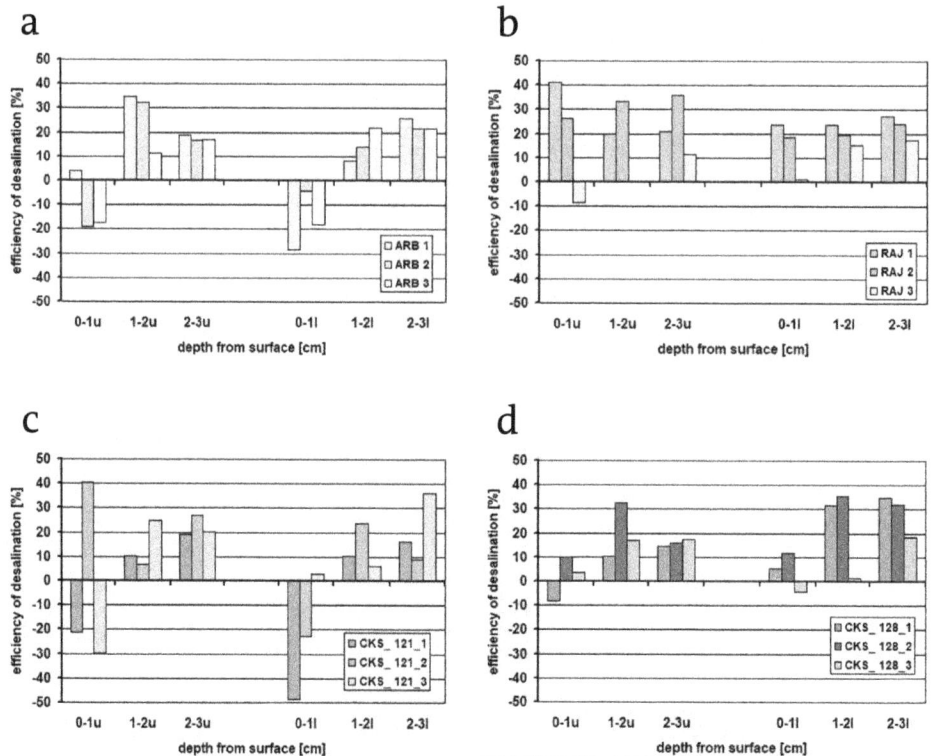

Figure 7: Efficiency of desalination displayed for different profile depths in each of the three sandstone specimens for a) Arbocel® mixture, b) Rajasil ready-made mixture, c) CKS_121 mixture and d) CKS_128 mixture. Positive values indicate salt reduction, negative ones increase of salt content after desalination, related to the original salt content in every specimen (u = upper part, l = lower part of the stone specimen)

(Table 2). Therefore another poultice with cellulose, kaolin and a higher amount of sand (1:2:8 by weight, shrinkage = 0) was additionally tested in the desalination experiment on artificially salt-loaded stone specimens, which also involved a customary Rajasil (RAJ) poultice and an Arbocel® mixture (ARB) often used by restorers. The results showed comparable amounts of extracted salts (75-80 g/m²) for RAJ and CKS_128 poultices, whereas the CKS_121 (61 g/m²) and the ARB mixture (47 g/m²) brought worse results. Visual observations showed an early loss of adhesion of both CKS_121 and ARB, leading to the concentration of salts activated by wetting beneath the stone surface (0-1 cm depth, *Figure 7*). Higher salt contents in the CKS_121 poultice compared to ARB indicate that the mixture with high clay content must be very active in the short period immediately after application until loss of adhesion. This is in accordance with[4], where good results were obtained on materials comparable to Cotta type sandstone with respect to their pore size distribution by repeated treatments with CKS_121 mixtures and very short desalination cycles of only 4-5 days. However, the long-term activity (and the total efficiency) of the mixtures with higher sand contents and no shrinkage is much better, although their pore size distributions do not fit best the theoretical demands for the pore structure of the Cotta type sandstone. The results clearly demonstrate that in desalination practice of stones with fine pores compromises have to be found between the competitive parameters of "ideal" pore structure and low shrinkage of the poultice, both influenced by clay contents in the mixture. Even if the shrinking of kaolin clay is significantly lower than that of bentonite, all poultice mixtures with high clay contents will shrink while drying and therefore raise problems with adhesion. Loss of adhesion before the system substrate-poultice has comple-

tely dried to moisture equilibrium with the surrounding air might lead to concentration of the activated salts near the stone surface *(cf. Figure 7 c)*, which could be dangerous to the stone substrate. Alternatively, more frequently repeated short application periods of only a few days (until adhesion gets lost) with the CKS_121 mixture could be tested. This might save time but will increase costs for material and personnel. More practical experiments are needed to work out strategies balanced between appropriate poultice materials, workability and costs.

Acknowledgements

Thanks are due to Annett Willomitzer (Institute of Construction Materials, TU Dresden) for performance of the MIP measurements and to Frank Hoferick (Zwingerbauhütte Dresden) for fruitful discussions and practical help.

References

[1] Vergès-Belmin, V. & Siedel, H.: Desalination of masonries and and monumental sculptures by poulticing: a review. - Int. Journal for Restoration of Buildings and Monuments 11 (2005) 6, 391-407.

[2] Pel, L. Sawdy, A. & Voronina, V.: Physical principles and efficiency of salt extraction by poulticing. - J. Cult. Her. 11 (2010), 59-67.

[3] Lubelli, B. & van Hees, R.P.J.: Desalination of masonry structures: Fine tuning of pore size distribution of poultices to substrate properties. - J. Cult. Her. 11 (2010), 10-18.

[4] Lubelli, B., van Hees, R.P.J. & De Clerq, H.: Fine tuning of desalination poultices: try-outs in practice. - In: Ioannou, I. & Theodoridou, M. (eds.): Proc. 2nd SWBSS Conference, Limassol 2011, 381-386.

[5] Kröner, S., Mañas Alcaide, B. & Mas-Barberà, X.: Influence of substrate pore size distribution, poultice type, and application technique on the desalination of medium-pore stones. - Studies in Conservation 61 (2016) 5, 286-296.

[5] Störr, M.: Die Kaolinlagerstätten der Deutschen Demokratischen Republik. - Schriftenr. Geol. Wiss. 18 (1983), 1-226.

[6] Bourgès, A. & Vergès-Belmin, V.: Application of fresh mortar tests to poultices used for the desalination of historical masonry. - Materials and Structures 44 (2011), 1233-1240.

Tracing back the origins of sodium sulfate formation on limestone as a consequence of a cleaning campaign: the case study on Charité and Espérance sculptures of Chartres cathedral

Sara Benkhalifa[2], Véronique Vergès-Belmin[1]*, Olivier Rolland[3] and Lise Leroux[1]
[1] *Laboratoire de Recherche des Monuments Historiques, CRC-LRMH-CNRS USR 3224, France*
[2] *Conservatrice - restauratrice de sculptures, France*
[3] *Conservateur - restaurateur de sculptures, France*
*veronique.verges-belmin@culture.gouv.fr

Abstract

In 2012, five years after cleaning by Mora paste, sodium sulfate related deterioration was noticed on two XVIIIth century limestone sculptures of Chartres cathedral. This paper presents the results of the diagnostic study and of the conservation intervention performed on these sculptures, with a focus on trials to determine the salt phases responsible for their deterioration. The methodology chosen includes mineralogical and chemical analyses of samples collected on the sculptures during the diagnostic phase and after kaolin-based poultice desalination (XRD, quantification of soluble salts, SEM-EDS), but also mineralogical analyses of the salts extracted from the poultices.

Mora paste chemicals (and in particular EDTA disodium salt) are probably responsible for the quick deterioration of the sculptures through complex and unidentified chemical pathways : at least two of the mineral phases found in this case study are not present in the JCPDS data base on mineral phases. It is proposed to set up an inventory of unidentified phases found in case studies similar to this one, in order to gather the knowledge spread in different institutions, and later to build scientific projects on the topic of EDTA-related deterioration.

Keywords: thenardite, EDTA, Mora, AB57, cleaning, desalination, limestone

1. Introduction

At Chartres cathedral, four limestone sculptures representing the virtues allegories Charity, Esperance, Faith and Humility and originating from its last Jube built between 1765 and 1770, are displayed on the ground floor of the southern Tower. The statues were separated into two groups after the dismantling of the rood-screen in 1866. Charity and Esperance were fixed to a wall located on the ground floor at Chartres Musée des Beaux Arts, in a situation sheltered from rain but exposed to outdoor environment, and probably in contact with the ground *(Figure 1)*. Faith and Humility were placed outdoors, some meters above the ground, against a monumental doorway at Hotel Dieu and later at Chambre de Commerce. No restoration is documented between 1866 and the 2000's.

In 2007, the four sculptures reintegrated the cathedral, where they were placed on pedestals ca.1 m height, back to the walls but without any contact with the masonries *(Figure 1)*. Before being reinstalled, Charity and Esperance were cleaned using "Mora" paste, while Faith and Humility did not undergo any intervention, apart the application of a limewash, clearly visible to the naked eye on their surface. It is believed that this intervention was performed to give the two sculptures an homogeneous colour, as this paint layer covers pollution black crusts.

In 2012, five years after their restoration, strong scaling, granular disintegration and salt efflorescence were noticed on Charity and Esperance sculptures *(Figure 2)*. This led Irène Jourd'heuil, the curator responsible for their preservation to ask for a condition survey, aiming at understanding the reasons for their quick degradation, and at setting up a conservation project.

This paper presents the results of the diagnostic study and conservation of these sculptures, with a focus on trials to determine the salt phases responsible for their deterioration.[1,2]

2. Analytical methodology

The nature and quarry area of the stone were identified through macroscopic in situ observations and observation, under binocular lens, of centimetric samples which were compared with samples from the LRMH collection of 6000 stone samples originating from 700 French quarries and 500 monuments. The pore size distribution was determined with a Micromeritics mercury intrusion porosimeter. Scales and efflorescence collected on the sculptures were characterized with a powder diffractometer Brucker D8 advance, fitted with a copper anticathode. Some surface samples were analyzed on a JEOL JSM T30, fitted with an Oxford energy dispersive spectrometer (SEM-EDS). A fragment of the poultice relicts

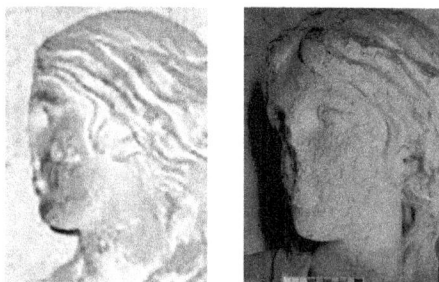

Figure 1: Charity and Esperance sculptures: presentation at Chartres Musée des beaux Arts in 1998 (left), and into the cathedral in 2012 (right)

Figure 2: Face of the statue Espérance: blackening in 1998 (detail of figure 1), scaling and granular disintegration in 2012.

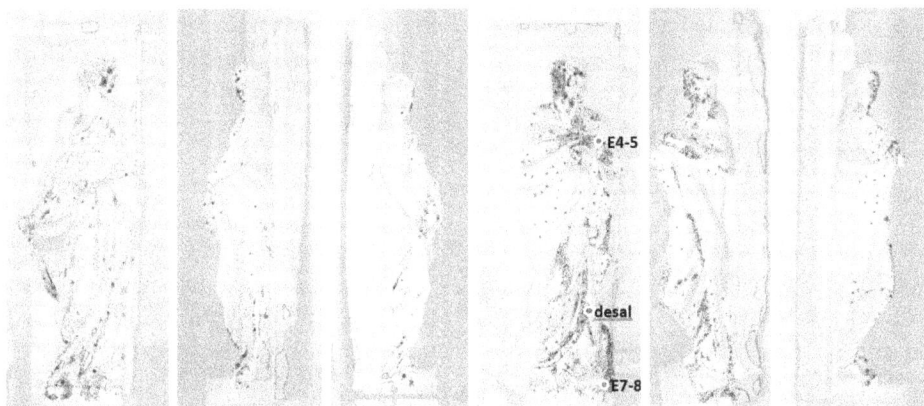

Figure 3: Charity (three images on the left) and Esperance (three images on the right): distribution of stone losses due to granular disintegration + scaling (blue spots), and to mechanical impacts (red lines). Sampling locations are shown as red spots.

found on the sculpture Esperance was analyzed in ATR mode with a Perkin-Elmer SP 100 IRTF spectrometer fitted with a SPOTLIGHT 400 microscope.

The conservation intervention consisted essentially into five successive applications of poultices based on 0.5 – 1 mm sand (Aqua gravel from Aquasand), kaolin (Speswhite from Imerys Mineral) and micron-size calcite (SoCal31 from Solvay). The first set of poultices removed from each sculpture was collected and put in distilled water (6 to 9 Kg / 12.5 L). After setting of the poultice material, the supernatant was collected, filtered and left to evaporate, using a fan to accelerate the process. Pizza-like light brown evaporates were obtained after one month. Crystals of different shapes were collected in the evaporates and characterized by X-ray diffraction

3. Condition survey and analytical results

The sculptures are 1.5 life size and are made of a fined grained light beige and very homogeneous oo-pelletoïdal and micritic limestone originating from quarries near Tonnerre, France. Mercury

intrusion porosimetry (MIP) revealed that the stone has a unimodal pore size distribution, centered on 1μm. Its porosity, as measured by MIP, is equal to 20%.

Many millimetric relicts of a paint layer and of a clayish material were found on the sculptures. SEM-EDS analyses performed on the paint relicts revealed that barium sulfate is present locally, suggesting that the paint layer might contain barite as a pigment.

The mineralogical characterization of the clayish material by X-ray diffraction revealed the presence of palygorskite $(Mg_5Si_8O_2O(OH).2,8H_2O)$ as the main phase, quartz (SiO_2) and calcite $(CaCO_3)$. This suggests that the material is an attapulgite poultice, most probably relictual from the 2007 cleaning campaign. These data do not seem at a first sight in line with the testimony of the conservator who worked on the sculptures. He indicated that he had applied "MORA" pads to clean them, after "many unsuccessful trials". He indicated that he had applied "MORA paste" to clean them, after "many unsuccessful trials". The original MORA paste recipe, also known as "AB 57 paste", was proposed in 1972 by Pr. Paulo Mora and his wife Laura Mora from Instituto central per il Restauro, Rome, for cleaning stone buildings. It comprised Water cc.1000; Sodium Bicarbonate g.50; EDTA (disodium salt) g.25; Desogen, a quaternary ammonium salt, cc.10; and carboxymethylcellulose (CMC) g.60. No clay component entered in the original recipe. However, its formulation has considerably changed in France since the original publication by Mora.[3] For instance on Amiens cathedral in 1995, cellulose powder (instead of CMC) impregnated with a water solution of biammonium carbonate was used as a first step of sculptures cleaning. CMC gels are quite difficult to remove from carved substrates; the operation often necessitates much water to be performed. This is probably the reason why other water retentive agents such as cellulose powder or clays, have replaced CMC gels in a number of workshops in France. It is thus possible that the conservator in 2007 considered using a modified MORA paste based on clay to clean the sculptures.

The sculptures were strongly deteriora-

ted: salt efflorescence, granular disintegration and scaling developed on a high proportion of Esperance and Charity surface *(Figure 3)*.

All surface samples collected in deteriorated parts of the sculptures contained calcite ($CaCO_3$) and sodium sulfate crystallized as thenardite (Na_2SO_4) *(Figure 4)*. Some samples contained also gypsum as a minor phase. Another crystallized phase was put in evidence on the diffractograms, but unfortunately it could not be identified *(Figure 4 and Table 2)*. Stone powder samples were collected at different depths through drilling, and their anionic and cationic contents were measured after salt extraction[4], following the recommendation of the Italian Standard Normal 13/83 : dosaggio dei Sali solubili *(Table 1)*.

Excessive sodium and sulfate quantities are put in evidence in the water extracts of the samples collected at 0-0,5 cm depth *(Table 1)*. This analytical result is in compliance with mineralogical data (pre-

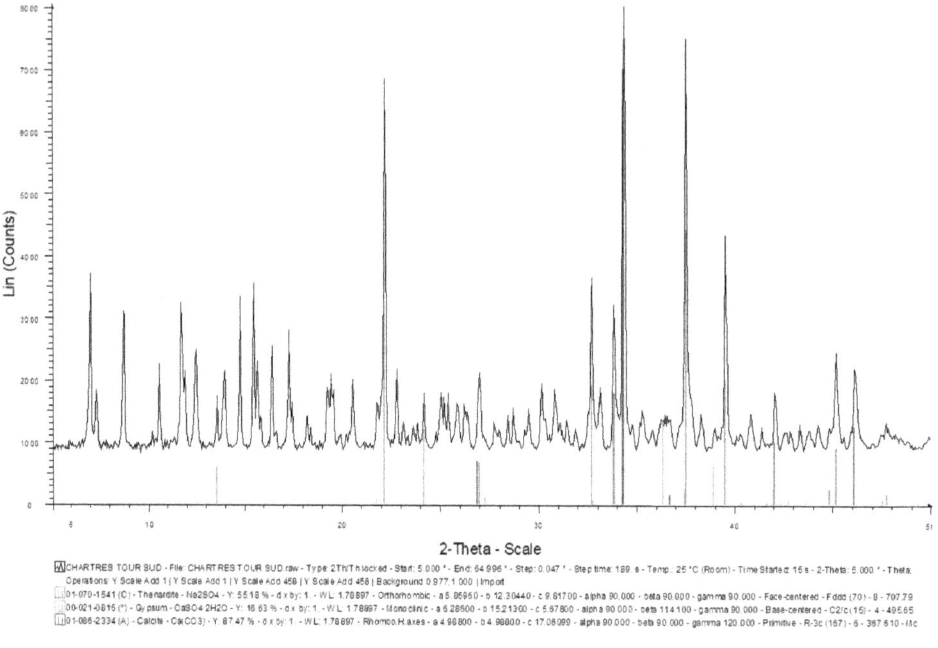

Figure 4: Diffractogram of a typical efflorescence sample collected on the sculpture Esperance. Thenardite (red), gypsum (green) and calcite (blue) are present; most peaks cannot be attributed to an identified mineral phase.

Location	Sample	Depth (cm)	Chloride (%)	Nitrate (%)	Sulfate (%)	Sodium (%)	Potassium (%)	Calcium (%)
Left arm	E4	0.0-0.5	0,07	0.06	0,50	0,59	0,01	0,69
	E5	4.0-5.0	0,01	0,01	0,08	0,04	0,01	0,72
Left foot	E7	0.0-0.5	0,09	0,24	0,22	0,64	0,03	0,83
	E8	4.0-5.0	0,06	0,15	0,06	0,08	0,02	0,73

Table 1: Results of the soluble salts quantification at two places of the sculpture "Esperance" Excessive ionic contents are indicated by bold letters. See Figure 3 for samples location.

sence of thenardite). At 4-5 cm depth, sulfates quantities are below 0.1% on both sampling locations, suggesting that the bulk of the sculpture may be considered as not contaminated by this ionic species.[4] Nitrates contents are rather high both close to the surface and in depth in the sampling location located close to the foot of the sculpture (E7-E8).

The same tendency, to a lower extent, is observed for chlorides contents. Calcium contents cannot be interpreted with confidence as some of the calcium carbonate of the substrate most probably got into solution together with salt related sources of calcium during water extraction. These results suggest, considering the location of the sample where nitrates and chlorides were found, that during its stay on the ground at Chartres Musée des Beaux Arts from 1866 to 2006, the sculpture has been contaminated by ground salts through capillary rise. The most probable kick-off event that determined the strong thenardite related decay seems to be the 2007 cleaning campaign, since no heavy stone loss can be observed on the picture taken before

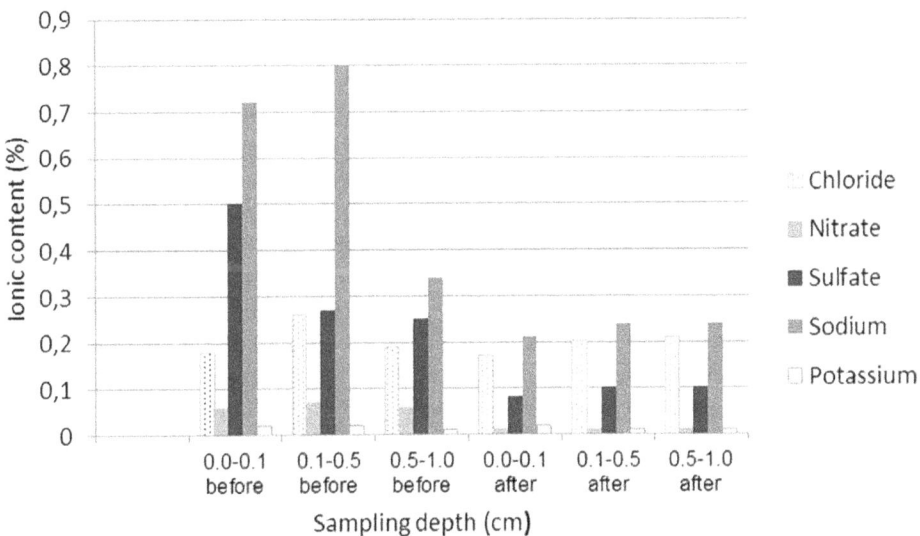

Figure 5: Results of the soluble salts quantification on the sculpture "Esperance. Location of sampling to be found on Figure 3 (ref. "desal")

the 2007 restoration. Interestingly scaling and granular disintegration have appeared preferentially on areas where blackening was present before 2006 (Figure 3), suggesting that some of the chemicals used for cleaning purposes may be responsible for the salt decay. This may be put in relation with the comments of the conservator in 2007: "MORA paste" was used to clean the sculptures, after "many unsuccessful trials".

Considering the nature and depth of the contamination on the one hand, and the pore size distribution of the substrate on the other hand, it was considered a reasonable solution to desalinate the sculptures with kaolin-based poultices.[5, 6]

4. Poultice desalination

After preliminary trials with poultices containing mixtures of various proportions of kaolin, cellulose powder, micron-sized calcite and sand, the following recipes were selected: Charity received five applications of a kaolin:sand:water 1:5:1,5 (vol) poultice. Esperance received one application of a mixture kaolin:nanocalcite:sand:water 0,7:0,3:5:1,5 (vol), followed by four applications of the mixture kaolin:sand:water 1:5:1,5 (vol) on the body parts of the sculpture, while its face received five applications of the mixture kaolin:nanocalcite:sand:water 0,7:0,3:5:1,5 (vol). The reason for a limited application of the mixture containing nanocalcite was that this poultice disintegrated too easily on removal, inducing a strong dust production (health issues and dust deposition on artworks around), while the kaolin/sand mixtures detached more easily into big and coherent pieces. The addition of nanocalcite was a trial to reduce the stone stripping problems during the removal of dried poultices.

Kaolin based poultices being much adhesive, a number of millimetric scales were stripped from the substrate on poultice removal, but the overall results were considered positive, as salt contamination was strongly reduced by the conservation intervention (Figure 4). Kaolin relics adhering to the surface were gently removed from the surface using a soft brush. Tonnerre limestone having a very light creamy colour, the slight and local whitening due to the presence of minute proportions of kaolin left after brushing did not impair the overall aspect of the sculptures.

The poultice desalination was evaluated on the basis of quantification of soluble ions from zero to 1 cm depth.[7] It appears that the main ions sodium and sulfates have clearly decreased while chlorides contents, relatively high before desalination, remained unchanged as a result of poulticing.

5. Salts extracted from the sculptures

The salts extracted from the two sculptures looked pretty much the same (Figure 6): a light brown colour, and a ring shape sequence of crystals. Samples collected from the outer to the inner ring were characterized by X-ray diffraction (Table 2).

The evaporates contain mainly thenardite. The other major phase is an unidentified phase (ref "unident phase b"), different from the unidentified phase (ref unident phase "a") found on surface samples collected on the sculptures before the desalination intervention. Three other phases are locally abundant: gypsum and halite, and the unidentified phase "a". Calcic EDTA is found in four samples. Interestingly, the sample containing the higher amount of calcic EDTA also contains high amounts of the two unidentified phases. The sample Charity 3 was collected on big flat quadrangular white crystals visible in the two evaporates, suggesting that they are thenardite (or rather mirabilite dehydrated into

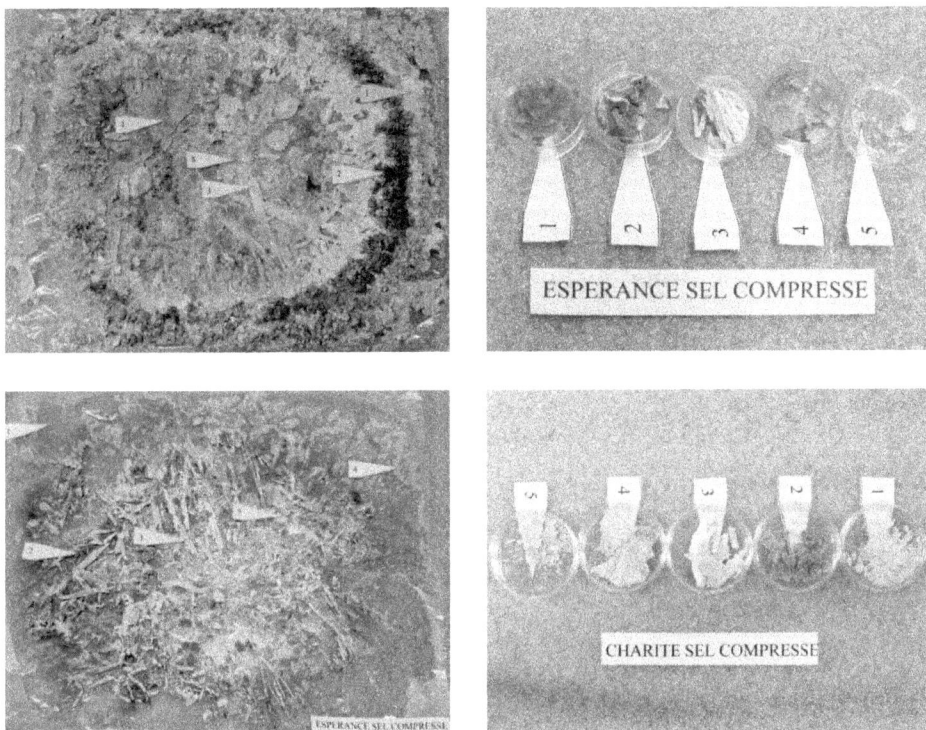

Figure 6: General aspect of the evaporates, and overview of the sampling.

Sample	Phases					
	thenardite Na_2SO_4	gypsum $CaSO_4.2H_2O$	calcic EDTA	unident. phase b	halite NaCl	unident. phase a
Charity 1	++++	+	+	++	(+)	++
Charity 2	++	++	tr	++++	++	tr
Charity 3	++++	-	-	-	-	-
Charity 4a	+++	++	tr	++	+	-
Charity 4b	++	++	-	+++	+	-
Charity 5	++	+	++	+++	(+)	+++
Esperance 1	++++	-	-	+	+++	-
Esperance 2	+++	++	-	++++	+++	-
Esperance 3	++++	+	?	++	+	-
Esperance 4	+++	++	-	++	++	+
Esperance 5	++++	+	?	++	++	+
Esperance efflorescence	+++	-	?	-	(+)	+++

Table 2 : Overview of the phases found in the evaporates.

thenardite). Other crystal shapes cannot be attributed clearly to a single phase, although it is clear that the cauliflower-like brown crystals (ref. Charity 2 and Esperance 2) in both evaporates are essentially made of the unidentified phase "b".

Halite and gypsum were probably present in the sculptures before the 2007 cleaning campaign: halite often originates from capillary rise in places where bodies have been buried[1], which is generally the case around churches. Gypsum contaminates very often limestone sculptures which have been exposed to industrial pollution, but can also originate from repairs: in France, plaster of Paris (calcium hemihydrate) has been used a lot for this kind of use. In the present case study, gypsum more probably originates from SO_2 atmospheric pollution, as plaster of Paris repairs are quite localized and absent close to the places where samples were taken.

Sodium sulfate formation is for us related to the 2007 cleaning campaign, in particular the use of a modified Mora paste. Sodium is present in the Mora formula as di-sodium EDTA salt, and as sodium bicarbonate. Sulfates were present as gypsum. Although we are not able to determine the exact sequence of reactions that took place into the stone as a result of Mora paste application, we think that its chemical cocktail is responsible for the formation of thenardite on the artworks. The strong deterioration

	d(nm)	
	Unidentified phase b	Unidentified phase a
14.	-	75287
	-	10283
12.	77839	-
11.	-	87934
10.	-	12769
09.	-	76787
08.	-	80737
	-	66869
07.	59332	39003
	-	98898
	36658	67528
06.	05857	58439
	-	52884
	-	26473

Unidentified phase a: found in samples collected on the sculptures before desalination and in the salt mixture extracted from the desalination poultices

Unidentified phase b: found in the salt mixture extracted from the desalination poultices

Table 3: Inter-reticular distances of unidentified phases

	d(nm)	
	Unidentified phase b	Unidentified phase a
	-	95659
	-	66108
05.	42659	60388
	-	35667
	-	31276
	-	26957
	-	01822
	-	53432
	-	45963
	17940	37221
04.	-	33351
	-	17657
	-	12826
	-	10238
	-	06840
	-	92234
	87233	73648
03.	54945	65018
	50377	61350
	41063	54237

is also most probably linked to the fact that the chemicals could penetrate the stone down to 1cm instead of staying at its very surface: In the original Mora formula the chemicals were kept by CMC, while in this case study attapulgite was used instead of CMC. Attapulgite ability to retain water is far lower: this clay easily changes from non-plastic to fluid over water addition, and in its fluid state, it cannot hold water. We believe that in such a condition, the Mora paste chemicals solubilized in water may have more easily penetrated the stone.

The unidentified phase "a" found in efflorescence may also have played a role in the deterioration process. The unidentified phase b has not been detected in efflorescence, but appears to be a major phase in the salts extracted by the poultice: this phase either was already present in the sculptures and could not be identified because absent in the deteriorated parts we sampled, or was formed via some chemical reaction between the salts present in the sculptures, the relicts of Mora components and the kaolin-based desalination poultice.

6. Conclusion

This case study of a sodium sulfate related deterioration following the application of chemical products is interesting as it points out one of the sources of a quite frequent secondary effect of stone cleaning: the formation of secondary salts, i.e. salts not present originally nor applied during conservation, and far more noxious than the original ones. The newly formed salts may modify the equilibrium relative humidity of the salts present in the substrate, or simply be more dangerous that the original ones (gypsum trapping soiling particles). Here the worst of all has formed: sodium sulfate.

Mora paste chemicals (and in particular EDTA disodium salt) are probably responsible for the quick deterioration of the sculptures through complex and still unidentified chemical pathways: at least two of the mineral phases found in this case study are not present in the JCPDS data base on mineral phases. The literature on stone conservation being very poor with that respect, we propose to build an international data base on mineral phases formed as a result of stone cleaning, and due to the use of EDTA. As a first step, an inventory of stone conservation issues related to the use of EDTA is to be completed.

Acknowledgements

The authors wish to thank Irène Jourd'heuil, Curator at Chartres Cathedral for her constant support during this project, and Susana Ramirez from ERM laboratory for her reactivity and professionalism concerning the salt analyses.

References

[1] Benkhalifa S.. Rapport d'étude pour quatre Vertus, par Berruer 1765-1770, cathédrale de Chartres, Chartres, Pierre calcaire. S. Benkhalifa, restauratrice de sculptures, 36 rue Damrémont, 75018 Paris. Report, 2012, 56p.

[2] Benkhalifa S. Rapport d'intervention dessalement après tests et recherche de composition de compresse de dessalement adaptée ; Vertus en pierre calcaire de Berruer : la Charité, l'Espérance, l'Humilité , la Foi, 1765-1770. Cathédrale de Chartres ; S. Benkhalifa, restauratrice de sculptures, 36 rue Damrémont, 75018 Paris. Report, 2014, 17p.

[3] Mora P., Mora L., Une méthode d'élimination des incrustations sur les pierres calcaires et les peintures murales, ICOM, In „Réunion du Comité de l'ICOM pour la conservation, Madrid,2-7 octobre 1972", 4p.

[4] Morin C., Analyses quantitative de sels extractibles à l'eau. Rapport ERM 12249CM342, 2012, 11 pp.

[5] Lubelli B., van Hees R.P.J., Desalination of masonry structures: Fine tuning of pore size distribution of poultices to substrate properties. Journal of Cultural heritage, Vol 11, Issue 1, Jan–March 2010, p. 10–18

[6] Bourgès A., Vergès-Belmin V., Comparison and optimization of five desalination systems on the inner walls of Saint-Philibert church in Dijon, France. SWBSS proceedings, Copenhagen, 2008, p. 29-40.

[7] Ramirez S., Morin C., Analyses quantitative de sels extractibles à l'eau. Rapport ERM 17103SR139, 2017, 4pp.

[8] Arnold A., Zehnder K., Salt weathering on monuments (Bari): Grafo, In „ the conservation of monuments in the Mediterranean Basin: proceedings / of the 1st International Symposium, Bari, 7-10 June 1989 ; scientific editor Fulvio Zezza", p. 31-58.

Granite and schist masonry desalination by poultices at Jacobine Church in Morlaix, France

Barbara Brunet-Imbault*, Benjamin Reidiboym and Clément Guinamard
Studiolo: engineering office for cultural heritage conservation and restoration
* barbara.brunet@cabinet-studiolo.com

Abstract

The Jacobine Church of Morlaix in France has been deconsecrated in the XIXth century. Afterwards, it has been used as a stable and then turned into a market which included salt storage for food preservation. Moreover, the church is built downhill of a significant slope next to the Jarlot river. The church, built with granite and schist, presents important salt resurgences, granular disintegrations and moisture areas on joint mortars. The project of the city is to restore the church and turn it into an exhibition space included in a new museum. According to this project, the architects want to apply new lime plasters on the walls, which is not possible because of the high levels of salts. In order to evaluate the salt contamination, samples have been taken in different heights and depths in stones and mortars. Contamination profiles have been obtained and poultices based on cellulose fibers, clays and fine quartz sand have been formulated. Stone and poultice porosimetries have been measured in order to adjust the poultice formulation. The goal of this adjustment was to obtain the appropriate pore size distributions to improve the moisture transport mechanisms between stones and poultices. Three poultices formulations containing variable ratios of cellulose, sand and clays have been selected. Test areas included join removal as much as possible and vacuum cleaning of the wall surface before the first poultice application. These on-site trials allowed us to study desalination efficiency, including poultice formulation and desalination practice (number of poultices applications, wall surface preliminary preparation). Granite cut stone and schist blocks masonries have distinct properties, so desalination practice must be adjusted to obtain an optimized efficiency, including a different poultice formulation and a different masonry preparation. Nevertheless, the desalination of such contaminated masonries is a challenge to preserve lime plaster from future degradations.

Keywords: desalination, practice, poultice, efficiency

1. Introduction

The church is built next to the Jacobine convent of Morlaix, turned into a museum in 1887. In the restructuration project of the museum, the church will be integrated in the museum and turned into an exhibition room. This church has a complex story, as it's been used as a stable and a market with salt storage. Moreover, the natural slope above the northern elevation and the proximity of the Jarlot river foster capillary rise in the masonries. During the XXth century, cement plasters have been applied on all the interior walls of the church, made of schist blocks; only the granite cut stones have been preserved from plastering. Cement has also been used for the pillars and arch's joints.

2. Materials and methods

2.1. Building stones and degradation patterns

The church's masonries are constituted of granite and schist. Salt efflorescences, granular disintegration and moist areas can be observed at several places on the stones and on the cement plasters.

2.2. Sampling

The sampling for the contamination diagnosis before the desalination tests had to be representative of the different materials in place (granite and schist stones, mortar) and of the different degradation patterns. In a first approach, we focused sampling on the most impressive patterns in order to analyse the feasibility of the desalination process in these areas, especially in the capillary rise zones.

Stones and mortar powders were sampled by drilling at two heights (around 20cm high and 2m high) and 3 depths (0-5, 10-15 and 30-40mm).; 30 powder samples were analysed before desalination tests for analysis of the soluble salts : sodium, potassium, chlorides, sulfates and nitrates were quantified.

2.3. Analysis

Salt analyses were performed by ionic chromatography by the BPE laboratory with the Thermo Scientific Dionex Aquion. Results correspond to the mass percentage of these ions inside the sample. Samples were prepared by aqueous extraction according to the standard NF EN 16455. Results are obtained with a relative precision of 5%.

Porosity and porometry measurement were performed on cores of granite and schist stones sampled on site. The same measurements have been also perfor-

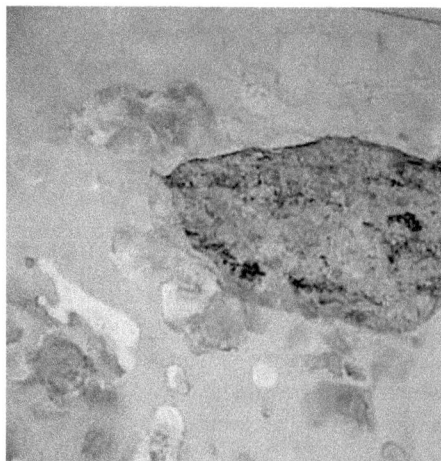

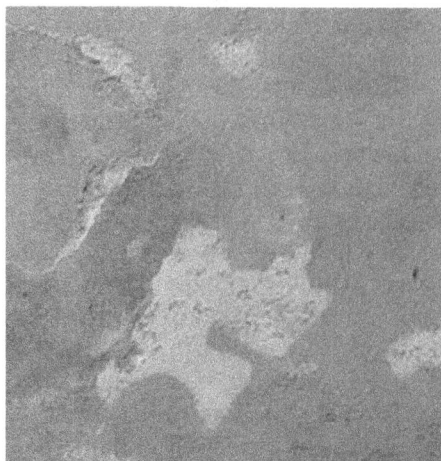

Figure 1: Overviews and details of salt efflorescences and moisture areas on granite and cement plaster covering the schist masonry

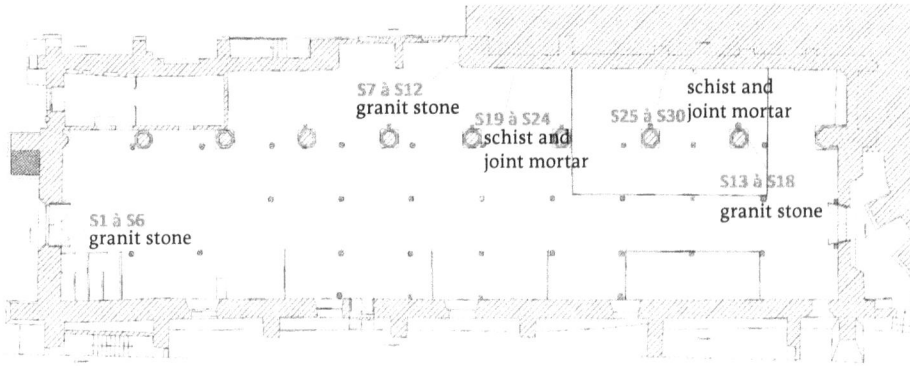

Figure 2: Location map of the samples. The nature of the analysed materials is indicated for each location

med on square samples of each poultices formulations, in order to optimise the poultices in terms of desalination efficiency. The open porosity was measured according to the NF EN 1936 standard. The mercury porosimetry was performed with a porosimeter Micrometrics Autopore IV 9500 V1.09 at mercury pressures ranging from 0.034 to 206 MPa.

Furthermore, water absorption of granite and shist stones was determined in situ using the sponge test method.

2.4. Desalination tests

2.4.1. Poultice formulations

Three formulations of poultices were performed with cellulose fibers, silica sand and attapulgite clay. To obtain poultices permitting advection on stone with a very low porosity, poultices with a high ratio of clay were studied.

Formulations are indicated for dry components. The best rheology is obtained by adding 30g of water for 100g of dry components.

2.4.2. Masonry preparation

Two surfaces were defined to perform the desalination tests: granite masonry (zone 1) and schist masonry (zone 2). Cement plaster was firstly removed from the schist masonry. Each zone were separated in two sub-zones according to the surface preparation:

- type a methodology: surface brushed and cleaned with vacuum cleaner and joint removal up to 4 to 6cm depth

- type b methodology: joint removal up to 2cm depth

	BWW 40 Cellulose	0-2mm Sand	Attapulgite clay
Poultice T1	6 volumes	2 volumes	2 volumes
Poultice T2	3 volumes	3 volumes	4 volumes
Poultice T3	3 volumes	2 volumes	5 volumes

Table 1: Poultice formulations

Sample	Location	Depth (mm)	Height (cm)	Chlorides (%)	Sulfates (%)	Nitrates (%)	Sodium (%)	Potassium (%)
S1	Granite South jamb of the west door	0-5	20	0.053	7.02	0.081	0.139	0.114
S2		10-15		0.052	0.109	0.065	0.241	0.075
S3		30-40		0.053	2.01	0.076	0.261	0.132
S4		0-5	150	0.242	3.22	0.457	0.173	0.114
S5		10-15		0.114	0.089	0.173	0.093	0.072
S6		30-40		0.136	0.091	0.201	0.140	0.099
S7	Granite Enfeu in the northern wall	0-5	100	0.164	0.555	0.845	0.235	0.354
S8		10-15		0.085	≤0.05	0.148	0.127	0.186
S9		30-40		0.101	0.057	0.215	0.186	0.274
S10		0-5	300	0.197	1.12	1.36	0.290	0.207
S11		10-15		0.096	0.453	0.342	0.192	0.213
S12		30-40		0.081	0.172	0.197	0.211	0.270
S13	Granit 1st north-east pillar	0-5	35	0.054	0.713	0.086	0.144	0.116
S14		10-15		0.055	0.113	0.077	0.165	0.057
S15		30-40		0.056	0.114	0.077	0.262	0.079
S16		0-5	230	0.128	0.081	0.251	0.219	0.103
S17		10-15		0.079	0.061	0.121	0.187	0.112
S18		30-40		0.083	0.062	0.128	0.207	0.056
S19	Schist northern wall	0-5	25	0.192	0.420	0.419	0.072	0.125
S20		10-15		0.082	0.081	0.137	≤0.05	0.088
S21		30-40		0.086	0.112	0.157	≤0.05	0.090
S22	Mortar northern wall	0-10	25	0.095	2.71	0.332	0.051	0.063
S23		10-20		0.182	1.78	0.561	0.079	≤0.05
S24		20-30		0.089	2.02	0.284	≤0.05	0.053
S25	Schist northern wall	0-5	300	0.107	9.19	0.465	0.199	0.108
S26		10-15		0.060	0.532	0.118	0.132	0.393
S27		30-40		0.066	1.98	0.161	0.127	≤0.05
S28	Mortar northern wall	0-5	300	0.193	2.17	0.624	0.390	≤0.05
S29		10-15		0.217	2.15	0.780	0.541	0.362
S30		30-40		0.065	0.916	0.127	0.071	0.099

Figure 3: Ion quantifications on granites, schists and mortars before desalination

2.4.3. Poultice applications

Desalination poultices were applied without previous water imbibition of the wall. Poultice was applied by projection then smoothed with a trowel. The protocol consists in 3 consecutive applications for each zones and methodology so 4 poultices were applied 3 times.

Poultices were removed after 3, 6 and 8 days respectively for the 1st, 2nd and 3rd applications. Very high levels of rain and humidity at that time slowed down the poultices drying, so the 2nd and 3rd applications remained in place longer. Even after 8 days, poultices were not totally dry.

3. Results

3.1. Salt contamination

The results of the soluble salt analyses are shown in *figure 3*. The french thresholds for stone conservation are the following ones :

- Chloride < 0.10%
- Sulfate < 0.10% if it's associated with sodium or potassium, <5% if associated with calcium,
- Nitrates < 0.5%
- Sodium <0.04% if associated with chloride
- Potassium <0.05% if associated with chloride or 0.20% if associated with nitrates

In the granite stone, chlorides and sulfates are mainly present in subsurface f while sodium and potassium does not really decrease down to 4cm depth. Schist contamination is extremely important, especially in the capillary rise area, and mortars are also strongly contaminated.

In the granite stone, chlorides and sulfates are mainly present in subsurface f while sodium and potassium does not really decrease down to 4cm depth. Schist contamination is extremely important, especially in the capillary rise area, and mortars are also strongly contaminated.

3.2. Porosity and porometry of stones and poultices

The results of the porosimetric measurements are presented in *table 3*. The poultice porosity significantly decreases when clay is added 80% of the schist pores are smaller than 1µm while for the granite stone, only 38% of the pores are smaller than 1µm, and 53% have a diameter /radius ranging from 1 to 10µm. Porometry of the poultices indicates that the clay clearly plays a role in the presence of very small pores (<0.1µm). The volume of this category of pores increases when the clay proportion increases in the poultice recipe. On contrary, the use of more cellulose fibers increases the proportion of pores bigger than 10µm. Porometry curves indicate that the poultices havea category of pore space smaller thatn the one of the granite and schist stones.

	Schist M1	Granite M2	Poultice T1	Poultice T2	Poultice T3
Developed pore surface (m2/g)	1.671	0.205	15.1	20.5	30.9
Porosity (%)	2.59	3.07	55.6	46.5	43.9
Pore grades					
Access radius (µm)	Porosity of the different pore grades (%)				
0.01 µm – 0.1 µm	59.9	11.5	17.4	31.5	41.6
0.1 µm – 1µm	19.5	27.4	5.03	9.52	10.2
1µm – 10µm	10.1	53.2	65.7	54.1	45.3
10µm – 100µm	10.5	7.9	11.8	4.91	2.86

Table 3: Porosity and porometry of the poultices

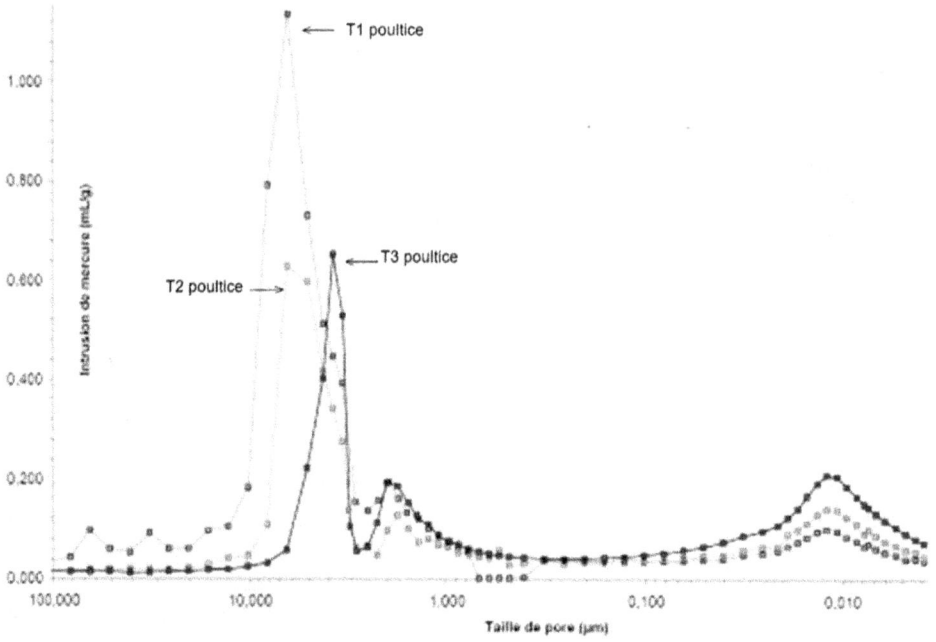

Figure 4: Pore size distribution curves of poultices formulations

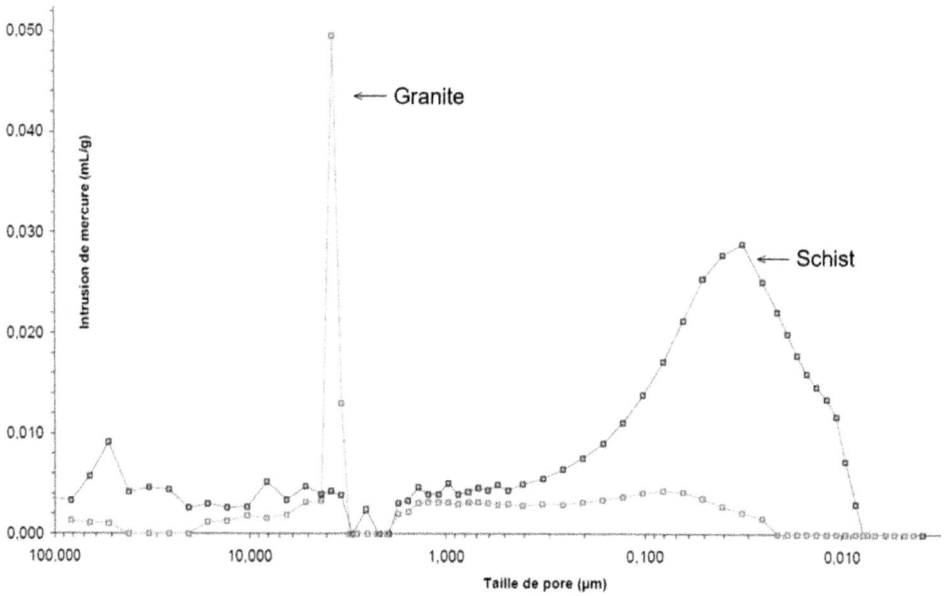

Figure 5: Pore size distribution curves of granite and schist

We can conclude that, the 3 poultice formulations may permit advective extraction on these granite and schist stones. According to the proportion of volume of the different pore categories, T2 formulation poultice is the most suitable formulation in relation with the granite pore size distribution while T3 formulation poultice is the most suitable for the schist in place.

3.3. Water rising measures

Sponge tests were performed on granite and schist stones. The results indicate that water absorption of the shist stone is 44% more important than the one of granite.

3.4. Poultice efficiency

The following tables indicate the results obtained on the granite and schist zones with the type a b desalination methodologies all over the desalination process. On granite, the desalination with methodology "a" may be considered as satisfactory as chloride, nitrate and sulfate rates reach values below recommended thresholds. With the b methodology, sulfates, remain over the thresholds. On schist stones, results indicate a very important reduction of the sulfate rates in both cases, with a and b methodology.

Tables 6 and 7 show that the b methodology allows a better extraction of anions and sodium while sulfates and sodiums remain higher with the a methodology.

Tables 8 and 9 indicating results on mortars shows a significant efficiency for chloride and nitrates but sulfates, sodium and potassium remain over thresholds for the mortars of granite masonry (zone1). For the mortars of schist masonry (zone 2), all ions except sulfates decrease under thresholds.

N°	Time of sampling	Depth of sampling	Chlorides (%)	Sulfates (%)	Nitrates (%)	Sodium (%)	Potassium (%)
S10		0 à 5 mm	0.197	1.12	1.36	0.290	0.207
S11	Before desalination	10 à 15 mm	0.096	0.453	0.342	0.192	0.213
S12		30 à 40 mm	0.081	0.172	0.197	0.211	0.270
P7	After opening the joints, brushing the surface and 1 poultice application	0 à 5 mm	< 0,05	0,06	< 0,05	0,05	0,09
P8		10 à 15 mm	< 0,05	0,01	0.09	0,07	0,09
P9		30 à 40 mm	< 0,05	0,01	0.09	0,09	0,13
P19		0 à 5 mm	< 0,05	0,05	< 0,05	0,06	0,10
P20	After 2 poultices applications	10 à 15 mm	< 0,05	< 0,05	0,06	0,10	0,16
P21		30 à 40 mm	< 0,05	< 0,05	0,07	< 0,05	< 0,05
P31		0 à 5 mm	< 0,05	0,06	< 0,05	0,08	0,09
P32	After 3 poultices applications	10 à 15 mm	< 0,05	< 0,05	< 0,05	0,14	0,19
P33		30 à 40 mm	< 0,05	< 0,05	< 0,05	0,14	0,14

Table 4: Poultice efficiency on granite in zone 1 with type a methodology

N°	Time of sampling	Depth of sampling	Chlorides (%)	Sulfates (%)	Nitrates (%)	Sodium (%)	Potassium (%)
S10		0 à 5 mm	0.197	1.12	1.36	0.290	0.207
S11	Before desalination	10 à 15 mm	0.096	0.453	0.342	0.192	0.213
S12		30 à 40 mm	0.081	0.172	0.197	0.211	0.270
P10	After opening the joints for 1 cm and 1 poultice application	0 à 5 mm	< 0,05	0,09	0,05	0,05	0,07
P11		10 à 15 mm	< 0,05	0,0	0,09	0,08	0,08
P12		30 à 40 mm	< 0,05	0,0	0,10	0,12	0,17
P22	After 2 poultices applications	0 à 5 mm	< 0,05	0,10	< 0,05	0,08	0,13
P23		10 à 15 mm	< 0,05	< 0,05	< 0,05	0,08	0,16
P24		30 à 40 mm	< 0,05	< 0,05	< 0,05	0,10	0,14
P34	After 3 poultices applications	0 à 5 mm	< 0,05	< 0,05	< 0,05	0,09	0,11
P35		10 à 15 mm	< 0,05	< 0,05	< 0,05	0,14	0,20
P36		30 à 40 mm	< 0,05	< 0,05	< 0,05	< 0,05	< 0,05

Table 5: Poultice efficiency on granite in zone 1 with type b methodology

N°	Time of sampling	Depth of sampling	Chlorides (%)	Sulfates (%)	Nitrates (%)	Sodium (%)	Potassium (%)
S25		0 à 5 mm	0.107	9.19	0.465	0.199	0.108
S26	Before desalination	10 à 15 mm	0.060	0.532	0.118	0.132	0.393
S27		30 à 40 mm	0.066	1.98	0.161	0.127	≤0.05
P1	After opening the joints, brushing the surface and 1 poultice application	0 à 5 mm	< 0,05	0,18	0,08	0,08	0,15
P2		10 à 15 mm	< 0,05	0,09	0,09	0,06	0,18
P3		30 à 40 mm	< 0,05	< 0,05	0,05	< 0,05	0,11
P13	After 2 poultices applications	0 à 5 mm	< 0,05	0,19	0,05	0,05	0,17
P14		10 à 15 mm	< 0,05	0,07	0,07	< 0,05	0,18
P15		30 à 40 mm	< 0,05	< 0,05	< 0,05	< 0,05	0,14
P25	After 3 poultices applications	0 à 5 mm	< 0,05	0,27	0,05	< 0,05	0,17
P26		10 à 15 mm	< 0,05	0,13	< 0,05	< 0,05	< 0,05
P27		30 à 40 mm	< 0,05	< 0,05	< 0,05	< 0,05	0,20

Table 6: Poultice efficiency on schist in zone 2 with type a methodology

N°	Time of sampling	Depth of sampling	Chlorides (%)	Sulfates (%)	Nitrates (%)	Sodium (%)	Potassium (%)
S25	Before desalination	0 à 5 mm	0.107	9.19	0.465	0.199	0.108
S26		10 à 15 mm	0.060	0.532	0.118	0.132	0.393
S27		30 à 40 mm	0.066	1.98	0.161	0.127	≤0.05
P4	After opening the joints for 1 cm and 1 poultice application	0 à 5 mm	< 0,05	< 0,05	< 0,05	< 0,05	0,11
P5		10 à 15 mm	< 0,05	< 0,05	< 0,05	< 0,05	0,10
P6		30 à 40 mm	< 0,05	0,07	< 0,05	< 0,05	< 0,05
P16	After 2 poultices applications	0 à 5 mm	0,05	0,14	< 0,05	< 0,05	0,15
P17		10 à 15 mm	< 0,05	< 0,05	< 0,05	< 0,05	0,13
P18		30 à 40 mm	< 0,05	< 0,05	< 0,05	< 0,05	0,11
P28	After 3 poultices applications	0 à 5 mm	0,05	0,10	< 0,05	< 0,05	0,16
P29		10 à 15 mm	< 0,05	< 0,05	< 0,05	< 0,05	0,23
P30		30 à 40 mm	< 0,05	< 0,05	< 0,05	< 0,05	0,16

Table 7: Poultice efficiency on schist in zone 2 with type b methodology

N°	Time of sampling	Depth of sampling	Chlorides (%)	Sulfates (%)	Nitrates (%)	Sodium (%)	Potassium (%)
J3	Before desalination	Hard grey joint (0-1cm)	0,17	0,38	1,36	0,11	0,07
J4		Beige mortar (2-3 cm)	< 0,05	< 0,05	0,12	0,06	0,06
J8	After opening the joints for 1cm, and 1 poultice application	Hard grey joint(1cm)	< 0,05	0,32	0,26	< 0,05	< 0,05
J9		Beige mortar (2-3 cm)	0,05	< 0,05	0,17	0,06	0,06
J13	After 2 poultices applications	Hard grey joint (1cm)	0,05	0,43	0,30	0,24	0,14
J14		Beige mortar (2-3 cm)	< 0,05	< 0,05	0,11	< 0,05	< 0,05
J18	After 3 poultices applications	Hard grey joint(1cm)	0,06	1,04	0,19	0,16	0,08
J19		Beige mortar (2-3 cm)	0,05	0,26	0,21	0,11	0,07

Table 8: Poultice efficiency on mortar in zone 1 with type b methodology

N°	Time of sampling	Depth of sampling	Chlorides (%)	Sulfates (%)	Nitrates (%)	Sodium (%)	Potassium (%)
J1	Before desalination	Thick beige joint (0-1cm)	0,07	0,50	0,23	0,21	0,24
J2		Beige mortar (2-3 cm)	< 0,05	< 0,05	< 0,05	< 0,05	< 0,05
J5	After opening the joints for 1cm, and 1 poultice application	Thick beige joint (1cm)	0,05	0,49	< 0,05	< 0,05	0,05
J6		Beige mortar (2-3 cm)	< 0,05	< 0,05	< 0,05	< 0,05	0,07
J10	After 2 poultices applications	Thick beige joint (1cm)	< 0,05	< 0,05	< 0,05	< 0,05	< 0,05
J11		Beige mortar (2-3 cm)	< 0,05	< 0,05	< 0,05	< 0,05	0,07
J15	After 3 poultices applications	Thick beige joint (1cm)	< 0,05	1,88	< 0,05	< 0,05	< 0,05
J16		Beige mortar (2-3 cm)	< 0,05	< 0,05	< 0,05	< 0,05	< 0,05

Table 9: Poultice efficiency on mortar in zone 2 with type b methodology

4. Result analysis

Initial contamination histograms are representative of an important and global contamination, especially schists which are strongly contaminated due to their thin sheet microstructure.

Tests for granite desalination indicate a better efficiency of the T2 poultice after previous brushing and cleaning with a vacuum cleaner and deep joint removal; anions are all under thresholds. Two T2 poultice applications enable significant salt extraction on granites but the 3rd application causes a cations increase, maybe because repeated applications bring to the surface some very deep salts.

For the schist masonry desalination, a single T3 poultice application after opening the joints for only 1cm gives the best results. The test including brushing, cleaning and deep joint removal doesn't permit anion reduction under threshold. This could be caused by large poultice cracks in case of deep joint removal. The poultice dries way faster on the joint removal zones and cracks appear rapidly.

Concerning the mortar contamination, results show an efficiency for schist masonry with two T3 poultice applications, especially for the cations, but sulfates comes up with the 3rd application, which is significant of a stock of sulfates deep in the stone. On the granite masonry, T2 poultice applications doesn't permit good extraction. Anions and cations remain in levels over thresholds.

5. Discussion

Results are significant of good poultice efficiency. The adaptation of the poultice porometry to the different stone porometries allows to perform efficient salt extractions, even for schist porometry, which is very thin. In this particular case, we obtain astonishing results. Nevertheless, sodium and potassium remain over threshold, which can, with the sulfate rates, provide salt cristallisations, even if anions are very low in several cases after desalination. The preliminary preparation of the masonry including brushing, cleaning with a vacuum cleaner and deep joint removal has a significant effect to the desalination efficiency. Results show that in case of a large amount of mortar, like in schist masonry, deep joint removal induces poultice craqueles while in granite cut stone masonry, with thin joints, deep joint removal fosters salt extraction. In both cases, for schist blocks masonry and granite cut stone masonry, salt extraction on mortar is not satisfying due to the high initial contamination of the mortars and the microstructure with large pores.

This study shows the feasibility of the poultice adjustment to the stone porometry, even with a very thin porous microstructure. Granite cut stone and schist blocks masonries of this church have distinct properties; this study shows that desalination practice must be adjusted to obtain an optimized efficiency, including different poultice formulations and different masonry preparations. Deep joint removal fosters salt extraction, avoiding their diffusion, but this deep joint removal leads to the cracking of the poultice, because of the large mortar microstructure inducing a high water capillary rising. Large microstructural mortar used with very thin microstructural schists complexifies the poultice adjustment. In this case study, we focused on the salt extraction in the schists due to their very high rate of contamination but the adjusted poultice microstructure for the very thin schist pores enhances the water capillary action in the large mortar pores, leading to this cracking phenomenon. In conclusion, this study permitted to obtain a significant salt contamination decrease. Nevertheless, desalinate such contaminated masonries is a challenge to preserve durable lime plaster in the future.

Bibliography

Bourgès A. and Vergès-Belmin V., Application of fresh mortar tests to poultices used for the desalination of historical masonry. Materials and structures, vol 44, n°7, 2011, 1233-1240.

Bourgès A. and Vergès-Belmin V., Comparison and optimization of five desalination systems on inner walls of Saint-Philibert Church in Dijon, France. Salt Weathering on Buildings and Stone Sculptures: 22-24 October 2008, The National Museum Copenhagen, Denmark: [Proceedings from the International Conference], 2008, 29-40.

De Clercq H., Godts S., Debailleux L., Vanhellemont Y., Vanwynsberghe N., derammelaere L. and de Swael V., Electrophoresis as a tool to remove salts from stone building materials – Results from lab experiments and on-site application. International congress on the deterioration and conservation of stone – Glasgow 2016, 289-297.

Lubelli B. and Van Hees R., Desalination of masonry structures : Fine tuning of pore size distribution of poultices to substrate properties. Journal of Cultural Heritage 11 (1), 2010, 10-18.

Case studies

Salt content of dust and its impact on the wall paintings of the church St. Georg at the UNESCO World Heritage site Monastic Island of Reichenau in Germany

Jürgen Frick*, Manuela Reichert and Harald Garrecht
Materials Testing Institute, University of Stuttgart, Germany
* juergen.frick@mpa.uni-stuttgart.de

Abstract

St. Georg is one of the three Romanesque churches on the island of Reichenau at Lake Constance, Germany, built between the 9th and 11th century. UNESCO inscribed the monastic island of Reichenau in the World Heritage List in 2000. St. Georg has meticulously restored wall paintings which are exposed to a very humid indoor environment. Anthropogenic risks and preventive mitigation measures to reduce the environmental stress were identified within a research project. One aspect of research was the impact of salt in dust collected from time to time from different areas in the nave and the crypt. The paper identifies possible sources and analyses the harmfulness in relation to the indoor climate by salt mixture simulation with ECOS/RUNSALT. The data were compared with drilling samples of walls and passive dust sampling data. An experimental analysis by dynamic vapour sorption (DVS) of dust samples is planned in near fu-ture.

Keywords: Salts in dust, climate assessment, DVS, source identification, salt mixture analysis by ECOS/RUNSALT

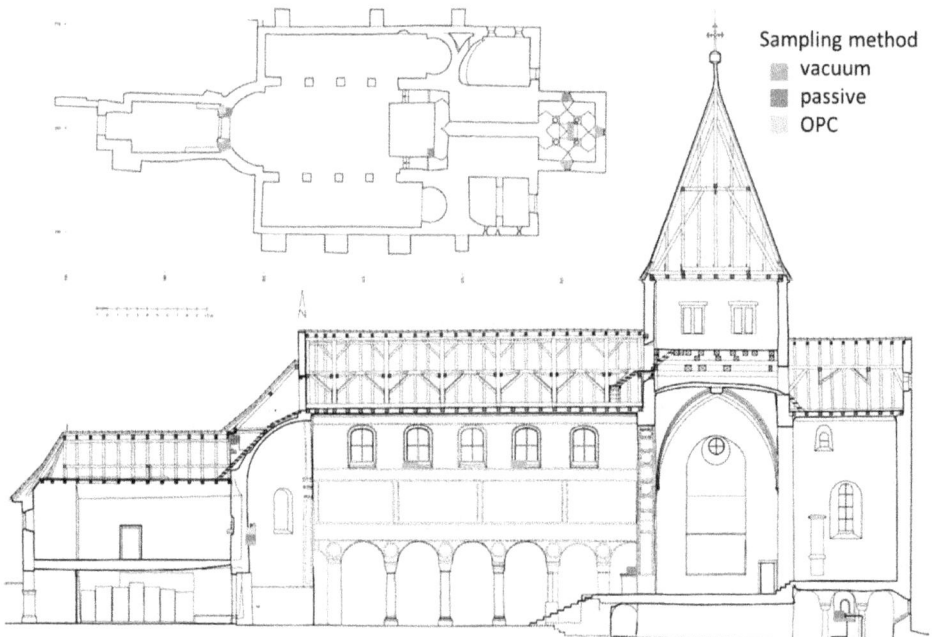

Figure 1: Sampling sites at St. Georg at the isle of Reichenau.

1. Introduction

In co-operation with the State Office for Monument Conservation Baden Württemberg, the Materials Testing Institute University of Stuttgart (MPA) initiated a national research project funded by the German Federal Envi-ronmental Foundation (DBU). The Project has started in 2015 with the aim of the identification of anthropogenic risks and preventive mitigation measures to reduce the environmental stress. One part was the collection and analysis of dust and fine-dust samples from time to time. Additional microbial volatile organic compounds (MVOC) sampling, fine-dust, CO_2 and climate monitoring and passive dust sampling was performed. See[1, 2, 3] for an overview. This paper will focus on the analysis of dust and fine-dust samples and the possible impact on the salt content on the wall paintings and walls in relation to the indoor climate.

2. Experimental

Dust samples were collected by vacuum cleaning on a quartz filter from several horizontal and inclined surfaces in nave, crypt and outside, mostly from window sills. Sampling intervals were monthly or longer. *Figure 1* shows the sampling sites. The samples were analysed by optical micros-copy and partly by ion chromatography and x-ray diffraction. Passive dust sampling on boron or carbon substrates and fine dust measurements with an optical particle counter (OPC, Grimm 1.109) were performed as well.

3. Results

3.1. Dust accumulation

The main series of dust samples was taken at the window sills of crypt and arcade windows of the west apsis. The accumulation differs with season with maximum values in autumn, see *Figure 1*. Additional samples from outside (crypt and cellar window sills) and nave (south and north upper window sills) were taken. *Table 1* lists the average values of dust accumu-lation for all sites.

The average daily accumulation of 6.1 to 6.8 mg/(m²d) is comparable for nave (west apsis) and crypt, except for the

Period	crypt east	crypt south	crypt north	west apsis north	west apsis south	crypt south exterior	crypt north exterior
	Average within period respectively in mg/(m²d)						
6.7.15-25.7.16	6,1	3,6	6,2	6,5	6,8	-	-
25.7.16-2.9.16	-	-	-	-	-	17	47,7
	nave south wall		nave north wall				
	west	east	west	east			
13.10.15-15.3.16	7,6	-	3,9	-			
15.10.15-16.3.16	-	3,7	-	2,9			

Table 1: Average values of dust accumulation for all sampled sites. The respective periods are given.

crypt south window with 3.6 mg/(m²d). The exterior samples (only one sampling period) showed higher values of 17.0 and 47.7 mg/(m2d). At the north and south wall of the nave an increase from east to west is visible *(Table 1)*.

3.2. Salt analyses

The salt content of the dust samples was analysed by ion chromatography. In *Table 2* the results for the two main series in crypt (east window sill) and nave

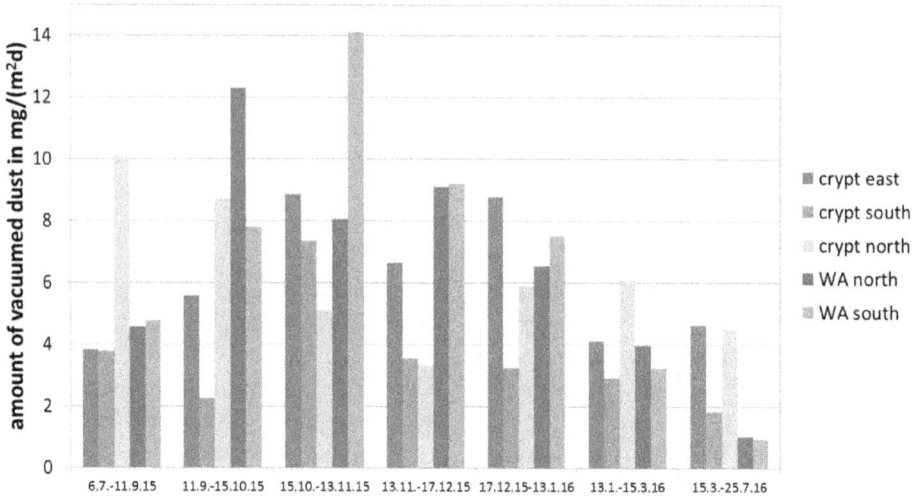

Figure 1: Dust accumulation of the main series in crypt (window sills) and nave (west apsis arcade windows sills).

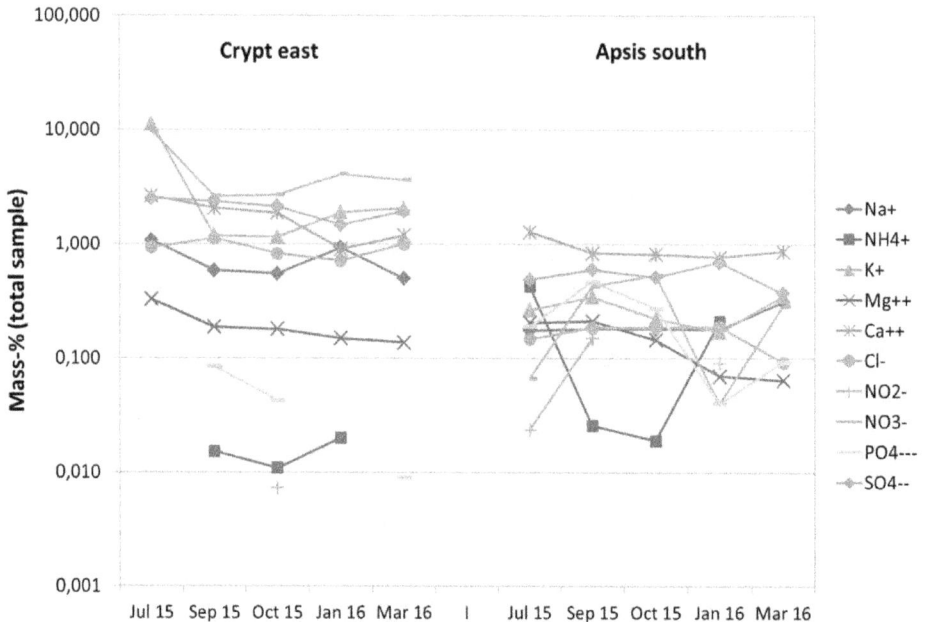

Figure 2: Salt analyses by ion chromatography of the two main series of dust collection.

(apsis south window sill) are given. A graphical interpretation of these results is shown in *Figure 2*. The first samples from July 2015 had a large collection time compared to the following samples, but the amount was comparable for most of the ions. Therefore within a short period of one or two months the salts were accumulated again.

In the crypt main ions were potassium, nitrate, as well as sodium, chloride and sulphate, whereas in the apsis higher concentrations of ammonia and phosphate occurred. Additional samples from window sills in the nave (north and south wall) showed nearly comparable results as the apsis sills *(see Table 4 in [1])*, but with higher amounts of ammonium (around 0.2 to 0.9 mass-%). Especially samples from the north wall showed high amounts of nitrate (up to 0.9 mass-%) and oxalate (up to 0.3 mass-%).

Some dust samples from crypt and apsis were analysed by x-ray diffraction. As expected calcite, gypsum and quartz occurred in both samples and some unspecific mixed crystals. *Figure 3* shows an analysis from the crypt with additional occurrence of niter. A sample from the apsis con-tained calcium oxalate hydrate.

Due to the lime plaster of the sills and walls it was not clear if the amount of calcium is due to the substrate. Additionally, some crystallized salts from the substrate could be in the collected dust samples. Therefore the fine dust collected in the filters of the optical particle counters (OPC) and dust samples from quasi neutral surfaces (organ, crypt altar) were collected and analysed. The results are shown in *Table 3* together with two samples from the outside window sills of the crypt. Due to the low amount of dust (1.2 to 1.7 mg) from the OPC filters, these values are less accurate.

On the filters high amounts of sodium, ammonium, sulphate and nitrite occurred compared to the main series *(Table 2)*. The amounts of calcium, potassium, magnesium and nitrate are comparable to the samples from the crypt series. Therefore salts from aerosol are a possible source.

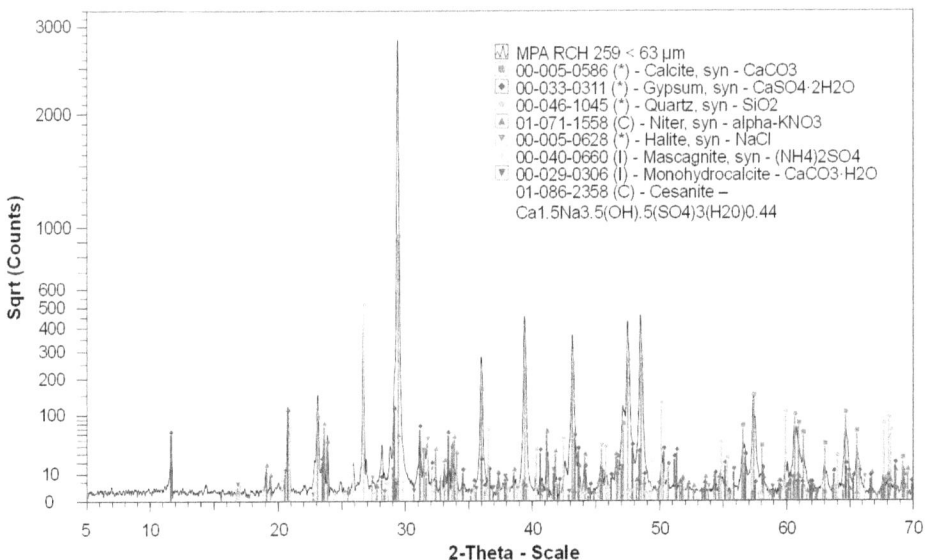

Figure 3: X-ray diffractometry of a dust sample from the crypt east window sill.

Positon/ Period/ Date	Na$^+$	NH$_4$$^+$	K$^+$	Mg$_2$$^+$	Ca$_2$$^+$
	mass-%	mass-%	mass-%	mass-%	mass-%
Crypt east					
06-07-15	1.07	< 0.01	11.30	0.33	2.64
06-07-15 – 11-09-15	0.58	0.02	1.19	0.19	2.07
11-09-15 – 15-10-15	0.55	0.01	1.16	0.18	1.87
17-12-15 – 13-01-16	0.93	0.02	1.90	0.15	0.90
13-01-16 – 15-03-16	0.50	< 0.01	2.07	0.14	1.19
Apsis south					
06-07-15	0.17	0.43	0.27	0.20	1.28
06-07-15 – 11-09-15	0.18	0.03	0.34	0.21	0.83
11-09-15 – 15-10-15	0.18	0.02	0.22	0.15	0.81
17-12-15 – 13-01-16	0.18	0.21	0.17	0.07	0.77
13-01-16 – 15-03-16	0.31	< 0.01	0.35	0.06	0.86

Positon/ Period/ Date	Cl$^-$	NO$_2$$^-$	NO$_3$$^-$	PO$_4$$^{3-}$	SO$_4$$^{2-}$	C$_2$O$_4$$^{2-}$
	mass-%	mass-%	mass-%	mass-%	mass-%	
Crypt east						
06-07-15	0.93	< 0.01	9.78	n.d.	2.52	n.d.
06-07-15 – 11-09-15	1.12	< 0.01	2.64	0.08	2.38	n.d.
11-09-15 – 15-10-15	0.82	0.01	2.70	0.04	2.13	n.d.
17-12-15 – 13-01-16	0.71	< 0.01	4.10	< 0.01	1.47	n.d.
13-01-16 – 15-03-16	1.00	n.d.	3.64	0.01	1.94	0.04
Apsis south						
06-07-15	0.15	0.02	0.07	0.19	0.49	n.d.
06-07-15 – 11-09-15	0.19	0.15	0.43	0.47	0.59	n.d.
11-09-15 – 15-10-15	0.19	< 0.01	0.53	0.27	0.51	n.d.
17-12-15 – 13-01-16	0.19	0.09	0.04	0.04	0.69	n.d.
13-01-16 – 15-03-16	0.09	n.d.	0.29	0.09	0.37	0.16

Table 2: Salt analyses by ion chromatography of the two main series of dust collec-tion. Top: Cations. Bottom: Anions. The end of period or the date represents the sampling date.

The quasi neutral sites are comparable to their counterparts in nave (organ and apsis) and crypt (altar and east window sill). Therefore crystallised salts from the walls could not be the only contribution to the salt content. The outside samples are comparable to the apsis samples with amounts of ammonium and phosphate but less sulphate and chloride. In general these sites are not protected from rain; therefore some dissolution of salts could occur.

4. Discussion

The discussion will focus on the risk potential for the precious wall paintings and possible sources of the salts.

Positon/ Period/ Date	Na^+	NH_4^+	K^+	Mg_2^+	Ca_2^+
	mass-%	mass-%	mass-%	mass-%	mass-%
OPC-Filter					
Crypt 01-09-16 – 20 10-16	1.9	3.1	1.2	0.3	3.3
Apsis north L 01-09-16 – 20-10-16	1.4	2.1	0.8	0.2	2.2
Apsis north D 01-09-16 – 20-02-17	2.7	1.6	2.2	0.2	1.8
Organ 01-09-16 – 20-10-16	1.6	2.4	1.0	0.2	2.4
Extra sites					
Organ 02-09-16 – 24-03-17	0.79	< 0.01	0.21	0.07	1.29
Crypt altar 24-03-17	0.34	< 0.01	0.30	0.09	1.68
Crypt south outside 02-09-16	0.16	0.50	0.13	0.05	0.64
Crypt north outside 02-09-16	0.19	0.76	0.27	0.06	2.15

Positon/ Period/ Date	Cl^-	NO_2^-	NO_3^-	PO_4^{3-}	SO_4^{2-}	$C_2O_4^{2-}$
	mass-%	mass-%	mass-%	mass-%	mass-%	
OPC-Filter						
Crypt 01-09-16 – 20 10-16	0.4	0.6	3.9	0.1	10.2	0.9
Apsis north L 01-09-16 – 20-10-16	0.3	0.2	2.1	0.1	6.8	0.5
Apsis north D 01-09-16 – 20-02-17	0.2	0.2	2.4	0.0	6.3	0.0
Organ 01-09-16 – 20-10-16	0.2	0.5	2.6	0.1	8.5	0.6
Extra sites						
Organ 02-09-16 – 24-03-17	0.06	n.d.	0.21	0.02	0.57	0.10
Crypt altar 24-03-17	0.58	n.d.	1.77	< 0.01	2.52	0.02
Crypt south outside 02-09-16	0.09	0.02	0.03	0.22	0.26	< 0.01
Crypt north outside 02-09-16	0.07	0.17	0.03	0.25	0.30	< 0.01

Table 3: Salt analyses by ion chromatography of dust samples from OPC filters, quasi neutral sites and from outside.
Top: Cations. Bottom: Anions.

4.1. Risk potential

The measured salt content of the dust is relevant and due to the accumula-tion of around 6 mg/(m²d) it is a continuous source for salts. The indoor climate is relatively humid therefore a solution of salts is possible which could be collected by the porous wall structure. To estima-te the risk poten-tial a simulation of the possible salt mixtures was performed with the soft-ware ECOS/RUNSALT.[4] Two samples from crypt and apsis of known sampling periods and a third sample of drill dust (0 to 0.2 cm depth) from the east wall of the crypt near to window sill (analysis taken from[5]) were chosen. RUN-SALT works with a limited number of ions and gypsum should be removed in advance due to the low solubility. There-fore the fol-lowing steps were performed to generate ion balanced data for the different samples:

1) Calculation of mole data out of the analyses.

2) Treatment of nitrite, ammonium and phosphate: In case of the crypt samples the amount could be neglected. In case of the apsis sample this was not pos-sible. In a first step nitrite was added to nitrate which keeps the charge and the particle number but changes the mass. The next step was the calculati-on of the charge sum of am-monium and phosphate because ammonium phosphate is one of the main salts for these ions. In our case the amount of ammonium was higher therefore nitrate and sulphate (as well cations of major ammonium salts) were reduced accordingly. The results are sets of ECOS/RUNSALT compatible ions.

3) Adjustment of charge sum to zero. Car-bonate could not be detect-ed by ion chromatography because for the anion analysis a hydro-gen carbonate/ car-bonate solution is used as eluent. The assump-tion was that the charge dif-ference is mainly due to calcium and

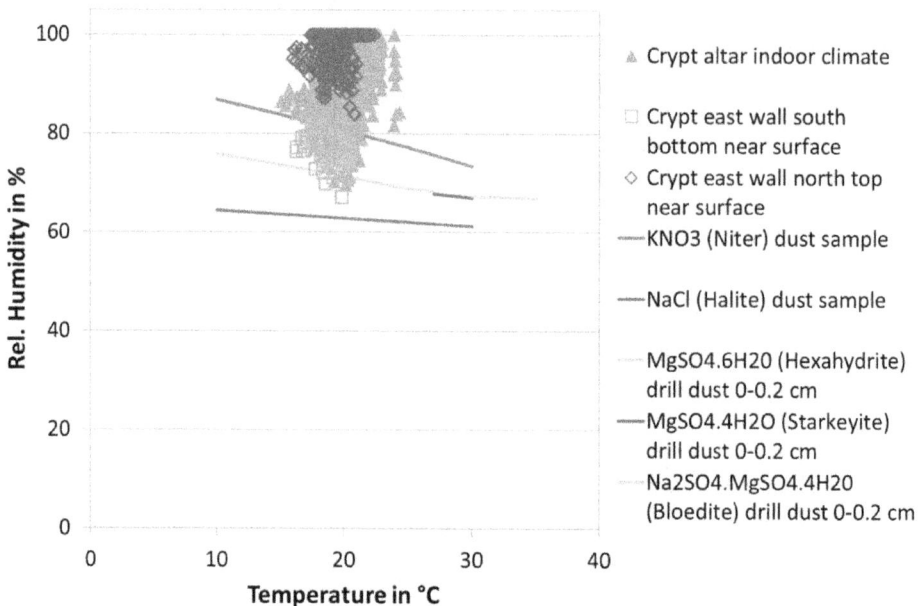

- ▲ Crypt altar indoor climate
- ☐ Crypt east wall south bottom near surface
- ◇ Crypt east wall north top near surface
- —— KNO3 (Niter) dust sample
- —— NaCl (Halite) dust sample
- ---- MgSO4.6H20 (Hexahydrite) drill dust 0-0.2 cm
- —— MgSO4.4H2O (Starkeyite) drill dust 0-0.2 cm
- ---- Na2SO4.MgSO4.4H20 (Bloedite) drill dust 0-0.2 cm

Figure 4: Comparison of simulated deliquescence humidities with climate data of the dust sampling period (6ᵗʰ July 2015 to 11ᵗʰ September 2015) in the crypt.

therefore the calcium content was re-duced accordingly.

4) The last step was the removal of gypsum from the data due to the low solubility. The result is a set of data with either calcium (apsis, crypt drill dust) or sulphate (crypt) set to zero.

Figure 4 shows the deliquescence hu-midities for the different simulated salt mixtures for the sampling period from 6th July 2015 to 11th September 2015 in the crypt in comparison with the climate data of the same period. It could be clear-ly seen that in most of the time all salts in the dust are in so-lution and a lot of phase changes occurred for KNO_3 (Niter). The drill dust sample which represents the wall surface would have some phase changes for hexahydrite in that period.

In *Figure 5* the results of simulati-on for the apsis sample are shown. The sampling period was from 17th December 2015 to 13th January 2016.

As for the crypt sample the salts are mostly dissolved. Phase changes oc-cur-red for the three salt mixtures mirabilite, aphthitalite and picromerite.

The simulations show that salt so-lutions occur in the crypt as well as in the nave which could penetrate into the porous walls. Therefore there is a risk of continuous salt accumulation in the who-le church. The risk is higher in the crypt due to the higher salt contents in the dust there. Two samples from nave and crypt were treated by dynamic vapour sorption (DVS), but the subsequent analy-sis by ion chromatography is in prepara-tion. This will give information about the viability of simulated data.

4.2. Salt sources

The relatively high amounts of salts in the filter of the OPCs indicate that aero-sols could contribute to the salt content in the dust. In the region of Baden-Würt-temberg[6] fine dust collected outside

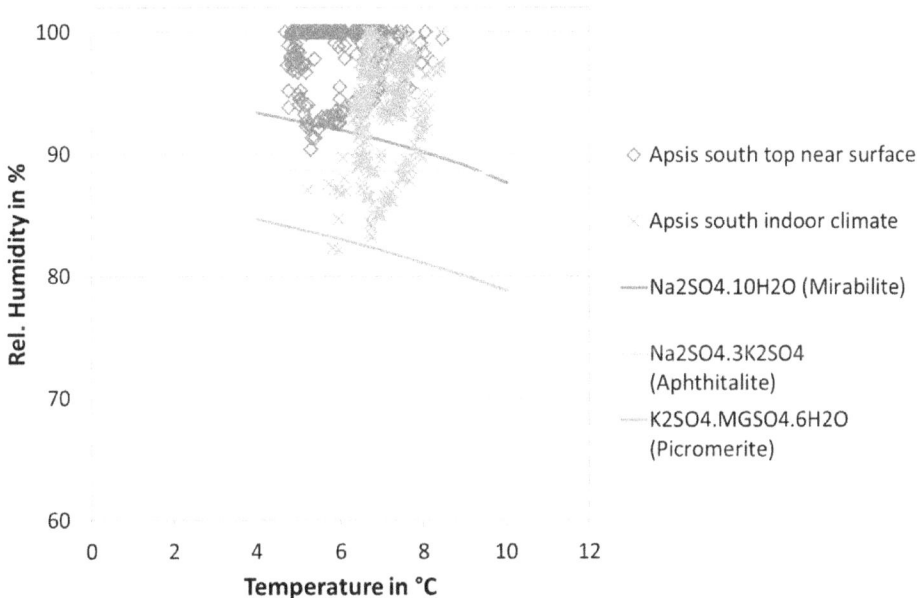

Figure 5: Comparison of simulated deliquescence humidities with climate data of the dust sampling period (17th December 2015 to 13th January 2016) in the ap-sis.

at different sites and periods was examined. Ammonium salts (nitrates and sulphates) occurred in percentages from 25% (high traffic street in Stuttgart) to 48% (Mann-heim urban background) and 51% (Black forest rural clean area). The sources are secondary aerosols formed from gaseous NO_x and ammonia in the atmosphere. In [1] the deposition rates in St. George were compared with data from a sheltered outside position (clock tower of Christchurch in Mainz).[7] The outside rates are 5 to 100 higher than in St. George which is not amazing due to the traffic surrounded site. This are hints that secondary aerosols are one of the sources for salts in the dust.

Another source could be salts from agricultural use. The Reichenau island is intensively used by agriculture. In [1] a calculation was performed by one of the authors (Weinzierl) which showed that nitrates from agriculture could contribute only to a minor part. The calculated amount of nitrate of 0.013 mass-% is small compared to inside values and only comparable to the outdoor samples *(Table 2 and 3)*. On the other hand the loose bound ground contains fine silt which could easily blow off by wind erosion. Sili-cate rich particles are one of the major compounds detected with passive sampling.[1, 2]

The higher amount of nitrates and sulphates in dust samples from the crypt *(Table 2)* might be due to the uptake of crystallised salts from the wall dur-ing sampling. There is as well a difference on quasi neutral surfaces (organ and crypt altar) visible, but on wall samples the difference is bigger *(compare Table 3)*.

There are several ways for aerosols to enter the church. One is the auto-matic opening of windows to control the climate in nave and crypt. Another source is the visitor traffic (only the nave could be visited). During the summer period (May to September) the visitor traffic is restricted to guided tours. The entrance

hall serves as sluice, only when the front door is closed the groups enter the nave by a second door. But due to the high interest as UNESCO world heritage the church has large amounts of visitors. Typical dust from visitors is textile and biological particles as well as dust from the surrounding area from the shoes. The lower amounts of salts in the dust samples of the nave compared to the crypt *(Table 2)* might have their ex-planation in the additional dilution by non-mineral dust from visitor traffic.

5. Conclusion

The salt content in the dust contributes to the salt content of the walls due to solution processes according to the climate conditions. Due to the con-tinuous accumulation of dust (around 6 mg/(m²d), *see Table 1*) it could increase the salt content in the walls over time and could be harmful to the precious wall-paintings. The ECOS/RUNSALT simulations show the poten-tial risks for the walls. Main sources of the salts in the dust are aerosols and input from visitor traffic. In the crypt some recrystallized salts from the walls could contribute as well.

It was suggested to keep the regulations for the visitor traffic in the summer period. Further research is needed to estimate the accumulation of dust by the automatic window opening for climatisation. It is planned to evaluate sedimentation rates and flow conditions.

Acknowledgements

We thank the "Deutsche Bundestif-tung Umwelt" (German Federal Envi-ronmental Foundation – DBU) for funding. We greatly acknowledge the help and support of our project and cooperation partners "Landesamt für Denkmalpfle-ge im Regierungspräsidium-Stuttgart", Institute of Applied Ge-osciences (Group of Environmental Mineralogy) Technical University of Darmstadt, "Landesamt für

Geologie, Rohstoffe und Bergbau (LGRB) im Regierungspräsidium Freiburg", Robert Lung, Corinna Luz, tourism office and catholic congregation of Reichenau.

References

[1] Reichert, M.; Frick, J.; Scheuvens, D.; Zapf, J.M.; Weinzierl, W., "Raumluft Monitoring in St. Georg auf der Reichenau: Leichtflüchtige organische Verbindungen, Staubpartikel und deren Quellen". In "UNESCO-Weltkulturerbe Reichenau – Die Wandmalereien in der Kirche St. Georg als Schlüssel zu einer nachhaltigen Denkmal-pflege", ed.: Jakobs, D. Garrecht, H., Arbeitsheft 33 Regier-ungspräsidium Stuttgart Landesamt für Denkmalpflege, Fraunhofer IRB-Verlag, Stuttgart, 2017, pp. 157 – 176.

[2] Zapf, J.M., "Analyse von Feinstaub-Depositionsproben aus der Kir-che St. Georg (Reichenau) mittels Rasterelektronenmikroskopie". Bachelor thesis, Institut für Angewandte Geowissenschaften, TU Darmstadt 2017.

[3] Weinbruch S.; Scheuvens, D., "Möglichkeiten und Herausforderun-gen der Einzelpartikelanalyse in denkmalgeschützten Innenräumen". In "UNESCO-Weltkulturerbe Reichenau – Die Wandmalereien in der Kirche St. Georg als Schlüssel zu einer nachhaltigen Denkmalpfle-ge", ed.: Jakobs, D. Garrecht, H., Arbeitsheft 33 Regierungspräsidi-um Stuttgart Landesamt für Denkmalpflege, Fraunhofer IRB-Verlag, Stuttgart, 2017, pp. 177 – 180.

[4] ECOS/RUNSALT 2005. RUNSALT (Copyright (c) 2002-2005 by Davide Bionda, Ph.D.) is a graphical user interface to the ECOS thermodynamic model for the prediction of the behaviour of salt mix-tures under changing climate conditions, as described in Price (2000).
http://science.sdf-eu.org/runsalt/
Price, C. A. (Ed.), 2000. An expert chemical model for determining the environmental conditions needed to prevent salt damage in po-rous materials. European Commission Research Report No 11, (Pro-tection and Conservation of European Cultural Heritage). Archetype Publications, London.

[5] Zöldföldi, J., "Zerstörungsfreie und mini-malinvasive Untersuchungen zu Feuch-te- und Salzbelastungen in der Krypta St. Georg auf der Reichenau". In "UN-ESCO-Weltkulturerbe Reichenau – Die Wand-malereien in der Kirche St. Georg als Schlüssel zu einer nachhaltigen Denkmalpflege", ed.: Jakobs, D. Garrecht, H., Arbeitsheft 33 Regie-rungspräsidium Stuttgart Landesamt für Denkmalpfle-ge, Fraunhofer IRB-Verlag, Stuttgart, 2017, pp. 111 – 124.

[6] Ahrens, D.; Anke, K.; Drechsler, S.; Gro-mes, B.; Holst, J.; Lumpp, R; Sähn, E.; Scheu-Hachtel, H.; Scholz, W., "Einfluss-größen auf die zeitliche und räumliche Struktur der Feinstaubkonzentratio-nen". Re-port LUBW Landesanstalt für Umwelt, Messungen und Naturschutz Baden-Württemberg, 2007, chapter 3.10. Online available at: http://www4.lubw.baden-wuerttemberg.de/servlet/is/36153/.

[7] Bundschuh, P.; Auras, M.; Kirchner, D.; Scheuvens, D.; Seelos, K., "Expo-sitionsprogramm zur Wirkung ver-kehrsbedingter Immissionen auf Na-tursteinoberflächen". In "Baudenkmäler unter dem Einfluss verkehrsbedingter Immissionen", ed.: Institut für Steinkon-servierung e.V., Bayerisches Landesamt für Denkmalpflege; IFS Bericht 49/2015, pp. 53 – 77.

Investigation of salts souces at the Karadjordje's Gate on the Belgrade Fortress

Maja Franković[1]*, Nevenka Novaković[2], Suzana Erić[3], Predrag Vulić[3] and Vesna Matović[3]
[1] *Central Institute for Conservation in Belgrade, Serbia*
[2] *Cultural Heritage Preservation Institute of Belgrade, Serbia*
[3] *University of Belgrade, Faculty of Mining and Geology, Belgrade, Serbia*
*maja.frankovic@cik.org.rs

Abstract

The Karadjordje's gate is a monument of culture, part of the historical complex of the Belgrade Fortress. The gate, dated from 1740 to 1791, is made of autochthonous limestone of Miocene age. After years of exposure to environmental conditions and different anthropogenic influences, the stone blocks showed a wide range of decay forms. The gate was subject to a restoration campaign in 2007 using cement based materials. Today, the gate shows renewed signs of degradation: detachment of "artificial stone" used for restoration; scaling and disaggregation of the original stone blocks. Salts efflorescence is present around the joints of stone blocks in the upper part of the gate and under the reconstructed rosettes. Characterization of salts was carried out by SEM-EDS and XRPD analyses. Results showed the presence of the following salts: syngenite, gypsum, thenardite, darapskite, bassanite, niter, aphthitalite and witzkeite. The paper concludes that there is an influence of restoration materials to salt contamination and to the decay of stone and "artificial stone".

Keywords: limestone, salt, restoration materials, decay

1. Introduction

Construction materials in monuments of culture are subject to a number of environmental factors that, acting together in different combinations, influence their degradation. Monuments are exposed to climate change, pollution, use demands, lack of maintenance, as well as inappropriate conservation treatments. Incompatibility in phisical and chemical properties of building materials, in combination with environmental factors, often causes damage of the built material.

A significant part of the damage of monuments is due to salt crystallization in the pores of stones and bricks.[1] The cristallization or hidration pressure of a particular salt or, in short „presence of salts" are often attributes of deterioration of materials.[2] The deteriorating effect of salts is grouped under the detachment category of the ICOMOS-ISCS Glossary[3] but deterioration induced by salts forms a continuum between the granular disintegration and scaling, delamination, and blistering patterns.[4] Many mechanisms are involved in the deterioration of porous materials by salts. For damage to occur, salts must move into and within porous bodies, a process that requires the presence of water or moisture.[2]

Some of main control factors of salt induced damages are: the rock fabric elements, especially porous network[5, 6], the type of salts and their amount[7], the climatic conditions, especially evaporation rate, water supply rate and mechanical resistance of the material.[8, 9]

Salts may accumulate in natural stone over time in several ways. Some of the primary sources of salt contamination include capillary uptake of ground and surface water, interaction between building materials[10], or deposition of acidity from the atmosphere.[11] Salts may also appear as a result of the interaction of aerosol pollutants with certain minerals, as in the case of gypsum[12], or may originate from mortar in contact with the stone, or even from the stone itself.[13] Other important sources of salts include biological activity[14] and polychrome, cleaning and conservation/restoration treatment.[15]

During restoration of heritage buildings, mortars are frequently used for the repointing of joints or for the "plastic" repair of stone. They consist of a binder, aggregates and sometimes additives or adjuvants.[16] Plastic repair mortars are often subdivided based on their binders, as repair mortars with cement, lime, or a combination of both.[17] Interpreting the philosophical and ethical guidelines of both the Venice Charter and the Nara Document, an ideal repair mortar for natural stone should be durable enough, but self-sacrificing on the long run.[18] Portland cement was often used in the past for repairing stone monuments. Due to the poor compatibility and low porosity of Portland cement compared to the stone types, use of this paste can have negative consequences. It forms low-permeable to the impermeable zone, which prevents circulation of moisture through the more permeable original materials and produces a accumulation of water in the more weathered original stone and to its freezing during winters.[15] The authors report that the Portland cement paste could also be a source of the water-soluble salts that contribute to the rapid disintegration of the more recently inserted slabs.

In the case of the Karađorđe's gate, on which this study was carried out, salts have emerged after the restoration treatment in which cement was used for restoration of stone blocks and joints. The study was carried out with the objective to characterize present salts and to determine their possible sources. The main hypothesis was that the cement based material used for restoration had significant influence on stone degradation that occurred after the restoration was carried out.

2. Site characteristics and historical background of the Karadjordje's Gate

The Karadjordje's Gate is located in the central and oldest part of the urban environment of Belgrade - the capital of Serbia. The climate in this area is temperate continental: average annual temperature 11.7 oC; average annual rainfall 669 mm/year; relative humidity 69.5% (data from the Republic Hydro-Meteorological Service of Serbia).

The gate, dated from 1740 to 1791, was not used for a long time – plans from the end of the 18th century already depict it walled up, with no bridge. It was put to use again only after World War II, during the rearrangement of Kalemegdan park and the fortress. Today, Karadjordje's gate lies in the second communication line approaching Belgrade fortress, connecting it with a part of the town along Sava River.

The gate consists of frontal (south) facade (investigated area in this paper), central aisle with two side rooms for guards and north facades built of bricks and stone. Frontal facade consists of double stone arches and it is finished with decorative cornice *(Fig. 1)*. Facade is built entirely of limestone blocks with simple decorations representing two simetrical rosettes. The gate was subject to the restoration campaign carried out in 2007. The cleaning was done by water jet and restoration of damaged stone blocks was carried out, without prior consolidation of the stone, with so called „artificial stone". Joints were restored with Portland cement mortar.

Figure 1: The view of the Karadjordje's Gate in Belgrade (Serbia) and position of the taken salt samples

3. Methodology

Salt samples were taken from stone blocks built in exterior walls of the Karadjordje's Gate in November 2016. The main criteria for the sampling positions were: significant quantities of salt in the form of efflorescence on the surface and the originality of the blocks (natural and artificial) where salts appear. In order to analyse of salt composition by X-ray powder diffraction (XRPD) and SEM/EDS, five samples of salts were taken from the substrate in following order: S1 – beneath the broken rosette made of "artificial stone"; S2 – from the surface of the cement mortar joint above restored stone block; S3 – beneath a scale on the original stone block; S4 – from contact zone between rosette of "artificial stone" and original stone substrate; S5 – beneath a scale located on the original stone block surrounded by blocks reconstructed with "artificial stone" (Fig. 1).

In order to examine the built stone, structurally different samples of natural stone were taken from the monument. Petrographic analyses of the built stone were performed on thin sections using a Leica DMLSP microscope for polarised light that was connected to a Leica DC 300 digital camera. Classification of built stone was done according to textural characteristics. Determination of the weathering forms types was done according to ICOMOS-ISCS glossary on stone deterioration patterns.[3]

SEM-EDS was performed using a JEOL JSM-6610LV scanning electron microscope connected to an X-Max energy dispersive spectrometer to identify the morphology and chemical composition of the mineral phases present in the salts. The samples were covered with gold and carbon using a BALTEC-SCD-005 sputter coating device, and the results were recorded under high vacuum conditions.

X-ray diffraction analyses were used to determine the phase compositions of powdered salts and powdered limestone. The XRPD was performed using a Rigaku Smartlab diffractometer. The diffraction patterns were obtained from 4 to 700 2θ using $CuK\alpha 1,2$ radiation with a scan of 50/min.

4. Results

4.1. Characteristics of built materials

Petrographic analyses of natural stones imply that the most dominant material built into the structure are autochthonous, allochemical limestones, Miocene in age. Textural features and micro- and macro-faunal characteristics of the limestones identify them as Grainstone microfacies and Algal rudstone microfacies. These two microfacies types are limestones mineralogically composed of calcite $(CaCO_3)$, with very small amounts of silici-clastic input. The limestones are whitish, pale yellow or beige in colour *(Fig. 2a)*.

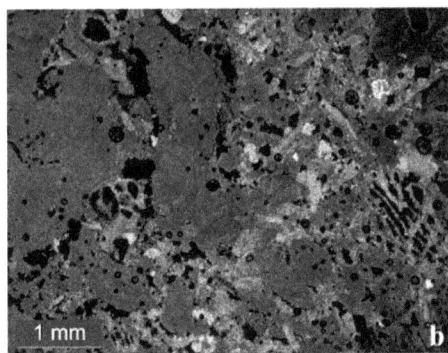

The prevailing allochemical compounds in all types are fossil-skeletal fragments of red algae (Lithotamnium ramisissimum; family Corallinaceae) *(Fig. 2b)*. The remains are micrite in composition with a well-developed and prominent mesh texture consisting of micron-sized rectangle chambers. Fossil remains of some other species (e.g., bryozoans, gastropods and different foraminifer species) were also recognised. They are recrystallised and filled with sparry calcite. The algal biosparrudite is extremely porous and weakly consolidated rock with sparite bounds with numerous vesicles, mesopores, rounded cavities and channels but fine/capillary pores, too. Grainstone microfacies is moderate porous rock with pores in range up to 0.1 mm. Water absorption of limestones used to build Belgrade fortress is about 15%, while the compressive strength vary from 4.5 to 13 MPa.[19]

The "artificial stone" used for restoration purposes was composed of 2 parts ground limestone aggregate, and binder that consisted of 1 part white cement and 0.1 parts slaked lime. Mineral pigments were added in order to match the colour of the original stone. Water absorption of the artificial stone was 1.6%, its compressive strength at 28 days is 43.3 MPa.[19] Portland cement mortar was used for restoration of the joints.

4.2. The decay forms of built stone

Mapping of the Gate's façades and registration of decay forms of stone degradation after the restoration was carried out. Damage occurs on the stone blocks at the upper part of the south façade. The most dominant decay forms are scaling and disaggregation of the original stone blocks, as well as detachment of "artificial stone" used for restoration from the rosettes *(Fig. 3a, b)*. Salts deposits occur around the joints on the stone blocks

Figure 2: The macroscopic appearance (a) and photomicrograph (XPL) of built Algal rudstone from the Karadjordje's Gate

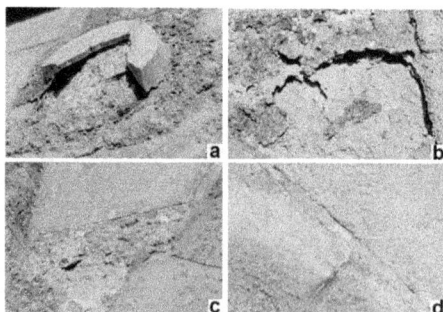

Figure 3: The damage forms on stone blocks. a) detachment of "artificial stone"; b) scaling and disaggregation of the original stone blocks; c-d) efflorescence on mortar joints and surface of artificial stone

gates and variable content (4.5-20.7%) ordinarily follows syngenite in the all samples. Although thenardite is present to a lesser extent (6.5-13.4%), its presence in the form of tabular aggregates or bipyramidal crystals, is evident on the left side of the façade. Darapskite and niter (crust-like aggregates) occur up to 4% in contents. Aphthitalite occurs only in sample S1 as trigonal crystals ranging from 3 to 10 µm in content of 8.3%, whereas basanite is present on right side of outer façade (3-5%).

in the upper part of the gate, as well as beneath the broken parts of rosette or beneath the scales on the original stone (Fig. 3c, d).

4. Identification and distribution of present salts

XRPD and SEM-EDS analyses of the samples showed the presence of the following salts: syngenite, gypsum, thenardite, darapskite, bassanite, niter, aphthitalite and witzkeite (Table 1, Fig. 4, 5).

Calcite prevails in the all samples but its presence origins of substrate. Syngenite occurs in prismatic or rarely elongated aggregates as dominant salt in all samples but vary in contents (8.5-13.4%). Gypsum in the form of tabular aggre-

5. Conclusion

The study of damage type and salts deposits on stone blocks on the Karadjordje's gate show that there are several factors which contribute to decay of restored stone blocks as well as to the occurrence of salts. First, cement mortars used for joint filling and "artificial stone" possess different properties from the original substrate. Its incompatibility is reflected in significantly lower water absorption and greater compressive strength comparing to original limestone substrate. Orginal soft stone blocks easily absorb and release water. That indicates that the contact zone between the original porous limestone and cement mortar used for restoration represents hygric impermeable zone. Longer moisture retention

Sample/stone type	Salt content (%)							
	G	S	T	N	D	B	A	W
S1/a-o	4	11	13	3	3	/	8	1
S2/a	11	9	7	4	1	/	/	/
S3/o	4	14	7	4	3	/	/	/
S4/a-o	21	11	/	/	2	3	/	/
S5/o	14	13	/	/	/	5	/	/

Legend: a – artificial stone; o – orginal limestone; G – gypsum; S – syngenite; T – thenardite; N – niter; D – darapskite; B – bassanite; A – aphthitalite; W – witzkeite

Table 1: Composition of investigated salt samples

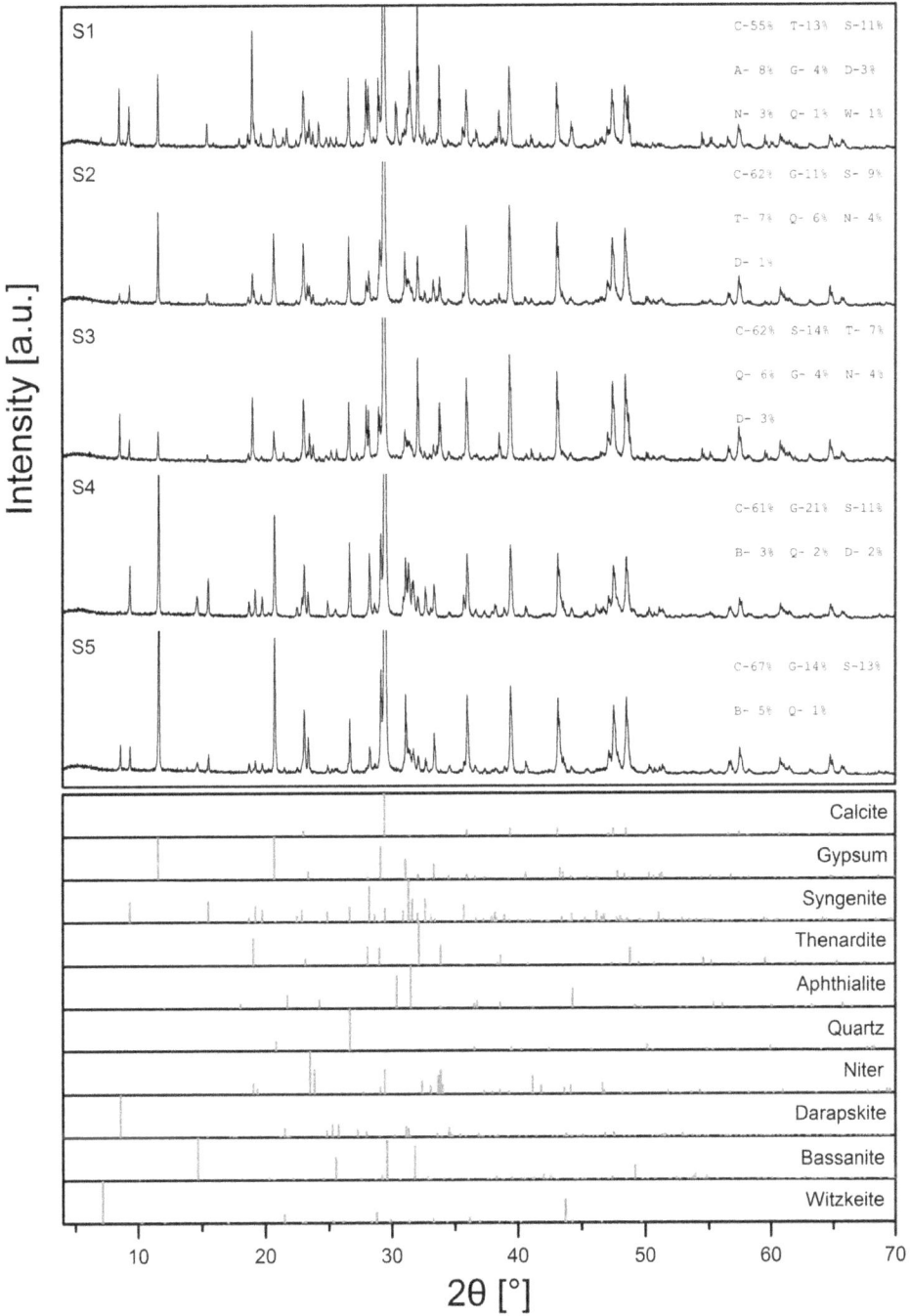

Figure 4: X-ray diffraction patterns of the salts formed on built limestones of Karadjordje's Gate with respective semi quantitative analyses results

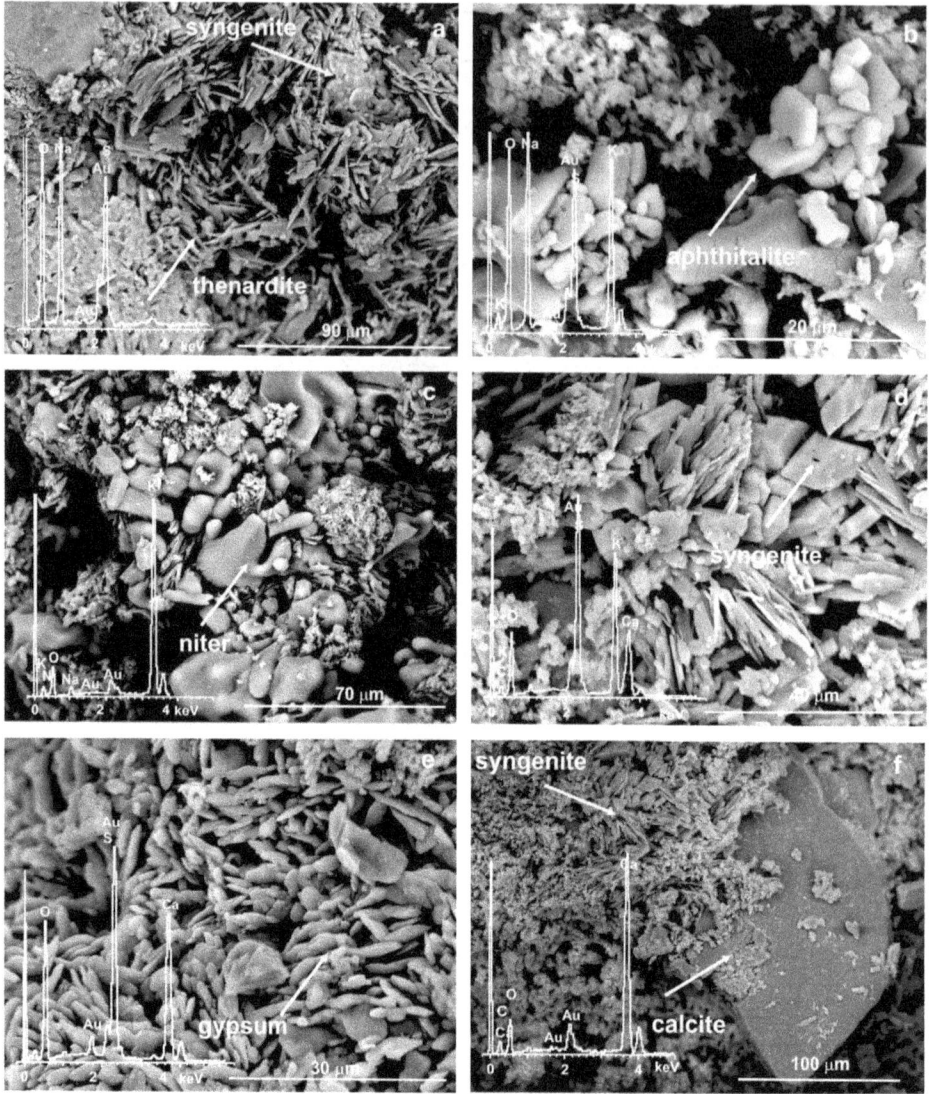

Figure 5: SEI images with EDS spectrums of salts from the analyzed samples S1-S5

in repair mortars can cause its chemical dissolution which could be potential source of sulphates. Water retention also causes physical damage due to freeze/thaw cycles. Gate's architectural features and its positioning within the rampart also contribute to described degradation mechanism. Slightly indrawn south facade creates semi-sheltered area at the upper part and prevents stone blocks to be washed by rain, which is pogodna environment for the forming of gypsum.

The thick earthen mound that covers the top part of the gate retains moisture which is drawn inside by capillary action and evaporates on the surface of stone blocks. The water percolation from the mound could also be the source of salts originating from grass fertilizers or decomposition of organic matter, such as niter, present in three of the samples.

This case study demonstrates the importance of compatibility of materials used in restoration with the original. If the properties of the „artificial stone" do not match those of the original, especially its hygric and mechanical properties, and if the Portland cement is used in restoration, damage is likely to occur and it will lead to the deterioration of the very stone that is meant to be protected.

References

[1] Goudie A, Viles H Salt weathering hazards. Wiley, Chichester (1997).

[2] Charola E.A. Salt in the deterioration of porous materials: an overview, JAIC (39), (2000) 327-343.

[3] Vergès-Belmin V. et al, ICOMOS-ISCS Illustrated glossary on stone deterioration patterns, Monuments and Sites XV, Ateliers 30 Impression, Champigny/Marne (2008).

[4] Snethlage R, Wendler E Moisture cycles and sandstone degradation. In: Baer NS, Snethlage R (eds) Saving our architectural heritage: conservation of historic stone structures, Wiley, Chichester, (1997) 7–24.

[5] Thomachot-Schneider C., Gommeaux M., Fronteau G., Oguchi C.T., Eyssautier S., Kartheuser B., A comparison of the properties and salt weathering susceptibility of natural and reconstituted stones of the Orval Abbey (Belgium), Environ Earth Sci (63) (2011) 1447–1461, DOI 10.1007/s12665-010-0743-8.

[6] Benavente D., Martínez-Martínez J., Cueto N., García-del-Cura M.A., Salt weathering in dual-porosity building dolostones, Engineering Geology 94 (2007) 215–226.

[7] Ruiz-Agudo E, Mees F, Jacobs P, Rodriguez-Navarro C The role of saline solution properties on porous limestone salt weathering by magnesium and sodium sulfates. Environ Geol (52) (2007) 269–281.

[8] Goudie AS, Parker AG Experimental simulation of rapid rock block disintegration by sodium chloride in a foggy coastal desert. J Arid Environ (40) (1998) 347–355.

[9] Diaz Gonçalves T.C., Salt crystallization in plastered or rendered walls, PhD thesis, Universidade Técnica de Lisboa Instituto Superior Técnico, (2007) 19-22.

[10] Price C.A., Stone conservation: An overview of current research. Santa Monica, Calif.: Getty Conservation Institute, J. Paul Getty Trust. (1996) 7-9.

[11] Steiger M., Charola A.E., Sterflinger K., Weathering and Deterioration. In: Siegesmund, Siegfried, Snethlage, Rolf (Eds.), Stone in Architecture Properties, Durability, Springer (2014) 265.

[12] Halsey, D.P., Dews, S.J., Mitchell, D.J., Harris, F.C., Real time measurements of sandstone deterioration: a microcatchment study. Build. Environ. (30) (1995) 411–417.

[13] McKinley, J.M., Curran, J.M., Turkington, A.V., Gypsum formation in non-calcareous building sandstone: a case study of Scrabo sandstone. Earth Surf. Process. Landf (26), (2001) 869–875.

[14] Gómez-Heras, M., Benavente, D., Álvarez de Buergo, M., Fort, R., Soluble salt minerals from pigeon droppings as potential contributors to the decay of stone based cultural heritage. Eur. J. Mineral. (16) (2004) 505–509.

[15] Přikryl, R., Novotná, M., Přikrylová Weishauptová, Z., Šťastná, A., Physical and mechanical properties of the repaired sandstone ashlars in the facing masonry of the Charles bridge in Prague (Czech Republic) and an analytical study for the causes of its rapid decay. Environ. Earth Sci. (63) (2011) 1623–1639.

[16] Isebaert A., Van Parys L., Cnudde V., Composition and compatibility requirements of mineral repair mortars for stone – A review, Construction and Building Materials (59) Elsevier (2014) 39–50.

[17] Feilden BM., Conservation of Historic Buildings. 3rd ed. Oxford: Architectural Press, Elsevier (2003).

[18] Schueremans L et al., Characterization of repair mortars for the assessment of their compatibility in restoration projects: research and practice. Constr Build Mater 25 (2011) 4338–50 http://dx.doi.org/10.1016/j.conbuildmat.

[19] Matović V., Condition report of the built stone in the King's gate of the Belgrade fortress, Institute for the protection of monuments of culture, Belgrade (2007) (unpublished document).

Investigation and examination of a degraded Egyptian painted limestone relief from Tell Hebua (Sinai)

Eid Mertah[1]*, Moamen Othman[2], Mohamed Abdelrahman[3], Mohamed Fatoh[4] and S. Connor[5]
[1] *Conservator of archaeological materials, Conservation Centre, Egyptian Museum of Cairo, Ministry of Antiquities, Egypt*
[2] *Director of conservation Centre, Egyptian Museum of Cairo, Ministry of Antiquities, Egypt*
[3] *Director of Scientific Research, Projects Sector, Ministry of Antiquities, Egypt*
[4] *Chief of stone Conservators, Conservation Department, Egyptian Museum of Cairo, Ministry of Antiquities, Egypt*
[5] *Research Fellow at The Metropolitan Museum of Art, New York*
* *eid.mertah@yahoo.com*

Abstract

This paper presents a study bas-relief from the New Kingdom (1550-1070 BC), which was found in Sinai in 2009 and came the Egyptian Museum, Cairo, in a quite challenging condition. The interest of this piece lies in the danger that the salt causes to the block's surface, and particularly to the pigments of its decoration. The block is sculpted on two opposite sides, which both need to be consolidated, in order to be safely displayed in the museum's galleries. The authors describe the investigation carried out to identify the types of salts and to select the possible techniques to remove them and preserve the surface and pigments.

Keywords: limestone relief, Sinai, salt weathering, crystallized salts, stabilization.

1. History

The piece which is the object of this study is made of limestone. The decoration is on two opposing sides. Its dimensions are: Height. 107; Width. 114; Thickness. 27 cm. It is currently in display in the gallery in the Egyptian Museum, Cairo, with the inventory number JE 100014 (= SR 4/15988).

The block was found in a New Kingdom military site, discovered in North-West Sinai on April 22, 2009, Tell Hebua II,[1] the ancient "khetem of Tjaru"[2], about 4 km from the city of al-Qantara East, on the East Bank of the Suez Canal, about 50 km north of Ismailia.

Recent excavation led by Mohamed Abd El-Maksoud for the Supreme Council of Antiquities revealed an impressive defence system, consisting of a series of mud brick fortresses, protecting temples, administrative buildings and domestic structures, dating back from the New Kingdom to the Late Period (ca. 1550-332 BC). These fortified cities were intended to protect Egypt's North-Eastern border.

The site of Tell Hebua II, which seems to have been used for military purposes essentially in the New Kingdom, continued to be occupied in later times. Indeed, the excavators found this relief, together with several other temple blocks decorated and inscribed, reused in the casing of later tombs set within earlier structures. According to the continuity of the ornamentation and the orientation of their inscriptions, these blocks most probably belonged to a tripartite sanctuary, the foundations of which could not be identified so far. These blocks attest two phases of decoration: the first one during the reign of Thutmosis II (1482-1480 BC) and a second during that of Ramesses II (1279-1213 BC).

The decoration of the studied relief presents a particular interest, since it displays offering scenes on either of its opposing sides, sculpted at two different times: on one side, carved in raised relief[3], Thutmosis II receives life from the god of war Montu, while on the other side, which is carved in sunken relief, Ramesses II makes an offering of bread to the god of earth Geb. Very few representations of Thutmosis II are known so far, which makes this relief, of very high quality, a precious testimony of his reign. Furthermore, this block is evidence of the restoration and reuse of stone structures during the reign of Ramesses II in the fortress and its temple, two centuries after its building.

Due to the quality of its reliefs and its poor state of preservation, the piece was brought to the Egyptian Museum, shortly after its discovery.

2. Conservation state

The block JE 100014 was not found in its original setting, but reused as the closing of the entrance to a tomb cut into the rock. Since it was then buried under soil and surrounded by mud, it was in relatively wet and closed environment

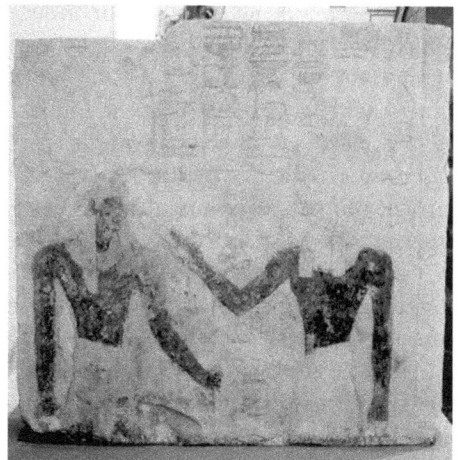

Figure 1: *Slab JE 100014 in its current display in the Egyptian Museum – face 1 [left image] (offering scene of Ramesses II) and face 2 [right image] (offering scene of Thutmosis II).*

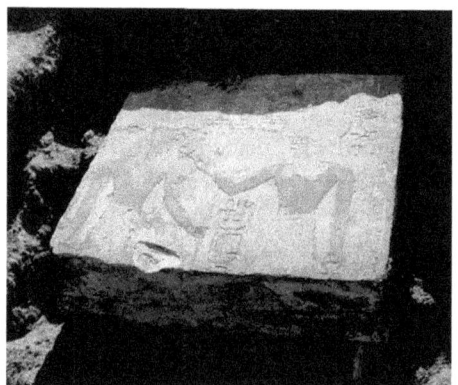

Figure 2: *Slab JE 100014 when discovered as closing for the entrance of a tomb*

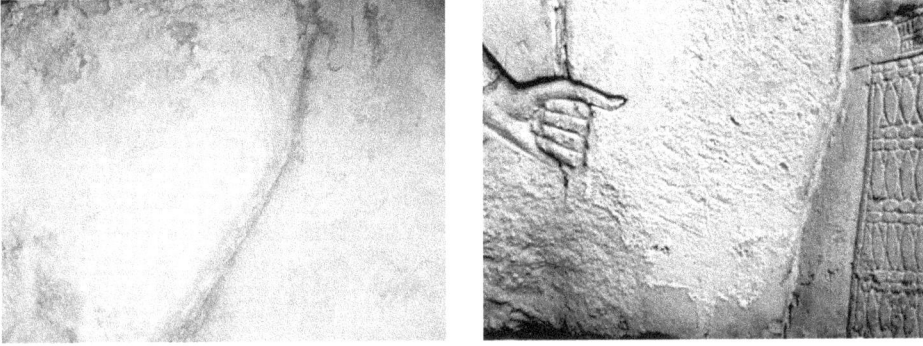

Figure 3: [left image] Crystallized salts in the upper area of the Ramesses II's side. [right image] Crystallized salts in the lower area of the same side.

for more than two thousand years. The photographs taken at the time of its discovery still show a good state of preservation of the upper side of the piece, which was that of Thutmosis II's offering scene.[4] According to the available records, and to the information provided by Gharib Sonbol, chief of the restoration team during the excavation, the lower face of the block, showing Ramesses II, had already suffered salt damage when the block was discovered. The removal of the limestone block from its wet atmosphere to a much dryer one, in the Egyptian Museum, pro-

bably intensified this process of salt damage since salts were moving to the surface of the block as drying occurred, and therefore accelerated the deterioration of its sculpted and painted decoration.

The majority of the painted areas became weak and needed urgent consolidation. In many zones, crystallized salts have appeared below the layer of colour, which might lead to its loss in the future. This situation is particularly visible on the side which contains the scene of Ramesses II (which was placed face down, when the piece was reused as a slab for closing

Figure 4: The circle shows the area where Microballoons filler was used to complete the missing parts of the pigment layer. This intervention should be reduced in order to become more discreet and suitable for the rest the surface.

the tomb). On that side indeed, crystallized salts are not limited to the painted layer, but also invade the whole surface, especially in the upper and lower areas, where the inscription is carved.

Tests and analyses were carried out on the painting (visual examination, examination using magnifying glasses and optical light microscope, portable X-ray fluorescence [PXRF], Fourier transform infrared spectroscopy [FTIR] Analysis). They revealed that the painting was previously restored; partly at least on the archaeological site itself, according to the information provided by the restoration team. As the analysis demonstrated *(see below, Fig. 7)*, it appeared that Paraloid B72 had been used in the consolidation process, and that microspheres known also as microballoons filler (i.e. a glass

Figure 5: Spots where Digital light microscope images were taken on the side of the block which shows the offering scene of Ramesses II.

Figure 7: Side of the offering scene of Thutmosis III, on the slab JE 100014. The photo shows the areas where Digital light microscope images were taken.

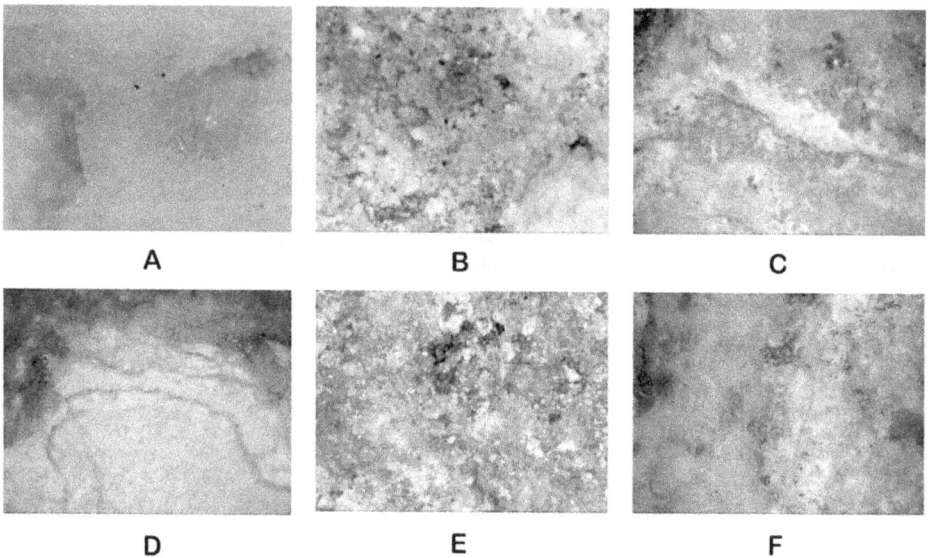

A

B

C

D

E

F

Figure 6: Digital light microscope images on Ramesses II's side of the slab. Images A and B show the crystallized salts on the surface of the relief. C and D show the crystallized salts above and below the pigments, which already led to the fall of some parts of the colour's layer. On image E the crystallized salts are captured inside the red colour, while F consists in a photograph of the crystallized salts above the blue colour. [A, B, E] 250x; [C, D, F] 50x.

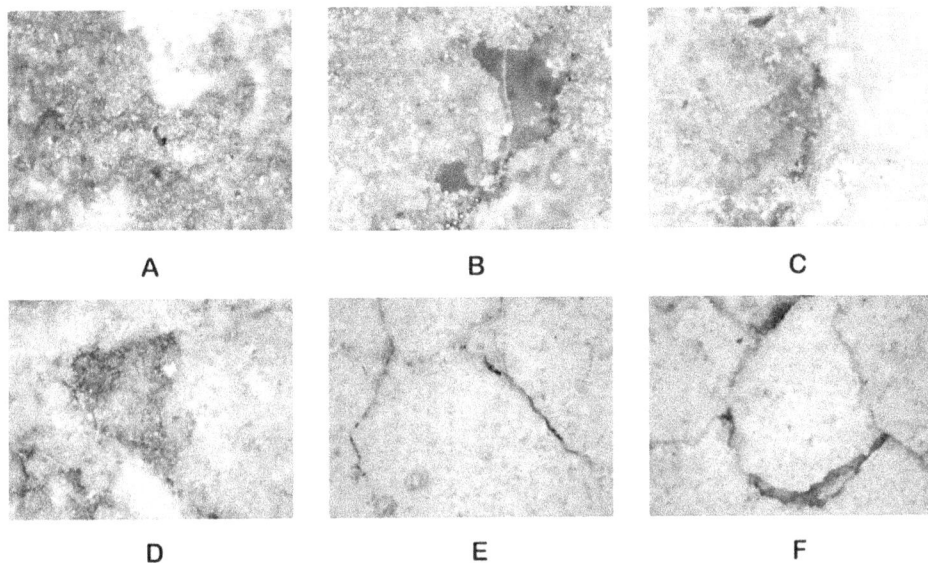

Figure 8: Digital light microscope images taken on the Thutmosis II's side. A shows the crystallized salts inside the red colour, B shows the crystallized salts above and below the pigments, which led to the fall of some parts of the colour's layer; on that photograph, we can see the flacking between the painted layer and the original surface. The crystallized salts inside the blue colour are visible on image C, while on D are shown the crystallized salts inside the eye of the figure of the king. E and F show the cracks and the micro-cracks on the surface of the relief. [A, C, D, E, F] 50x. [B] 250x.

fibre based on Silica [Si]) was also used to complete some parts, probably after the relief entered the Egyptian Museum.

The block needs a re-restoration, a stabilization of its fragile areas and a study of the possible ways to remove crystallized salts from the weak parts, and to preserve the painted layer.

In the framework of our study, different techniques, described in the following parts, have been used to identify the causes of rapid deterioration, salt weathering, which may have partially occurred in the external environment where the relief was lying before its discovery, but which has continued, since the relief has been installed in a poorly controlled interior environment, in the galleries of the Egyptian Museum.

3. Analytical Procedure

This section discusses the results of the different analyses and investigation techniques carried out to identify the salts emerging from the block and damaging its surface, and the nature and the state of preservation of the pigments on the decorated surface.

3.1. Optical light microscope (O.L.M.)

A portative optical light microscope was used in order to reveal details such as the porosity, grain morphology and micro-cracks. A series of twelve spots (six on each side) were analysed *(see fig. 5 and 7)*. The used device (Dino-Lite Digital Microscope) allows taking high quality photos (from 50 to 250x), even without direct contact with the surface of the object – the fragile state of preservation of the relief JE 100014 required a distance of a few millimetres to be kept. This first examination allowed determination of the different lithotypes of the block: it provided information about the damaged layers, such as the sequence of layers, the particle size, as well as the colour and texture of these layers. It also showed the cracks and the micro-cracks on the surface of the relief.

Furthermore, the microscope examination revealed that the salts are not only above the layer of colour, but also under it and inside it and thus putting it in particular danger *(see fig. 6 and 8)*.

3.2. Portable X-Ray Fluorescence [PXRF]

X-Ray Fluorescence is widely used in the field of archaeometry due to its non-destructive properties, its high sensitivity and its applicability to a wide range of situations. The Egyptian Museum's lab owns a portable device (Elio Device

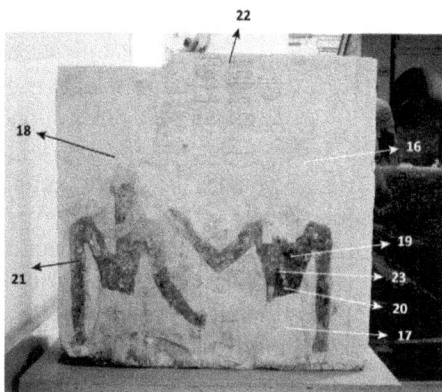

Figures 9 and 10: Slab JE 100014, with numbers showing the areas where PXRF analysis was carried out – face 1 [Fig. 9] (offering scene of Ramesses II) and face 2 [Fig. 10] (offering scene of Thutmosis II).

SN 177), which we used to determine the specific elements that are present within the pigments and in the stone block itself (including the salts).

We selected a series of twenty-two points (fifteen on the Ramesside size and eight on the Thutmoside one) in order to cover all the different areas of the surface of the block, including the different painted layers, as well as the various parts covered with salt.

The analysis revealed a widely spread crystallization of chlorine salts on Ramesses II's side of the slab, and what appears to be calcium sulphate salts on the other side, due to the presence of sulphur *(see table 1)*. Both types of salt are also present within the red pigment (made of hematite, *see Table 2*), as well as in the blue and green pigments (both acquired from copper firing process, *see Tables 3 and 4*).

On the side (1), depicting Ramesses II (see figure 9-10), two analyses were made (14 and 15) on what seemed to be white colour. This revealed itself to be, through the PXRF, not to be a pigment but the microballoons used during the previous restoration (see above).

3.3. Fourier Transform Infrared Spectroscopy [FTIR] analysis

A sample of the red colour which had already fallen from the surface of the block was sent to the laboratory of the Ministry of Antiquities for a FTIR analysis, which allowed identifying an adhesive applied on the pigment layer, in order to fix it during the previous restoration: Paraloid B72.

4. Preliminary conclusion and salt-reduction treatment

This article discusses the investigation carried out on a limestone relief found in

	No.	Elements	Results
(1) Side with Ramesses II's offering scene	1	Ca, Cl	The stone itself is limestone (calcium carbonate). As revealed by the analysis, the side (1) suffers from a widely spread crystallization of chlorine salts on the surface. The side (2), however, contains sulphur, and may therefore contain calcium sulphate salts.
	2	Ca, Cl	
	3	Ca, Cl	
	4	Ca, Cl	
	5	Ca, Cl	
(2) Side with Thutmosis II's offering scene	16	Ca, S	
	17	Ca, S	
	18	Ca, S	

Table 1: Results of the [PXRF] analysis on the stone itself

	No.	Elements	Results
(1) Side with Ramesses II's offering scene	9	Ca, Cl, Si, Fe, S	The PXRF revealed the presence of Iron (Fe) inside all the red pigment, which means that it consisted of red hematite (iron oxide hematite Fe_2O_3). The presence of chlorine and calcium sulphate is due to the crystallization of salts inside the pigments (as well as under and above them).
	10	Ca, Cl, Si, Fe, S	
	11	Ca, Cl, Si, Fe, S	
(2) Side with Thutmosis II's offering scene	19	Ca, Cl, Fe, S	
	20	Ca, Cl, Fe, S	
	21	Ca, Cl, Si, Fe, S	
	23	Ca, Cl, Fe	

Table 2: Results of the [PXRF] analysis on the red colour

	No.	Elements	Results
(1) Side with Ramesses II's offering scene	6	Ca, Cl, Si, Cu, S	The presence of chlorine and calcium sulphate inside the green pigment is due, like in the case of the red one, to the crystallization of salts (see above). However, the colour was here acquired from copper.
	7	Ca, Cl, Si, Cu, S	
	8	Ca, Cl, Si, Cu, S	

Table 3: Results of the [PXRF] analysis on the Green colour (only on side 1)

	No.	Elements	Results
(1) Side with Ramesses II's offering scene	12	Ca, Cl, Si, S, Cu	The elements are similar to those observed previously. Indeed, green and blue colours are both acquired from a firing process of copper. The presence of chlorine and calcium sulphate provides again the nature of the crystallized salt inside the pigment.
	13	Ca, Cl, Si, S, Cu	
(2) Side with Thutmosis II's offering scene	22	Ca, Cl, Si, S, Cu	

Table 4: Results of the [PXRF] analysis on the Blue colour

a wet environment ten years ago, currently in display in the Egyptian Museum, Cairo (inv. JE 100014). This block has been subject to different deterioration processes, and needs urgent consolidation and restoration. Atmosphere variations during the years following the block's removal from its original environment caused salt crystallization and numerous internal and external stresses.

Our team carried out various techniques of analysis, in order to establish the different causes of damage and to identify the most suitable methods of restoration. First of all, before carrying out any conservation work, the block was carefully examined on all its sides. Each of the alterations and degradations was mapped, documented and registered in the conservation records.

A first investigation with the Optical light microscope (O.L.M.) allowed identification of the nature of the stone and the different lithotypes present in the block, and drawing out a map of all cracks and micro-cracks. It revealed the presence of salts not only above, but also inside and beneath the layers of colour, on the decorated surface of the reliefs.

An examination with a Portable X-Ray Fluorescence [PXRF] tool established the nature of the different elements from which the stone and pigments were composed. The whole surface of the block and its painted decoration were found to suffer from a crystallization of chlorine and calcium sulphate salts.

Finally, a Fourier Transform Infrared Spectroscopy [FTIR] analysis applied on a sample of red pigment revealed the presence of adhesive Paraloid B72, applied on the pigment layer in order to fix it during a previous restoration.

These different methods demonstrate that the weathered surfaces of the block have undergone continual warming and cooling cycles, due to the semi-dry atmosphere of Tell Hebua. The surfaces now exposed are enduring physical disintegration, as assessed from macroscopic visual analysis. The block suffers from several deterioration of its structural coherence: cracks, microcracks, cavities, loosing of material, resulting of needles of salts between grains, and accumulation of different kinds of dirt.

We are currently in the process of identifying the nature of the salts and their

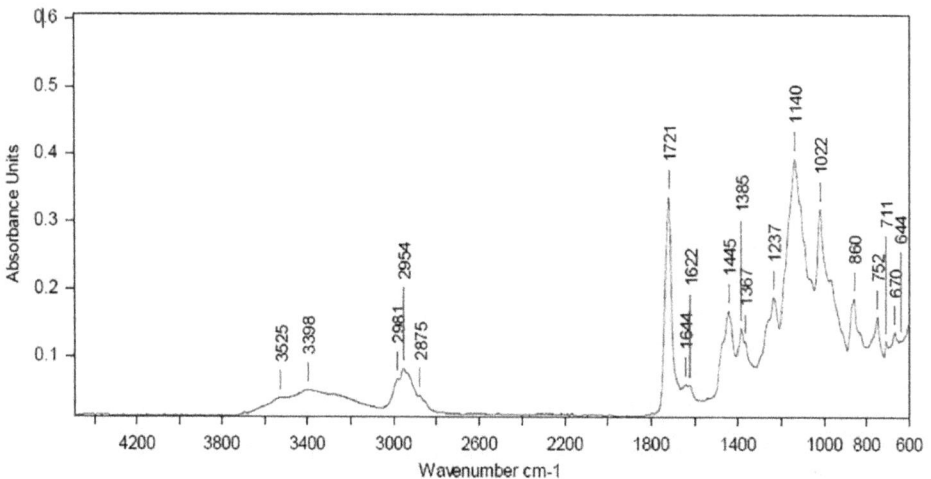

Figure 11: Spectrum FTIR of the sample of red pigment

degree of solubility, in order to choose the most appropriate method for removing them to prevent further desintegration of the stone due to their active presence. Some salts revealed themselves to be of calcium sulphate, which is very soluble and will be easily removable, but some other salts might be of unsoluble calcite, which would therefore require a mechanical cleaning. The chlorides identified on the side of Ramesses II (see table 1), they may show a variable level of solubility (either high solubility, if they consist of calcium or copper chlorides, or low solubility in the case of sodium chloride), and therefore require further investigation.[5]

Before any new restoration, we will first proceed to some de-restoration of the piece. The microballoon filler, applied at the arrival of the block in the museum, should be reduced to a smaller area, in order to produce a more suitable and sympathetic aesthetic appearance. In other situations, a de-restoration, although desirable, will probably be difficult: indeed, the presence of the materials used in early restoration treatment will have an impact on the ways of undertaking future consolidation work and must be taken into account. The Paraloid B72 applied on all the painted surfaces forms a film, covering large areas, which prevents the salts reaching the surface of the block during their migration. This creates a detachment between the stone and the painted layer, which is likely to continue to increase.

Due to the friable nature of the stone, and the fragility of the pigments[6] each consolidation and conservation intervention will be performed according to the needs of each of the individual problematic zones, after thoroughly testing all possible procedures.

In order to prevent further degradation of the relief, some precautions will be necessary to keep it on display in the Egyptian Museum. Controlling, to the maximum of possibilities, the relative humidity and temperature in the gallery would help prevent further salt damage; this is of course quite difficult in the current conditions of display, but will hopefully be improved in the near future.

References

[1] Concerning the discovery of the relief, see Abd El-Maksoud M. and Valbelle D., "Tell Héboua II: Rapport préliminaire sur le décor et l'épigraphie des éléments architectoniques découverts au cours des campagnes 2008-2009 dans la zone centrale du khétem de Tjarou", Revue d'Égyptologie 62 (2011), p. 1-39.

About the site of Tell Hebua, its defence system and its region, see: Abdel Maksoud M. and Valbelle D., "Tell Héboua-Tjarou: l'apport de l'épigraphie", Revue d'Égyptologie 56 (2005), p. 1-44; Id., Tell Heboua (1981-1991). Enquête archéologique sur la Deuxième Période Intermédiaire et le Nouvel Empire à l'extrémité orientale du Delta, Paris, 1998. Abdel Maksoud M., "Projet de sauvetage des sites antiques du Nord Sinaï", Discussions in Egyptology 24 (1992), p. 7-12; Oren E., "Migdol: A New Fortress on the Edge of the Eastern Nile Delta", BASOR 156 (1984), p. 7-4; Redford, D., "Report on the 1993 and 1997 Seasons at Tel Qedwa", Journal of the American Research Center in Egypt 35 (1998), p. 55.

[2] Concerning the name of the site in the antiquity and the nature of frontier fortresses, see Somaglino C., "Les 'portes' de l'Égypte de l'Ancien Empire à l'Époque Saïte", Égypte, Afrique & Orient 59 (2010), p. 10; Id., Du magasin au poste-frontière dans l'Égypte ancien-

ne : étude lexicographique du vocable khetem, (unpublished Phd dissertation, Paris IV University, under the direction of D. Valbelle, 2010, forthcoming in the Ifao series.

³ On the side of Thutmosis II, the figures and the hieroglyphs of are all in raised relief, except for a few signs in the middle of the inscription, which are in sunken relief. These signs, which give the name of "Montu, lord of Thebes", were most probably originally in raised relief too, but were erased during the iconoclastic phase of the Amarnian period (1353-1334 BC), and then restored, necessarily in sunken relief, probably in the time of Ramesses II, when the king commissioned a new ornamentation of the temple. Cf. Abdel Maksoud M. and Valbelle D. 2011 (cf. supra).

⁴ Our thanks go to Gharib Sonbol, who directed the team of restoration on the site during the excavation, and who kindly provided us information about the context of discovery and the state of preservation of the block in 2009, and allowed us to use the photos taken during the mission, in this article. Thanks are also due to Bianca Madden, for her coments and kind revision of English writing.

⁵ Nord A.G., "Effloresecence salts on weathered building stone in Sweden", in Geologiska foreningens I stockholm forhandlingar 114/4 (1992), p. 223-229: "The solubility of calcium sulphate salt is (2,4g/l at 20°C and PH7) is more than 200 times greater than that of calcite, but the solubility of sulfates is much less than that of chlorides."

⁶ For similar cases and valuable sources of inspiration for salt treatment, see: Mirabootalebi R., "Conservation of A Highly Degraded Egyptian Limestone Stele", e-conservation Journal 3 (2015), available online 06 November 2015 (http://www.e-conservation.org/issue-3/52-Egyptian-Limestone-Stele); Doehne E., "Salt Weathering: A Selective Review", Geological Society Special Publication. Natural Stone, Weathering Phenomena, Conservation Strategies and Case Studies 205 (2002), p. 51-64.

Technique for transportation of stone sculptures damaged by salt crystallization

Vinka Marinković
Croatian Conservatian Institute, Croatia
* vmarinkovic@h-r-z.hr

Abstract

The paper presents a technique for handling and transportation of stone sculptures damaged by salt crystallization. A limestone sculpture from the Cathedral of St. Lawrence in Trogir was studied and pre-consolidated with cyclododecane. First, test were performed in the lab. On 5 glass slides were applied powder of the Seget stone and then consolidated with 50% solution of CCD in Shellsol T, 50% solution of CCD in white spirit, 50% solution of CCD in toluene, with holt melted CCD and hot melted CCD with cotton gauze. After the tests were performed, a technique of pre-consolidation with melted cyclododecane and facing tissue was used in situ on the sculpture. Practical uses of the technique were considered in the paper.

Keywords: cyclododecane, stone, pre-consolidation, salts.

1. Introduction

The Cathedral of St. Lawrence in Trogir (Croatia) is a Roman Catholic triple-nave basilica constructed in the Romanesque-Gothic style. Trogir Cathedral was constructed over a period of four centuries (1200-1598), and it illustrates all the successions of styles in that time in Dalmatia. It is one of the most important monuments in Croatia, and was declared a World Heritage Site by UNESCO in 1987. The integral part of the cathedral is the richly decorated Renaissance Chapel of St. John (The Chapel of the Blessed Giovanni Orsini), created by the masters Nikola Firentinac and Ivan Duknović in the period from 1461 to 1497. The chapel was restored and cleaned in between 2000 and 2002 as part of the American foundation project - Venetian Heritage.

The Croatian Conservation Institute started the monitoring and preliminary investigations of the chapel in 2016 because some adverse changes were obsereved on the surface of sculptures and stone interior. In a few locations, it was possible to notice high levels of powdering, sanding and sugaring of the stone, which was caused by salt crystallization.

The Annunciation group (sculptures of the Angel and the Virgin) positioned on the triumphal arch of the chapel was in a terrible state. The bases and lower back zones of both sculptures were extremely damaged, and were in unstable conditions with a high loss of the substrate. First analyses of samples from the background zones of both sculptures showed presence of the following salts: chlorides and nitrates, in concentrations ranging from 0,13-0,17%. Both sculptures were made of local limestone (called Seget).

2. Methods and materials

The presence of water soluble salts in the samples was determined by chemical analyses and UV/VIS spectroscopy with a Perkin Elmer Lambda 25 instrument. Compositional salt analysis was performed using X-ray diffraction method with a Philips Vertical X-ray Goniometer

(X-Pert type). Water content measurement of the stone showed that the water coefficient values were within the reference values for the same type of stone collected from the Seget quarry.

Considering the extremely damaged state of the sculptures and their original position (5m above the floor level), we decided to dismantle them and carry out the conservation work of desalination in the workshop.

Before the dismantling and transportation of the sculptures it was necessary to pre-consolidate the most unstable powdery areas of the sculptures. Cyclododecane was chosen as consolidant.

Cyclododecane (abbreviated CDD) is a waxy 12-carbon cyclic hydrocarbon ($C_{12}H_{24}$) that sublimes from a solid to gas at room temperature.[1] This has made it an appropriate material in conservation as a temporary consolidant, adhesive or barrier layer during the last few decades.[2] When the object is highly damaged and contaminated with soluble salts, and when it needs to be transported, CDD is an appropriate solution. One of the main reasons is that harmful soluble salts are not locked in the material. Some successful cases have already been reported in literature.[3]

CDD can be applied as a melt or as a solution [Burton, 2009]. In order to choose an appropriate technique and method of application, some test were performed in the laboratory prior using it on the sculptures of Trogir. On glass slides we applied powder of the Seget stone, and then we applied 50% solution of CDD in Shellsol T *(glass slide 1, Fig. 1)*, a 50 % solution of CDD in white spirit *(glass slide 2, Fig. 1)*, a 50 % solution of CDD in toluene *(glass slide 3, Fig. 1)*, hot melted CDD *(glass slide 4, Fig. 1)* and hot melted CDD with cotton gauze *(glass slide 5, Fig. 1)*.

After a visual observation, it was possible to conclude that the hot melted CDD in both cases formed a dense film, and the stone powder was consolidated and packed (the particles of powder are interconnected). When CDD was dissolved in solvent, the layer was less dense and the powder was less consolidated and packed.

After a fingernail scratching test (after 24 hours) it was possible to conclude that the powder treated with hot melted CDD (in both cases, glass slide 4-5 with hot melted CCD and hot melted CCD with gauze, *Fig. 2*) was stable and still packed. Powder treated with CDD dissolved in the solution *(glass slide 1-3, Fig. 2)* after the

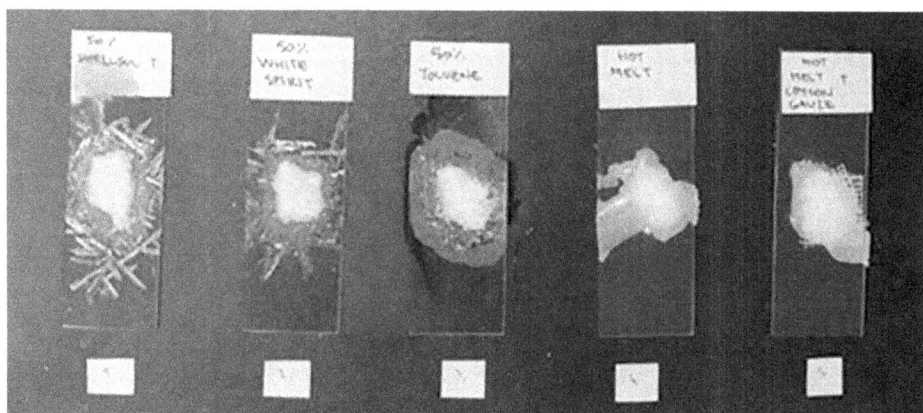

Figure 1: Glass slides with Seget powder and different types of application of CDD (from left to right: 50% solution of CDD in Shellsol T; 50 % solution of CDD in white spirit; 50 % solution of CDD in toluene; hot melted CDD; hot melted CDD with cotton gauze)

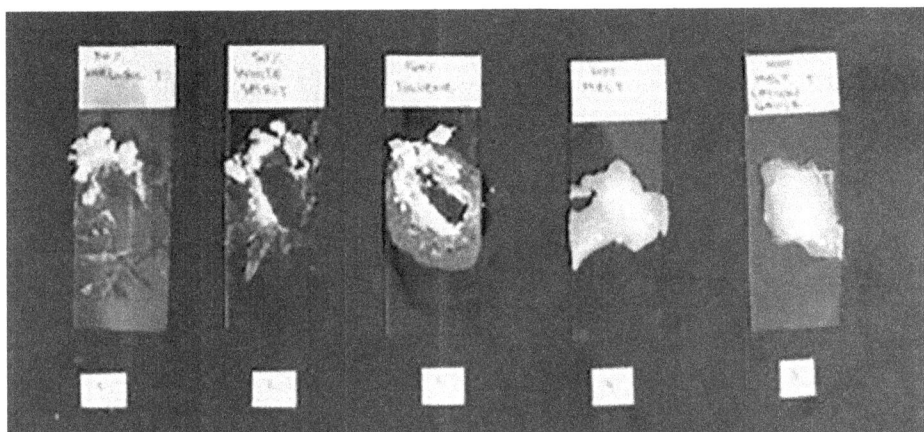

Figure 2: Glass slides with Seget powder and different types of application of CDD after fingernail scratching test (from left to right: 50% solution of CDD in Shellsol T, 50 % solution of CDD in white spirit; 50 % solution of CDD in toluene; hot melted CDD; hot melted CDD with cotton gauze)

test was unstable, and the consolidation effect was less effective.

After the preliminary laboratory tests, the technique of pre-consolidation with melted CDD and facing tissue was used in situ. CDD was reactivated with heat and applied on the surface of the sculpture of the Angel using a brush. Simultaneously, on the surface we applied two layers of cotton gauze. After the pre-consolidation, the sculpture was wrapped in foil and lowered to the ground using a crane. The facing tissue did not detach in 10 days in the lab environment (approximately 20°C with moderate airflow). CDD was reactivated with a heating gun and removed.

3. Results

Laboratory and field tests of pre-consolidation, handling and transportation of stone sculptures damaged by salt crystallization showed good results when the technique with hot melted cyclododecane and facing tissue was used. The sculpture of the Angel from Trogir Cathedral was successfully transported to the workshop using the aforementioned method. A minimum of unstable material was lost. Other negative changes were not recorded on the surface of the stone.

4. Disscusion

Only practical uses of the technique were considered in the paper. The investigations were limited because of technical and financial issues. Future analytical research should provide answers to the following questions:

- depth of the penetration of CCD on the Seget stone,
- sublimation rates,
- negative aspects in the presence of harmful soluble salts.

5. Conclusion

Cyclododecan (CDD) has been used increasingly as a temporary consolidant for a variety of weak or fragile materials. This contribution focuses on the practical use of cyclododecane as a temporary consolidant in cases of handling and transportation of stone sculptures damaged by salt crystallization. Melted

CDD, melted CCD with facing tissue and saturated solutions of consolidant in white spirit, shelsoll T and toluene were applied on glass slides with stone powder. By visual observation and using a fingernail scratching test, it was possible to conclude that the best effect of consolidation was obtained during the application of hot melt CDD. After the lab tests, the technique of pre-consolidation with melted cyclododecane and facing tissue was used in situ. The extremely damaged sculpture of the Angel from Trogir Cathedral was consolidated with melted CDD and cotton gauze. After the consolidation, it was transported to the workshop of the Croatian Conservation Institute. During the intervention, a minimum of unstable material was lost.

Acknowledgements

The project is supported by the Ministry of Culture of the Republic of Croatia.

References

[1] Watters, C. "Cyclododecane: A Closer Look at the Practical Issues", Anatolian Archaeological Studies (XVI), (2007), 195–204.

[2] Burton, D. J: An investigation into the efficacy of cyclododecane as a volatile exclusion layer providing isolation of an un-saturated surface during consolidation, Conservation Studies, 2009.

[3] Cleere, D. C.: „Cyclododecane re-investigated - An experimental study on using cyclododecane to secure a stable ceramic surfaces prior to transportation", Conservation News (94), (2005), 26-28.

Investigation and conservation of salt damaged epitaphs in the church of Werben (Saxony-Anhalt, Germany)

Steffen Laue*, Dörte Poerschke and Benjamin Hübner
University of Applied Sciences Potsdam, Department of Conservation and Restoration, Germany
*st.laue@fh-potsdam.de

Abstract

The case study of the St. Johannis church in Werben (Saxony-Anhalt) is one of those instances where conservators and restorers working in the field of building heritage have to find sustainable solutions for a cultural heritage site affected by high humidity levels and salt content for centuries. In the church of Werben, the ground floor of the steeple houses a remarkable collection of predominantly polychrome epitaphs from the 16th – 18th centuries. These epitaphs reflect in an unique way the story of important citizens of Werben and provide an extensive overview of the design and techniques used in the manufacturing of such monuments through various epochs. During an inventory of the epitaphs, several damages were attributed to salts and the high moisture content in the ground floor of the steeple. Salt analyses, climate measurements and calculations with the computer model ECOS/RUNSALT were used to understand the damage processes of the epitaphs. The results provide the basis for applying appropriate conservation methods and materials pertaining to the room and the polychromic epitaphs. As a case study one epitaph was removed and preserved in the workshop of the conservation and restoration department in Potsdam.

Keywords: Werben, polychromic epitaph, salt damage, climate, ECOS/RUNSALT, conservation concept

1. Introduction

Werben is located in the northern part of Saxony-Anhalt at the river Elbe in the administrative district of Stendal between Magdeburg and Wittenberge. Because of the proximity to the river, the town was and is continuously threatened by floods, especially in the past when no adequate flood prevention was in place.

St. Johannis church was built with bricks in the 15th century but includes older parts from the Romanesque period [1]. The church consists of a three axial nave with the choir in the East and a huge steeple in the West *(Fig. 1)*. In this paper, we focus on the ground floor of the steeple in the western part of the church

Figure 1: St. Johannis church in Werben, southwestern part with steeple

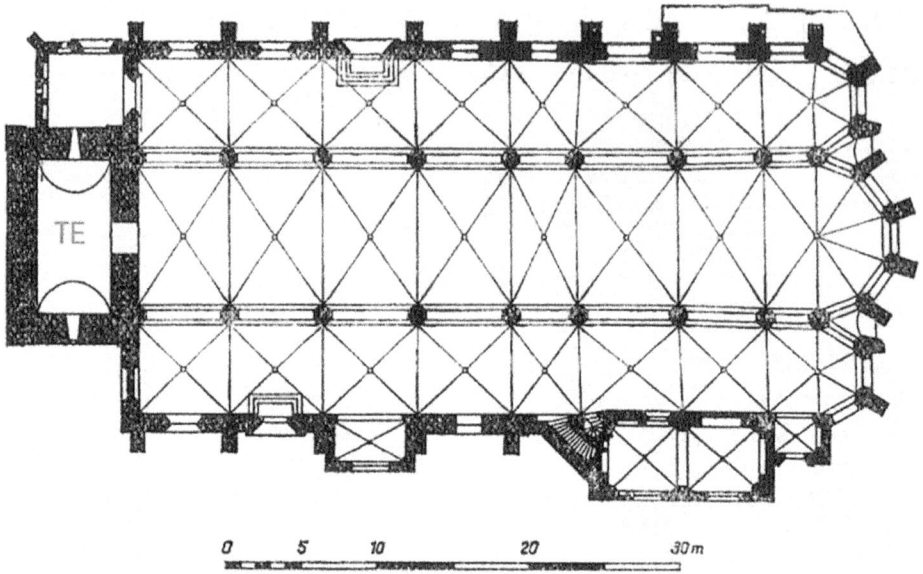

Figure 2: Ground map of St. Johannis 1 containing the ground floor of the steeple (TE)

with a floor area of 9,20 m x 4,20 m and a height of 4,82 m in the zenith of the barrel-vault. Today, the ground floor can only be reached through a wooden door from the nave *(Fig. 2)*. The masonry of the ground floor consists of 1,50 m – 1,80 m thick walls made of bricks and mortar that today, contain only one window to the south leading to a moist and dark room. The walls were plastered with a lime mortar that is now partly damaged by moisture and salts.

Since 1868, the ground floor has been used as a depository for 18 stone epitaphs from the 16th – 18th centuries. These epitaphs reflect in a unique way the story of important citizens of Werben and provide an extensive overview of the design and techniques used in manufacturing these monuments through various epochs. The epitaphs are partly polychrome and gilded and show the wealthy members of citizenry and clergy of the town *(Fig. 3)*. It is not quite clear where the epitaphs were located before 1868 but it has to be assumed that they were placed somewhere in or around the church and the cemetery.

Figure 3: polychrome epitaphs at the southern wall and the only window in the ground floor, in the middle the epitaph of Leonhardt Kämpfe

Today, each epitaph stands on a ca. 40 cm high pedestal made of bricks that are plastered *(Fig. 3)*. The epitaphs are fixated by iron anchors to the back. Additionally, bricks with mortars were applied irregularly to support the backward connection of the epitaphs to the walls. Probably later grey cement mortars were used to reinforce the connection of the epitaphs to the back.

In an inventory of the epitaphs made by students and in the M.A. thesis by Dörte Poerschke[2], materials and damages of the epitaphs were determined and mapped showing advanced degradation of the stones and colored surfaces obviously caused by moisture, salt crystallization and microbiological impact. Thus, the aim of the investigation was to understand the damage processes in the ground floor and to find solutions for the conservation and restoration of the epitaphs as well as the masonry.

2. Analytical approach

To find an appropriate conservation concept for the ground floor of the steeple as well as for the epitaphs the following investigations were completed.

During the inventory the epitaphs and the damage of the walls and epitaphs were documented. Salt efflorescences and crusts were sampled from different surfaces and analyzed by polarizing microscopy, microchemistry and XRD (Siemens D5005, CuKα-radiation) to detekt the crystallizing salts.

Drill powder samples were taken to investigate the distribution of salt ions concentration in two vertical profiles and into the depth at the southern and western wall. 1 g of sample material was mixed with 100 ml distilled water for about 24 hours and then filtered before the analyses of Na^+, K^+, Ca^{2+}, Mg^{2+}, Cl^-, NO_3^- and SO_4^{2-} were completed using ion chromatography (Dionex, IC90).

Permanent measurement of room temperature, surface temperature and relative humidity was implemented using RH/T sensors and a data logger.

To investigate under which climatic conditions new salt crystallization takes place and what kind of damage they produce, a monitoring process was implemented: Climate measurements were executed combined with the periodic observation of the crystallized salts at defined surface areas.[3]

The observations were compared using the computer model ECOS/RUNSALT in order to assess under which climate condition salts are crystallizing in the ground floor of the steeple.[4, 5]

The reduction of the salt ion content through cellulose poultices from the epitaph of Leonhardt Kämpfe were verified by taking 5 x 5 cm sized poultice samples from the surface, extracting the ions by deionized water and measuring the electrical conductivity which correlates to the total amount of extracted salt ions.

3. Damage

In the ground floor of the steeple a high rate of damage in bricks and plaster due to moisture and salts was discerned.

At the masonry the plaster shows deteriorations like lacunas, cavities, bulges, and partly chalking[6] whereas bricks partly decay by flaking and granular disintegration.

In sum, the strongest damages are visible up to a height of approximately 1.5 m. Between heights of about 1.5 and 3 m the damage is considerable but scattered and above 3 m strong damages of the plaster mainly show at the western walls including salt efflorescences.

On the epitaphs the loss of paint by flaking and peeling[7] is visible as well as the disintegration of the sandstone itself. Salt crystallization on the epitaphs can be observed in areas where the epitaphs are connected with the walls by bricks or mortar indicating the transport of ions through the pore system of the bricks and mortar into the epitaphs.

Salts

On the wall surfaces, crystallized salts consist of niter (KNO_3), epsomite ($MgSO_4 \cdot 7H_2O$) hexahydrite ($MgSO_4 \cdot 6H_2O$) and gypsum ($CaSO_4 \cdot 2H_2O$).

The distribution of salt ions was analysed in two vertical profiles at the western and at the southern wall. As an example, *Fig. 4* demonstrates the distribution of ions at the western wall in the heights of 0.9, 1.4, 2.0, 3.0, 3.7 m. *Table 1* shows the ion content (in weight percent) at each height and depth at the same wall. Although the distribution of salt ions in both walls is rather complex, some general tendencies can be noted:

- Sulphate and calcium are always accumulated in the first centimeter of each height, probably due to the conversion

of calcite to gypsum and the low solubility of gypsum.

- Nitrate, sodium and potassium are distributed in all heights and depths especially above 2.0 m of height. The concentration of sodium compared with potassium is mainly twice as much, which was a little surprise at first because no crystallized sodium salts could be detected on the ground floor surfaces.

- Slightly increased amounts of magnesium can be found in the lower parts of the wall which correlates with the magnesium sulfates found in this region of the walls.

Although hardly any tendency of enrichment of salt ions can be observed in different heights and depths, there is no clear explanation why the salt ions spread in the wall as pointed out. The ions might have been transported partly by rising damp but not into the height of 3.7 m. The infiltration of water through defects in the masonry could be a pos-

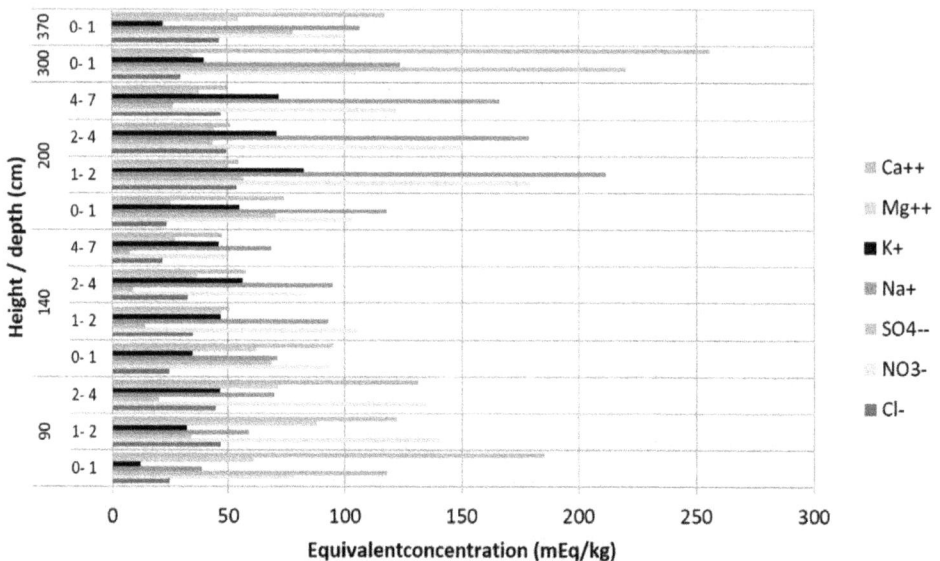

Figure 4: Salt ion distribution in the western wall expressed in Equivalentconcentration (mEq/kg)

sample	height (cm)	depth (cm)	Ion concentration (wt%)						
			Cl⁻	NO$_3$⁻	SO$_4$²⁻	Na⁺	K⁺	Mg²⁺	Ca²⁺
1	90	0- 1	0,09	0,48	0,57	0,09	0,05	0,07	0,37
2	90	1- 2	0,17	0,87	0,16	0,14	0,13	0,11	0,24
3	90	2- 4	0,16	0,84	0,10	0,16	0,18	0,09	0,26
4	140	0- 1	0,09	0,58	0,33	0,16	0,13	0,08	0,19
5	140	1- 2	0,12	0,66	0,07	0,21	0,18	0,06	0,10
6	140	2- 4	0,12	0,52	0,04	0,22	0,22	0,04	0,12
7	140	4- 7	0,08	0,31	0,04	0,16	0,18	0,03	0,10
8	200	0- 1	0,08	0,64	0,34	0,27	0,21	0,03	0,15
9	200	1- 2	0,19	1,11	0,27	0,49	0,32	0,06	0,11
10	200	2- 4	0,18	0,94	0,21	0,41	0,28	0,05	0,10
11	200	4- 7	0,17	0,76	0,13	0,38	0,28	0,05	0,10
12	300	0- 1	0,11	0,65	1,06	0,28	0,16	0,04	0,51
13	370	0- 1	0,16	0,62	0,37	0,24	0,09	0,07	0,23

Table 1: Salt ion distribution in the western wall expressed in weight percent (wt%)

sible explanation for the accumulation of water and ions in higher regions of the wall, but there is no evidence for that.

Nevertheless, it can be concluded that the ground floor of the steeple is affected by an accumulation and complex distribution of salt ions in the masonry consisting mainly of Na⁺, K⁺, Ca²⁺, Mg²⁺, NO$_3$⁻ and SO$_4$²⁻.

Climate

Permanent climate measurements of room temperatures, surface temperatures and relative humidity in the ground floor started in 2013. *Fig. 5* shows the daily average temperature and relative humidity from December 2013 until June 2014 demonstrating that the relative humidity in the ground floor consistently lies between 80 and 90% both in summer and in winter. After evaluation of the climate measurement results of two years it can be concluded that the temperature in the ground floor varies between 4°C and 18°C and the relative humidity bet-

ween 75% and 95% resulting in a permanently moist environment.

Simulation of the crystallization of salts using ECOS/RUNSALT

ECOS/RUNSALT is a computer model to predict the crystallization behavior of salts based on the thermodynamic model ECOS (Environmental Control of Salts).[4, 5] For this purpose, a set of representative salt ion contents is needed derived from the first cm of the surface of a monument. Then, the data have to be simplified (even ionic balance, removal of gypsum) before the model is used to calculate which soluble salts will crystallize under predetermined environmental conditions like temperature and relative humidity. Thus, the program makes it possible to predict under which environmental conditions salts will probably damage the surfaces of a monument.

As an example, in *Fig. 6* the data of sample 12 are calculated with ECOS/RUNSALT assuming a room temperature

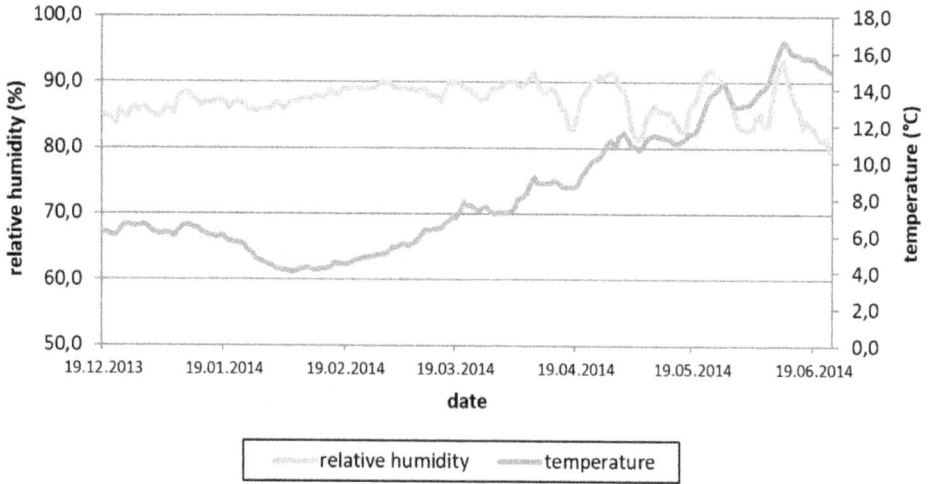

Figure 5: Daily average values of room temperatures (scale right) and relative humidity (scale left) in the ground floor

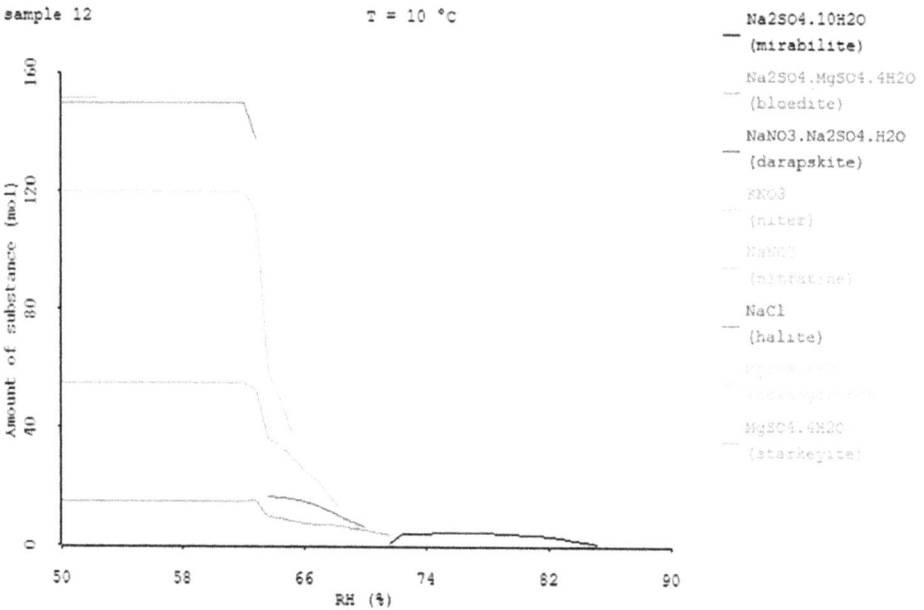

Figure 6: Image of ECOS/RUNSALT after calculation of sample 12 at 10°C showing the crystallized salt phases between 50% and 90% of relative humidity

of 10°C. It can be seen that the crystalliz-ation of salts starts below 86% of relative humidity with the salt mirabilite (Na-SO$_4$.10H$_2$O). Furthermore the chart de-monstrates that the main crystallization of salts will be below a relative humidity of 70% mainly consisting of the sodium and magnesium sulfates and nitrates. In this area of the wall, efflorescences of ni-ter (KNO$_3$), but no mirabilite were deter-mined.

Further ECOS/RUNSALT calculations with data from other surface areas in the ground floor yielded similar results with regard to the crystallization process of salts presuming that crystallization mostly happens below a relative humidity of 70% with much damage to the walls and epitaphs. Thus, the calculations give an important hint for conservation measures: To avoid further damages in the ground floor through salt crystallization, the relative humidity should not fall below 70% of relative humidity.

Conclusion of the weathering situation

In the ground floor of the steeple, the masonry and epitaphs suffer from high moisture content with a relative humidity of predominantly above 80%. Moisture sources are probably rising damp and infiltration from outside over the years. Only one door to the nave, which is mostly closed, and one little window to the south do not allow for enough air exchange with the outdoor climate resulting in this quite stable humid environment. The environmental condition keeps most of the analyzed salt ions in solution, but enables biological growth. With higher air exchange rates between the climate in the ground floor and the outdoor climate the relative humidity level would probably be lower than 70% at certain times resulting in more damage due to salt crystallization.

On the other hand, the high relative humidity levels produce damages to the painted epitaphs resulting especially in the degradation of the binder [8]. The paint layers of the epitaphs show in parts extreme damages like flaking. Keeping the epitaphs permanently in the ground floor, the polychromy of the epitaphs would be slowly destroyed forever.

Based on the climate situation in the ground floor today, it can be concluded that the high relative humidity levels are

Figure 7: Epitaph of Leonhardt Kämpfe in Werben before (left) and after conservation (right)

beneficial regarding salt damage but promote the decomposition of the paint layers on the epitaphs as well as biological growth everywhere in the room.

Conservation of the epitaph of Leonhardt Kämpfe

After the awareness of the weathering situation in the ground floor, the decision was made that one epitaph should be removed and transported to the workshop in Potsdam where a conservation of the epitaph was to be completed as a case study. The option of working on the epitaph under stable and relatively dry environmental conditions in the workshop was a decisive factor for removing the object from its original location where the preservation would have been more complicated due to the humid environment.

We chose the epitaph of Leonhardt Kämpfe *(Fig. 7)* dated to the 17th century and located at the southern wall of the ground floor of the steeple. The original polychromy of the epitaph must have been outstanding[9], but the polychromy now shows much damage and some parts are completely lost.[2]

The epitaph consists of four sandstone fragments which are piled up on top of each other. The stone fragments are also in a degraded condition showing numerous cracks, delamination and partly granular disintegration.[2]

To minimize the risk of the transport, the surface of the front of the epitaph was temporary protected by cyclododecane – a temporary measure of fixing the surface.[10]

The application of poultices is a common measure of salt reduction.[11] Therefore, after arriving in the workshop, a moist cellulose poultice was applied on the back of the epitaph to avoid salt crystallization at the surface of the stone which was expected due to the change to more arid climate conditions in the workshop. The front was still covered with cyclododecane; thus, no further damages or crystallization of salts to the polychromy were expected.

Additionally, a salt ion reduction of the epitaph was realized by exchanging the poultice weekly until the salt ion load in the poultice was significantly reduced verified by electrical conductivity measurements. Selected poultice samples were analysed by ion chromatography to determine which soluble ions could have been extracted. The results proved that nitrates together with chlorides, magnesium and sodium are the main extracted salt ions.[12]

After the salt reduction several conservation measures were implemented, among them dry cleaning with a paintbrush, structural consolidation of the stone with silicic acid ester and fixing of the paint layers by using the aqueous consolidant JunFunori®. Additionally, cracks were filled and repair mortars were applied with silica sol as a binder.

Fig. 7 shows the epitaph of Leonhardt Kämpfe before and after conservation.

The investigation and conservation measures were completed during a period of about one year and a half before the epitaph was returned to the ground floor of the steeple. There, the epitaph was mounted in the same area, but in future no rising damp or capillary suction from the back of the walls is possible: to the ground, the epitaph was sealed up with lead, and to the back wall, a ventilated space of 10 cm for air circulation was created. Thus, no salt ion transport into the epitaph by capillary uptake will take place.

Summary

The investigations in the ground floor of the steeple and the conservation of one epitaph as a case study could indicate how to preserve the epitaphs of St. Johannis in Werben. The epitaphs as well as the masonry suffer from extreme humidity and high salt contents. The ventilation of the room would decrease the relative humidity which would cause more salt damages as demonstrated by the calculations with ECOS/RUNSALT. Therefore, we recommend to remove the epitaphs from the ground floor, and preserve them after salt reduction as shown in this paper. Of course, after a salt reduction and the completion of preservation measures the best solution for the epitaphs would be to find for them another location with moderate relative humidity around 50% or 60% and with low variation in relative humidity. But if they have to remain in the ground floor of the steeple, the preserved epitaphs should stand on platforms as suggested so that no new salt ion can penetrate into the epitaphs.

Additionally, sustainable measures for the whole church should be implemented, among them optimizing the water drain and removing salt ion sources like cement repair mortars.

However, in Werben, as in many other places, there is a lack of funding to realize the best possible solutions for the cultural heritage. Thus, step by step interventions in the surroundings and the monument itself can often improve the condition of monuments resulting in less degradation in the future.

Acknowledgements

We would like to thank for support and fruitful discussions: Gottfried Hauff, Peter Kozub, Katja Schmeikal, Catalina Schulz, and many other students who worked with us in Werben. We appreciate the support of the parish of St. Johannis in Werben.

References

[1] Dehio, G., Handbuch der deutschen Kunstdenkmäler, Der Bezirk Magdeburg, Berlin, (1975).

[2] Poerschke, D., Schadensanalyse und Auswertung des Schadenspotentials im Turmerdgeschoss der St. Johanniskirche in Werben.- unplublished Masterthesis at the University of Applied Sciences Potsdam, (2013).

[3] Laue, S., Salt Weathering of Porous Structures Related to Climate Changes, Restoration of Buildings and Monuments, 11, No. 6, (2005), 381-390.

[4] Bionda, D., Modelling indoor climate and salt behaviour in historical buildings: A case study, PhD thesis, Diss. Nr. 16567 ETH Zürich, (2006).

[5] Price, C (Ed.), An expert chemical model for determining the environmental conditions needed to prevent salt damage in porous materials. European Commission Research Report No 11, (Protection and Conservation of European Cultural Heritage). London: Archetype Publications, (2000).

[6] Weyer, A., Picazo, P.R., Pop, D., Cassar, J.A., Özköse, A., Vallet, J.M., Srša, I., EwaGlos European Illustrated Glossary of Conservation terms for Wall painting and Architectural Surfaces. Michael Imhof, Petersberg, (2015).

[7] ICOMOS – ISCS, Illustrated glossary on stone deterioration patterns.- Monument and Sites XV, Michael Imhof, Petersberg, (2010).

[8] Horie, C.V., Materials for Conservation - Organic Consolidants, adhesives and coatings.- 2. edition, Butterworths, London, (2010).

[9] Wollensen, E. & Haetge, E., Werben, Johanniskirche und Austattung. In: Der Kreis Osterburg. Die Kunstdenkmale der Provinz Sachsen Band 4, August Hopfer Verlag Burg b.M., Osterburg, (1938).

[10] Rowe, S., Rozeik, C., The uses of cyclododecane in conservation.- Reviews in Conservation, Volume 9, IIC, London, (2008), 17-31.

[11] Heritage, A., Heritage A. & Zezza, Desalination of historic buildings, stone and wall paintings.- Archetype Publications Ltd, London, (2013).

[12] Hübner, B. & Schulz, C., Documentation of the conservation measures on the epitaph of Leonhardt Kämpfe.- unpublished report, University of Applied Sciences Potsdam, (2016).

Salt-induced flaking of wall paintings at the Mogao Grottoes, China

Lori Wong[1]*, Su Bomin[2], Wang Xiaowei[2], Amarilli Rava[3] and Neville Agnew[1]
[1] The Getty Conservation Institute, USA
[2] Dunhuang Academy, Mogao Grottoes, Dunhuang, Gansu, China
[3] The Courtauld Institute of Art, UK
* lwong@getty.edu

Abstract

The Mogao Grottoes, a World Heritage Site in northwest China, is known for its surviving 492 painted Buddhist cave temples. Commissioned over a thousand year period, from the fourth to the fourteenth centuries, the caves were hewn into a 1.6 km long cliff face and the wall paintings executed on earthen plasters. Situated in a remote and arid desert landscape, these painted caves have endured throughout the centuries but many have also suffered from salt-related deterioration.

Repeat cycles of treatment for flaking on salt-damaged wall paintings have caused worsening of conditions resulting in significant loss of painted plaster. A research project to study this intractable problem and to develop and implement improved treatment methods was undertaken as part of a collaboration between the Getty Conservation Institute (GCI) and the Dunhuang Academy (DA) under China's State Administration for Cultural Heritage (SACH).

This case study looks at the activation mechanisms and deterioration processes of salt-induced flaking and highlights the development and implementation of remedial and preventive measures to prevent further loss from occurring. Topics investigated include the material composition of the paintings and plaster, previous treatments, salt identification and distribution, environmental conditions and the impact of increased humidity. Results show that past treatment of flaking wall painting with polyvinyl alcohol (PVA) and polyvinyl acetate (PVAc) created a film-like barrier that reduced permeability and trapped salts below the painted surface. This led to a build up of salts that when exposed to periods of high humidity caused disruption and powdering of the plaster from cycles of deliquescence and crystallization; the consolidated upper layer, then separated and lifted, in a new form of flaking, referred to as exfoliation.

The study also aimed to improve methods of condition monitoring to better assess when change due to salt activity occurs and to implement findings from the Visitor Carrying Capacity Study for the site. This included identifying caves at risk of salt-related deterioration and closing them to visitation during periods of high humidity.

Keywords: Mogao Grottoes, Dunhuang, wall painting, flaking, salt deterioration, previous treatment

1. Introduction

Salt-induced flaking of wall paintings is a condition phenomenon afflicting caves at the Mogao Grottoes, a World Heritage Site in northwest China, known for its 492 surviving painted Buddhist cave temples dating from the fourth to the fourteenth centuries *(Figure 1)*. Situated in a remote and arid desert landscape, these caves have endured throughout the centuries but their wall paintings have been subject to a particular form of loss related to both soluble salt deterioration

Figure 1: Cave 85, one of the 492 painted Buddhist cave temples at the Mogao Grottoes, was selected as the test cave for this project. Its wall paintings which date from the ninth century suffer from salt-induced flaking.

and paint flaking, known as exfoliation *(Figure 2)*. Repeat cycles of treatment in past decades have played a role in causing and exacerbating the deterioration phenomena, resulting in significant loss of the cave wall paintings. A joint research project to study this problem was undertaken as part of a collaboration between the Getty Conservation Institute (GCI) and the Dunhuang Academy (DA) under China's State Administration for Cultural Heritage (SACH) with the aim of developing and implementing improved treatment methods as well as measures to prevent this condition from recurring. Improvements in documentation and condition monitoring methods were also undertaken to better assess when changes to the wall paintings are occurring.

In order to understand the activation mechanisms and deterioration processes of exfoliation, the composition of the paintings and plaster was studied, previ-

Figure 2: Salt-induced flaking or exfoliation of the wall paintings in Cave 85

ous treatment materials and hygroscopic salts were identified, and environmental parameters and visitation history monitored. To provide a complete case study, Cave 85 was selected as a test cave for this project. This ninth century cave has been extensively studied and information was readily available [1].

Treatment of porous, inorganic substrates with synthetic polymers like polyvinyl alcohol (PVA) and polyvinyl acetate (PVAc) can decrease pore size with a consequent reduction of vapor and gas permeability.[2, 3] Upon exposure to high humidity the salts present undergo a phase change, from solid to liquid, and are free to move through the stratigraphy. The less permeable surface, where the synthetic organic materials were applied, can act as a barrier trapping the ions in solution; as the moisture evaporates, the solution becomes saturated, causing the salts to precipitate.[4, 5] Cycles of deliquescence and crystallization then causes gradual decohesion of the lower plaster layer; the upper layers are then separated from the powdering plaster below and are free to lift.

When caves are visited, doors are opened allowing outdoor humid air to intrude into cave interiors. In instances of high humidity, exterior damp air enters rapidly increasing humidity and resulting in damaging salt activity.[6, 7] This aspect of the collaborative project was also part of the implementation of findings from the Visitor Carrying Capacity Study for the Mogao Grottoes that developed visitor management solutions to close caves at risk of salt-related deterioration – including exfoliation – during periods of high humidity.[8] Environmental conditions such as rain and flood events are occurring with increasing frequency causing heightened concern over the long-term preservation of the site and its wall paintings.

2. Wall painting technique

Commissioned over a thousand year period, the caves were hewn into a 1.6 km long cliff face of conglomerate rock. Once

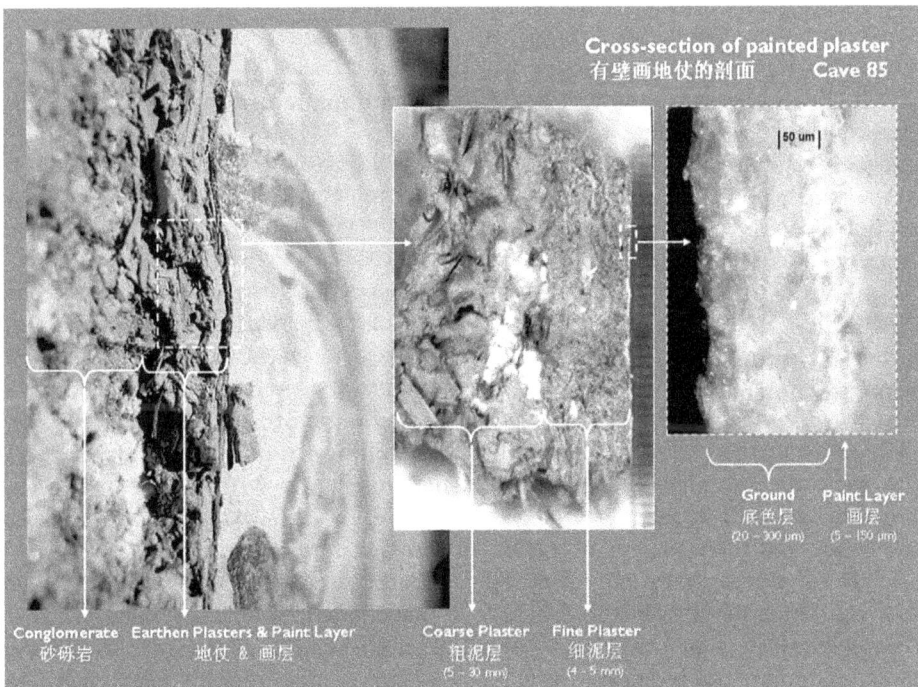

Figure 3: Stratigraphy of the painted plaster in Cave 85.

excavated, the roughly carved rock walls and ceilings were plastered and smoothed with a coarse leveling plaster of variable thickness (5-30 mm) and finished with a fine plaster (4-5 mm) *(Figure 3)*. The composition of the plasters were found to be a close match to the riverbed silt from the Daquan River that runs through the site with additional sand and plant fibers mixed in to prevent shrinkage and cracking. The plasters are bound with clay and contain 19% clay-sized fraction including illite, chlorite and the mixed-layer mineral, illite/smectite.[1]

A ground layer, composed of calcium carbonate, talc, and mica and bound with possibly a plant gum or mucilage, was applied as a thin wash to the fine plaster to prepare the surface for painting. Onto the ground, delicate line drawings were executed in red and black ink. The decorative scheme was then filled in with a rich palette of inorganic pigments and organic colorants applied singly or in combination, as thin washes or thick layers, and finished with colored glazes. Because of the variation in paint application, the overall thickness varied considerably (5-150 µm). Analysis found bone glue to be the most extensively used paint medium but samples also showed varying amounts of polysaccharides possibly indicating the presence of plant gums, honey, and mucilage [1].

3. Physical history of site and conservation history

Images from the early twentieth century show the site abandoned and the fronts of the cliff eroded leaving the caves open and their wall paintings exposed to the elements. Periodic flooding also occurred in the site's history leading to complete loss of painting up to one meter high in ground level caves and associated moisture and salt-related deterioration in the lower zone of painting.

With the founding of the Dunhuang National Art Research Institute in 1944, doors were installed on caves and beginning in the 1950s full-scale stabilization work on the grottoes commenced. Flood control measures were put in place in the 1960s which prevented occurrences of flooding until June 2011 when a major flood caused substantial damage to the site infrastructure. A subsequent flood occurred in 2012.

Past documentation records show a history of salt-related deterioration exhibited as both surface paint loss and detachment and collapse of painted plaster. In the 1960s, widespread paint flaking was observed in a number of caves and water-soluble synthetic polymers, polyvinyl acetate (PVAc) and polyvinyl alcohol (PVA), were used to treat the paintings, both to relay individual flakes and as a surface consolidant. Today, other adhesives such as gelatin are also being used to treat occurrences of flaking.

In the 1980s, the site was first opened to visitors.

4. Condition recording

Paint flaking has been the most prevalent problem at Mogao. During an assessment of 112 caves at the site, seventy-one were found to exhibit this condition making up 63% of the caves assessed.[9] The type of flaking varied considerably from localized to widespread and ranging in severity; sixteen caves exhibited serious and extensive flaking. Salt-related deterioration was found in thirty-four caves making up 30% of caves assessed. Twenty-eight of these are ground level caves with a history of flooding.[9]

In the past, types of deterioration have tended to be looked at individually rather than jointly and as a result salt-related flaking has only recently been identified as a unique phenomenon. A terminology was established to provide a clear and common language for the naming, recording and communicating of information regarding this condition. The terminology also reflects the specific technique of execution of the wall paintings in Cave 85 and focuses on the particular condition phenomena associated with salt-related flaking. Conditions were mapped, macrophotography undertaken and a detailed description of each produced.

4.1. Salt deterioration

Conditions related to salt deterioration included plaster disruption which exhibits itself as powdering plaster caused by repeat cycles of salt activity *(Figure 4)*; salt efflorescence seen as salt crystals on the painted surface; and, punctate loss, tiny, rounded losses, less than 1mm in diameter, in the paint, ground and/or fine plaster layer. Salt efflorescence and punctate loss are related conditions: individual salt crystals push through to the surface creating tiny losses in the paint layer. Over time, this gradually leads to complete loss of the paint layer in areas where salt problems are concentrated.

Figure 4: Powdering and decohesion of the plaster and punctate losses, visible in the red paint layer, are both forms of salt deterioration.

4.2. Paint flaking and exfoliation

Paint flaking occurs when the paint, or paint and ground layer, separate from the layer below and then lifts. This deterioration phenomenon is sometimes referred to as "pure flaking" to distinguish it from "exfoliation" or salt-induced flaking. Exfoliation occurs deeper within the stratigraphy of the painted plaster than pure flaking, typically within the fine plaster layer, and is defined as lifting of the fine plaster and/or ground and paint layers. Exfoliation was found to only occur in caves that have been previously treated for flaking with PVA and PVAc and have active salt deterioration. Exfoliation can be broken down into three stages of development:

1) Decohesion of the plaster due to salt activity causing volume expansion that pushes out the layers above leading to bulging and tenting of the paint, ground and some fine plaster layers. In some cases the protruding area can have associated cracking but otherwise the surface is still closed *(Figure 5a)*.

2) The protruding layers of paint, ground and some fine plaster breaks open along cracks and begins to lift further *(Figure 5b)*.

3) The breakage and opening of protruding areas progresses further to a point where loss of original material occurs *(Figure 5c)*.

Photography alone cannot always accurately describe the extent of lifting, as the way conditions are perceived by the viewer are affected by the angle of incidence of the light on the object. Other documentation techniques were employed to improve characterization of conditions such as Reflectance Transformation

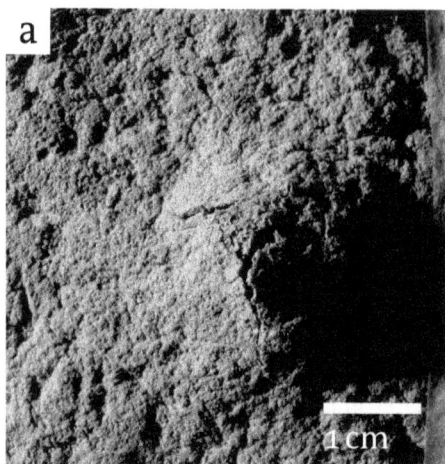

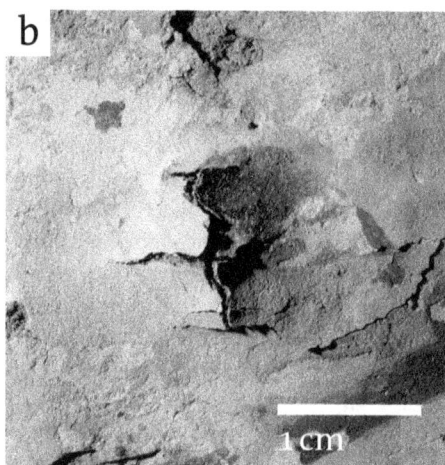

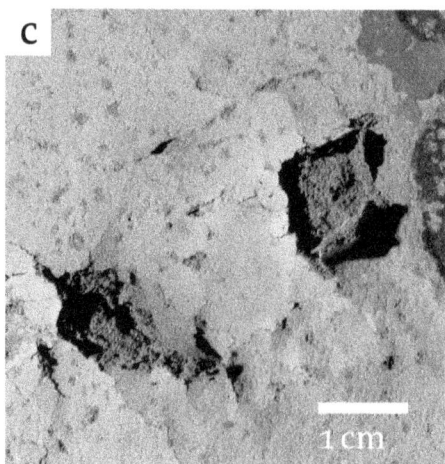

Figure 5a-c: The three stages of progression of exfoliation.

Imaging (RTI) which allows you to view an area under different lighting angles.

5. Analytical investigation

Sampling of exfoliating layers from Cave 85 was undertaken in areas where deterioration was concentrated. Analytical methods employed included ESEM-EDX on samples mounted in cross-section, FTIR to identify the presence of PVAc, XRD to identify salt species and ion analysis to quantify soluble ions.

5.1. Treatment materials

The presence of some treatment materials on the paintings could be seen with UV fluorescence in situ but in general FTIR was undertaken on unmounted samples using an acetone extraction. FTIR can give semi-quantitative information on the concentration of PVAc present on samples, with standardized weight, based on the strength of signal. Out of sixteen samples, twelve indicated the presence of PVAc, two had borderline PVAc present and two showed no positive results for PVAc. Attempts to quantify PVAc content more precisely were unsuccessful.

5.2. Soluble salts

The main salts present were identified by X-ray diffraction (XRD) as sodium chloride (NaCl) and sodium sulfate (Na$_2$SO$_4$).[1] Core sampling at different depths into the clay plaster was undertaken in three caves to provide topographic and stratigraphic distribution.[1] Samples were analyzed with ion chromatography and results indicated that salts were not evenly distributed between caves and within individual caves. Deterioration, however, was found to be correlated with salt cont-

ent: in Cave 85, areas with the most severe damage had salt content up to 2-3.9%
(expressed in weight percent of NaCl per
100gm of plaster) in contrast to less than
1% salt content in areas without visible
damage. Stratigraphic soluble ion content for chloride and sodium was highest
at the second increment (2-5mm) indicating enrichment of NaCl within the fine
plaster and at the interface between the
fine and coarse plaster layers.

6. Environmental monitoring

Environmental monitoring data was
collected from three locations within
Cave 85, measuring temperature and relative humidity. The moisture levels in
the cave are predominantly influenced

through natural ventilation which allows
cave air to freely exchange with the outside when cave doors are opened during
periods of visitation. Moisture produced
by visitors does not significantly contribute to moisture levels in the cave.[10] The
impact of the cave door being opened during high exterior humidity is dramatic,
causing a sharp and sudden spike in RH
(Figure 6). For the most part when a cave
is not visited and the cave entrance door
remains closed, a fairly stable climate is
maintained.[1] However, when high humidity persists for extended periods,
the RH inside the cave increases slowly
even when the cave door is not opened.
The louvered panels and gaps around the
door allow humid air to enter.

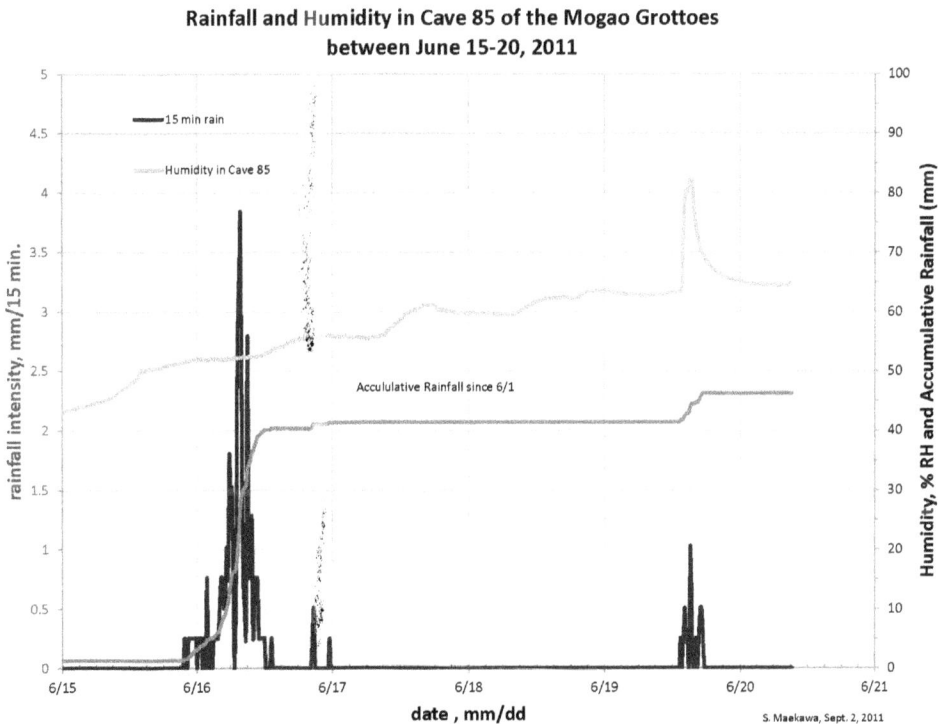

Figure 6: The humidity in Cave 85 rises even with the cave door closed after a prolonged period of rainfall on 6/16. The sharp peak in humidity that occurs between 6/19 and 6/20 is when the door is opened. In 2011, the RH exceeded 75% several times throughout the summer months.

7. Diagnostic investigation

Caves at the site with a history of treatment with PVA and PVAc were identified and examined. Of these, a few were also found to exhibit salt deterioration and exfoliation. However, there was no corresponding information regarding environmental data and visitation history that could be used to correlate with deterioration. Cave 85 was the only cave where complete information existed.

7.1. Linking visitation and environmental data

Visitation information for Cave 85 shows that prior to exfoliation being observed, the cave was opened on a limited basis to visitors. During summer months, several periods of high RH – some of which reached over 80% – were recorded. The repeat fluctuations of RH led to widespread plaster disruption and exfoliation in the parts of the cave where salts are concentrated.

7.2. Role of PVA and PVAc

Prior to the use of PVA and PVAc on the paintings, salts would have been able to move through the porous plaster before encountering the less permeable barrier constituted by the paint layer. This caused crystallization within or underneath the ground and paint layers, pushing off the paint layer and leading to punctate loss. Treatment with PVA and PVAc, penetrated into the paint, ground and fine plaster layers, affecting the permeability and porosity of the plaster, creating a less permeable layer. When humidity levels fluctuated, salts present in the plaster, unable to push through to the surface, underwent cycles of crystallization and deliquescence resulting in plaster disruption in the fine plaster. This caused

volumetric expansion and the bulging of areas that is characteristic of stage 1 of exfoliation. If conditions persist then stage 2 occurs, where the protrusions burst open; and, then stage 3, when the area continues to open and lift to the point when losses occur.

7.3. Humidity chamber

In order to understand the impact of fluctuating humidity on the surface and subsurface layers treated with PVA and PVAc, samples taken from exfoliated areas were placed in an environmental chamber to assess their response to changes in RH (possible moisture uptake and contraction) *(Figure 7)*. The RH in the chamber ranged from approximately 38-86%. A selection of samples were subjected to RH fluctuation and a few showed movement beginning around 60%, indi-

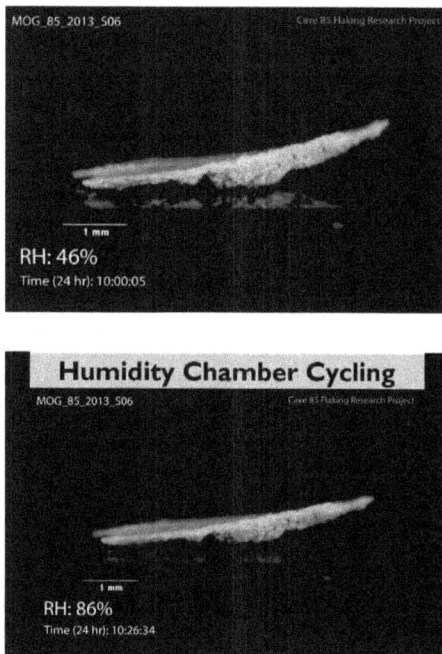

Figure 7: A sample from Cave 85 placed in an environmental chamber contracted at 46% RH (left) and relaxed at 86% RH (right) in response to changing humidity. Movement in the sample was observed around 60% RH.

cative of the lifting of layers that occurs with exfoliation. Time lapse images were put together into a short video to record the movement. However, given the complexity of the stratigraphy of the samples including paint, ground, fine plaster plus soluble salts and PVAc – each of which has a different moisture uptake rate – it was not possible to pinpoint the exact reasons for the lifting.[11]

7.4. Modeling salt behavior

A salt mixture collected from Cave 85 was tested using Dynamic Vapor Sorption (DVS) between 60% and 75% RH with increments of 0.5% RH.[1] The resulting isotherm showed mass increase starting at 67% RH, as the relative humidity increased, the slope increased exponentially. Salt samples were also examined with ECOS-RUNSALT, a program that uses a thermodynamic model to predict the behavior of salt mixtures under changing climatic conditions 1. However, as salt mixtures can vary depending on the cave and location and conditions of the sampling; and, as different salt species can form depending on the RH, maintaining stable RH and T within a cave was determined to be the best means of preventing future damaging salt activity.

8. Remedial treatment

Given that exfoliation is occurring within the fine earthen plaster and in order to avoid adding a film-forming material in the stratigraphy, treatment to relay flakes was undertaken using only water *(Figure 8)*. Water, can act to restore clay binding properties and proved to be an effective means of relaying areas of exfoliation. However, as water can further activate salts, causing damage, and there are water sensitive elements in the painting, treatment trials were undertaken

with reduced water content and non-aqueous systems. Water was used mixed with ethanol in varying percentages: 100% water, 50% water: 50% ethanol, 30% water: 70% ethanol, and 100% ethanol. Prewetting with ethanol or ethanol and water (1:1) was also undertaken which helped to soften the layers and acted to consolidate the powdering loose material behind the flakes prior to relaying.

Application methods also aimed to control the quantity of liquid used and to keep the treated area as contained as possible: small brushes were used and the liquid applied through two layers of paper tissue. The flakes were gently pressed back. A Preservation Pencil, an ultrasonic humidifier, was also tested in order to further reduce water content being introduced into the wall paintings but was found to be too slow to be practical.

9. Condition monitoring

Photographic condition monitoring of areas treated were established. Context and macro images were taken from different angles and under different lighting conditions both before and after treatment. Microscopy areas were also established to monitor the emergence of salts that may not be visible without magnification. Monitoring aims to be undertaken using the same equipment setup including the photographer, camera and lighting system.

10. Conclusions

In all treated areas, exfoliation has not recurred and the decohesive areas of plaster have remained stable. However, in an otherwise arid climate, rain and periods of high humidity are occurring with increasing frequency. Environmental data shows that humidity inside the caves increases even with doors closed. Though

Figure 8: Relaying of exfoliated areas was undertaken with water and water: ethanol mixes applied with brushes through paper tissue.

the aluminum door acts to buffer against the outside conditions, recommendations were made to improve the seal by closing gaps around the door and covering louvered panels with non-woven permeable fabric to slow air intrusion. This should be done in conjunction with continued condition and environmental monitoring in order to assess the stabilizing effects this has on the RH of the cave. Finally, visitation to the cave should be managed in a way that does not endanger the wall paintings with a system for closing the caves to visitors during periods of high RH.

11. Acknowledgements

The authors would like to acknowledge the significant contributions of the rest of the project team: from the Dunhuang Academy: Chai Bolong, Chen Bo, Fan Zaixuan, Li Yanfei, Qiao Hai, Sun Shengli and Zhang Guobin; from the GCI: Martha Demas, Shin Maekawa and Julie Chang (2014 GCI graduate intern); and, GCI consultants, Po-Ming Lin and Peter Barker.

References

[1] Wong, L. & Agnew, N. eds., The Conservation of Cave 85 at the Mogao Grottoes, Dunhuang: A collaborative project of the Getty Conservation Institute and the Dunhuang Academy, Los Angeles: Getty Conservation Institute, 2013.

[2] Camaiti, M., Bortolotti, V. and Fantazzini, P., 'Stone porosity, wettability changes and other features detected by MRI

and NMR relaxometry: a more than 15-year study, Magnetic Resonance in Chemistry 53 (1), (2015), 34–47.

3 Carretti, E. and Dei, L., 'Physiochemical Characterization of Acrylic Polymeric Resins Coating Porous Materials of Artistic Interest', Progress in Organic Coatings 49(3), (2004), 282–89.

4 Zhang, H., Qiang L., Ting L. and Bingjian Z., 'The preservation damage of hydrophobic polymer coating materials in conservation of stone relics', Progress in Organic Coatings 76(7-8), (2013), 1127–34.

5 Arnold, A., & Zehnder, K., 'Monitoring wall paintings affected by soluble salts', The Conservation of wall paintings: proceedings of a symposium organized by the Courtauld Institute of Art and the Getty Conservation Institute, London, July 13-16, 1987, S. Cather ed., Marina del Rey, CA: Getty Conservation Institute, 1991, 103–35.

6 Agnew, N., Maekawa, S. and Wei, S., Causes and mechanisms of deterioration and damage in Cave 85, Conservation of Ancient Sites on the Silk Road: Proceedings of the Second International Conference on the Conservation of Grotto Sites, Mogao Grottoes, Dunhuang, People's Republic of China, June 28-July 3, 2004, N. Agnew ed., Los Angeles: Getty Conservation Institute, 2010, 412–420.

7 Li, H., Wang, W., Zhan, H., Qiu, F., Guo, Q., Sun, S. and Zhang, G., 'The effects of atmospheric moisture on the mural paintings of the Mogao Grottoes', Studies in Conservation, 62(3-4) (2017), 229–239.

8 Demas, M., Agnew, N., and Fan, J., Strategies for Sustainable Tourism at the Mogao Grottoes of Dunhuang, China, Springer International Publishing, 2015.

9 Wong, L., Graves, K. and Wang, X., Assessment of Cave Condition and Visitation Potential: Undertaken as part of the Visitor Carrying Capacity Study at the Mogao Grottoes, China, unpublished report, Los Angeles: Getty Conservation Institute, 2012.

10 Maekawa, S., Zhan, Y., Wang, B., Fu, W., and Xue, P., Environmental monitoring at the Mogao Grottoes, Conservation of Ancient Sites on the Silk Road: Proceedings of an International Conference on the Conservation of Grotto Sites, N. Agnew, ed. Los Angeles: Getty Conservation Institute, 1997, 301–313.

11 He, X., Xu, M., Zhang, H., Zhang, B., and Su, B., 'An exploratory study of the deterioration mechanism of ancient wall paintings based on thermal and moisture expansion property analysis', Journal of Archaeological Science, 42, (2014), 194–200.

i This project builds upon a long lasting partnership between the Getty Conservation Institute and the Dunhuang Academy at the Mogao Grottoes. This includes the conservation of wall paintings in Cave 85 as well as master planning for the site and visitor management [1, 8]. For more information, see: http://www.getty.edu/conservation/our_projects/field_projects/china/app_mogao.html.

ii The assessment was undertaken as part of the Visitor Carrying Capacity Study at the Mogao Grottoes by the Getty Conservation Institute and the Dunhuang Academy [8, 9].

iii The acetone extraction process consisted of leaving the sample at room temperature for 3 hours then centrifuging it for 10 minutes. 0.2 µl of the solvent extract was then run under FTIR.

iv The source of the salts in the Cave 85 wall painting may have originated in the conglomerate substrate, in the original materials of the painted plaster or was brought in with flood events. For complete information on the salt investigation in Cave 85 see Chapter 16, Salt-induced Deterioration of Wong & Agnew 2013 [1].

v Copyright © 2002-2005 by Davide Bionda.

vi Manufactured by Preservation Equipment Ltd.

Development of a network-based climate monitoring system for climate assessment and regulation

Christian Leonhardt*, Sabine Leonhardt and Julika Heller
Werkstatt für Kunst und Denkmalpflege, Kiel, Gerrmany
* intonaco@gmx.de

Abstract

In the cloister (the so-called Schwahl) of the St. Petri Cathedral in the German town of Schleswig, the current situation of the historical wall paintings was investigated in the context of a research project funded by the German Federal Foundation for the Environment (German: "Deutscher Bund für Umwelt" – DBU). The project is focused on the investigation of the intense salt contami-nation and its connections with the climatic situation.

In this context, a network-based climate monitoring system was installed and tested in 2016. The data is accessible online, which is more convenient and advantageous than conventional data log-ger records for long-term climate measurement and evaluation, which can only be accessed in situ. It allows direct response to critical climatic values by connected alarm, control and regulation systems. This way, measurements can be taken promptly thus preventing damage and high resto-ration costs. Previously, climatic data had been collected in the course of the research project since 2007, the comprehensive evaluation of which served as a basis for the setting of the param-eters for the new system.

Measurement data are transferred directly to an open-source based server in-

Figure 1: The Schwahl in Schleswig, exterior walls with the chronological depictions of Christ's life cycle

frastructure, which is scalable and ready for future requirements. Due to the data being directly available on the internet, it is possible to provide a minute-accurate climate monitoring in real-time. Thus, the evaluation of an arbitrary number of measurement locations and the analysis of local causes (events, weather changes) can be put into practice.

The system was conceived and tested for the requirements of the Schleswig Cathedral. It can also be used in archives, depots and churches and thus constitutes a reliable basis for climate monitoring.

Keywords: Climate monitoring, climate data, networking-based monitoring, preventive conservation, regulation systems

1. The cloister in the St. Petri cathedral at Schleswig

The St. Petri cathedral in the north German town of Schleswig is one of the most important medieval monuments of the federal state of Schleswig-Holstein. The church, which was mentioned for the first time in 1134, was extended on the north side by a three-winged cloister between 1310 and 1320 by order of Bishop Johannes II von Bockholt. The cloister is called "Schwahl" (Danish: Svalen) which means "semi-open passage". The 24 yokes of the Schwahl are opened towards the courtyard by pointed-angled windows. Small slits are integrated into the windows, so that the cloister is a permanently unheated, slightly ventilated room under the partial influence of the external climate.

The Schwahl has been used as a walkway for processions and shows polychrome paintings on the arches of the 24 yokes. The paintings represent fabulous creatures and are largely well preserved. Moreover, the wall panels shown in the picture below are covered by paintings giving a chronological depiction of Christ's life cycle.

Figure 2: Application of desalination poultices

Figure 3: After the application of desalination poultices

2. Necessity for a network-based climate-monitoring system

Starting with a comprehensive restoration and conservation measure in 2007, the condition of the stock of the Schwahl had been continuously observed. Here, an accelerated accumulation of gypsum and hygroscopic salts was observed on the outer walls causing intense damage to the wall paintings. Consequently, several work campaigns have been carried out to address this problem since 2010. Among other measures, these included the reduction of the accumulations of gypsum and salt by the application of poultices (compare *figures 2 and 3*). The campaigns were accompanied by comprehensive analyses prior to and after the concerning measures.

Subsequently, the complex causes and conditions of the given damage processes are being in-vestigated within in the framework of a research project funded by the German Federal Environ-mental Foundation (German: Deutsche Bundes-stiftung Umwelt (DBU) since 2016 (compare [2]).

The primary goal of the project was to investigate the modification of the pro-perties of gypsum due to the interaction with highly soluble salts depending on the indoor and outdoor climate. Here, besides long term changes in clima-te conditions during a period of several days and weeks, short term changes in the indoor climate could have a relevant influence on the damage process. Thus, for instance, opening the exterior doors for a longer period of time under certain weather conditions could cause an unfa-vourable change in the interior climate thus causing a temporarily accel-erated damage progression on the wall pain-tings.

Since this is a common problem affec-ting a great number of architectural mo-numents, appropriate measures need to be taken to deal with this issue. Conclu-sively, this gave rise to the development of a modular and reusable solution for a network-based and real time-capable cli-mate monitoring system. Thus, a remote observation of the conditions in situ can be realised and proper advice can be gi-ven in real time. This involves automatic alarms based on guidelines derived for the individual object as well as a remote evaluation of critical situations. In doing so, the authors aim to provide a tool to si-

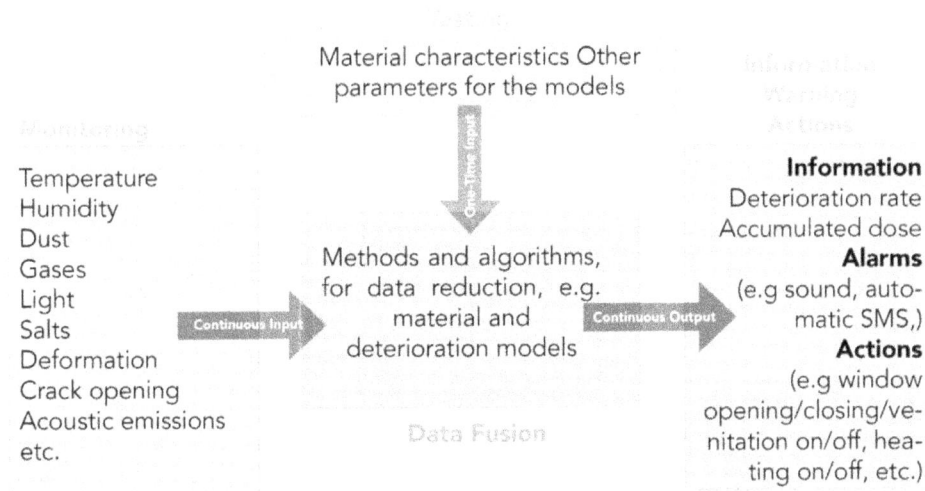

Material characteristics Other parameters for the models

Temperature
Humidity
Dust
Gases
Light
Salts
Deformation
Crack opening
Acoustic emissions
etc.

Continuous Input

Methods and algorithms, for data reduction, e.g. material and deterioration models

Continuous Output

Information
Deterioration rate
Accumulated dose
Alarms
(e.g sound, auto-matic SMS,)
Actions
(e.g window opening/closing/ve-nitation on/off, hea-ting on/off, etc.)

Figure 4: Principle sketch of intelligent monitoring and its integration in cultural heritage maintenance, based on [1]

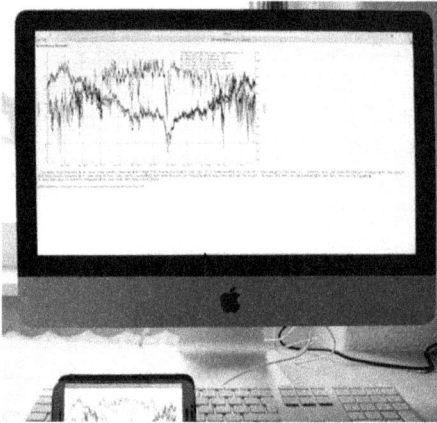

Figure 5: Read the climate data from the Schwahl, via Mobilephone and PC.

gnificantly reduce the amount of damage caused by temporarily unfavourable climate conditions.

3. Development of a network-based climate monitoring – IMMOMON

Within the scope of the DBU project it was possible to develop a system for measuring and moni-toring the climate in re-al-time. This network-based climate monitoring system is named IM-MOMON (www.immomon.net) and was created by collaboration of restorers involved in the pro-ject and IT system architect Dirk Hoffmeister.

As an outcome of this cooperation, proper measuring locations, measuring intervals, holding time of the data, threshold values were determined based on the restorers' expert knowledge. Also, appropriate measures were derived to deal with alarm messages caused by critical climate condi-tions due to e.g. open exterior doors. The control of external devices (e. g. ventilation) can also be initiated by the system.

The climate data are measured with set intervals by sensors distributed in the monument, collected in a local data hub and then transferred directly to a secured internet server. In addition to the commonly requested data on room temperature and relative humidity, sensors for outside climate, etc. can be easily integrated into the system. Via authorized access rights clients can access their data in real-time.

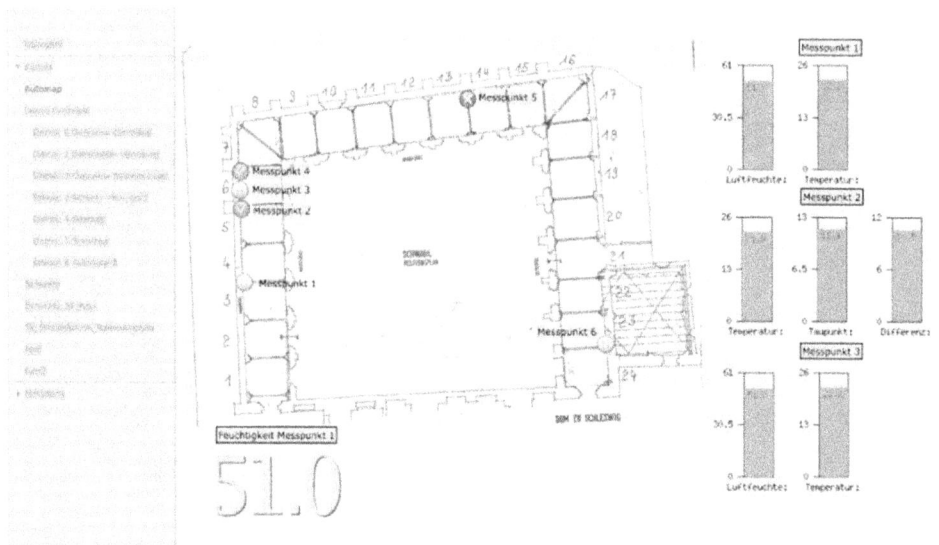

Figure 6: Visualization of the measuring points on site, defective devices are displayed in real time (meas-uring point 5).

Finally, the read-out and evaluation of the climate data was usually carried out in an annual cycle prior to the installation of the network-based climate monitoring. Accordingly, the process of data evaluation becomes cumbersome and laborious due to the huge amount of data to be analysed. Furthermore, the association of the data to specific events (e.g. concerts, exhibitions, etc) and their effects on the climate becomes difficult this way. This issue can now easily be dealt with by regular evaluation of the data available in real-time *(figures 4 and 5)* on the servers.

4. Conclusion and outlooks

The recording and evaluation of the interior climate of historic buildings and historic items or art objects in exhibition or storage is a fundamental, preventive measure for preservation. Even costintensive interventions due to climateinduced damage processes can be avoided by characterisation of the climate profile and development of corresponding restoration recommendations.

Here, IMMOMON can provide a significant contribution. Thus, the network-based measurement, control and warning system saves the necessity for regular control in-situ and allows for immedi-ate response to irregularities or unfavourable climatic situations.

More information on the network-based climate monitoring can be found on www.immomon.net.

References

[1] ReMonitoring of historic structures, project smoohs, p. 17, www.smoohs.eu

[2] Project 3216945: The historical paintings in the silt of St. Peter's Cathedral in Schleswig: investigations on the conversion, mobilization and recrystallization of gypsum on heavily salt-laden ground

SWBSS 2017

SALT WEATHERING OF BUILDINGS AND STONE SCULPTURES

SWBSS
2017
20 – 22 September 2017

FH;P UNIVERSITY OF APPLIED SCIENCES POTSDAM

all about salts: www.saltwiki.net